Edward A. Birge

Bacterial and Bacteriophage Genetics

An Introduction

Second Edition

With 150 Figures

Springer-Verlag
New York Berlin Heidelberg
London Paris Tokyo

Edward A. Birge
Department of Microbiology
Arizona State University
Tempe, Arizona 85287

LIBRARY OF CONGRESS
Library of Congress Cataloging-in-Publication Data
Birge, Edward A. (Edward Asahel) .
Bacterial and bacteriophage genetics : an introduction / Edward A. Birge.—2nd ed.
p. cm.—(Springer series in microbiology)
Includes bibliographies and index.
ISBN 0-387-96644-7
1. Bacterial genetics. 2. Bacteriophage—Genetics. I. Title. II. Series.
[DNLM: 1. Bacteria—genetics. 2. Bacteriophages—genetics. QW 51 B617b]
QH434.B57 1988
589.9'015—dc19
DNLM/DLC
for Library of Congress 88-12241
CIP

Typeset by David E. Seham Associates/Metuchen, New Jersey.
Printed and bound by Arcata Graphics/Halliday, West Hanover, Massachusetts.
Printed in the United States of America.

9 8 7 6 5 4 3 2 1

ISBN 0-387-96696-X Springer-Verlag New York Berlin Heidelberg
ISBN 3-540-96696-X Springer-Verlag Berlin Heidelberg New York

For Lori, Anna, and Colin

Preface to the Second Edition

It is with real pleasure that I offer some introductory remarks to this second edition of *Bacterial and Bacteriophage Genetics*. The reception of the first edition was very good, and most of the criticisms offered have only served to strengthen this new edition. The majority of the suggestions for revision of material included requests for more molecular detail. I have tried to provide this throughout the text but especially in the heavily revised Chapter 1 and completely rewritten Chapters 2 and 10. The former Chapter 2 has been repositioned as an appendix so as to offer an uninterrupted flow of material in the main body of the text.

The tremendous increase in our knowledge of the genetics of eukaryotic microorganisms has permitted another sort of change in coverage. Instead of general discussions about eukaryotic organisms, wherever possible I have tried to offer specific details about *Saccharomyces cerevisiae*, whose wealth of genetic detail will soon rival that of *Escherichia coli*. Although this material is not intended as a substitute for a course in yeast genetics, it is my hope that it will enable the beginning student to comprehend some of the similarities and differences between these two popular microorganisms.

As before this book is intended for the student who is taking a first course in bacterial and bacteriophage genetics and who brings to it some background in genetics. The best background would be an introductory course in general genetics, but extensive coverage of genetics in an introductory biology course might well prove sufficient. As an example, it is assumed that the student is familiar with the standard Watson and Crick model for DNA structure. A broad knowledge of microorganisms is helpful but not required to understand the material presented. In general the material from a good introductory biology course should be adequate.

Several books can be noted as being particularly useful sources of detailed supplementary information. David Freifelder's *Physical Biochemistry,* Second Edition, is particularly valuable as a resource for methods used in analyzing macromolecules. Additional information about *Saccharomyces* can be found in *Yeast Genetics: Fundamental and Applied Aspects* edited by J.F.T. Spencer, D.M. Spencer, and A.R.W. Smith. Cold Spring Harbor Laboratory constantly publishes many monographs dealing with various aspects of bacterial and viral genetics. Their current book list should be consulted for details. Probably the most important one for this book is *Genetic Maps,* Volume 4, edited by S.J. O'Brien. The American Society for Microbiology also is a good source for monographic literature. Their recent publications include *Escherichia coli and Salmonella typhimurium: Cell and Molecular Biology,* edited by F.W. Neidhardt. Most references to classic papers have been omitted, and the reader is referred to collections of papers that have been reprinted such as the volume edited by Abou-Sabe.

Acknowledgments

I am indebted to many people for helpful suggestions and comments about the first edition. I would particularly like to acknowledge the following individuals who assisted in correcting errors in the first edition: Dr. Paul A. Lemke, Auburn University; Dr. Margarita Salas, Universidad Autonoma de Madrid; and Dr. T.A. Trautner, Max Planck Institut, Berlin. Dr. W. Scott Champney's suggestions on reorganizing the material were most helpful. Dr. Martha Howe graciously volunteered some assistance with the material on phage Mu. However, any errors that remain must be attributed to me.

It has once again been a pleasure to work with the people at Springer-Verlag who have been very helpful during all phases of this revision.

Linkage Maps

Inside the front cover The illustrations on the front inside cover are linear scale drawings representing the circular linkage map of *E. coli* K-12. The time scale of 100 minutes, beginning arbitrarily with zero at the *thr* locus, is based on the results of interrupted-conjugation experiments. Major genetic symbols used in this figure are defined in Table 2-2. Parentheses around a gene symbol indicate that the position of that marker is not well known and may have been determined only within 5 to 10 minutes. An asterisk indicates that a marker has been mapped more precisely but that its position with respect to nearby markers is not known. The small vertical arrows indicate the directions of transcription of certain well-studied loci. Parentheses around an operon indicate that, although the direction of transcription of the genes in the operon is known, the orientation of the operon on the chromosome is not known. A similar map for *Salmonella typhimurium* may be found inside the back cover. From Bachmann, B.J. (1983) Linkage map of *E. coli* K-12, edition VII. Microbiological Reviews 47:180–230.

Inside the back cover The illustrations on the back inside cover are linear scale drawings representing the circular linkage map of *Salmonella typhimurium*. The scale of 100 units has been chosen to emphasize the similarities to the *E. coli* map (inside the front cover). A length of one unit represents the amount of DNA carried by P22, KB1, or ES18 transducing phage particles, while a length of two units represents the amount of DNA carried by a P1 transducing phage particle (see Chapter 7). The segmented lines to the right of the gene symbols indicate genes that are jointly transduced and the linear distances determined from these data. Major genetic symbols used in this figure are defined in Table 2-2. Parentheses around a gene symbol indicate that the location of the gene is known only approximately, usually from conjugation studies. An asterisk indicates that a marker has been mapped more precisely, usually by phage-mediated transduction, but that its position relative to adjacent markers is not known. Arrows to the extreme right of operons indicate the direction of mRNA transcription at these loci. From Sanderson, K.E., Roth, J.R. (1983). Linkage map of *Salmonella typhimurium,* edition VI. Microbiological Reviews 47:410–453.

Contents

CHAPTER 13
Repair and Recombination of DNA Molecules

CHAPTER 14
Applied Bacterial Genetic Principles

Appendix

The Laws of Probability and Their Application to Prokaryote Cultures

Chapter 1
Prokaryote Molecular Biology

When beginning a study of the genetics of bacteria and bacteriophages, it is important to have clearly in mind the ways in which these prokaryotes and their viruses organize their genetic and molecular processes, and the ways in which these processes differ from those used by eukaryotic organisms. This chapter provides a brief overview of major cell activities with comparisons specifically between the eubacteria, primarily represented by *Escherichia coli*, and the "lower" eukaryotes, primarily represented by *Saccharomyces cerevisiae*. The focus in this chapter is on the major molecular biologic processes of DNA replication, transcription, and translation. Specific genetic considerations are the subject of the next chapter. Some of the material in these chapters may be familiar from introductory biology classes, but all of it forms a necessary foundation for the topics to be presented later. The many varied replication and transcription mechanisms found among the bacteria, their plasmids, and their viruses refer back to this material.

Prokaryotic Cells and Eukaryotic Cells

Structure

The key feature that distinguishes prokaryotic organisms from eukaryotic organisms is their lack of an organized nucleus. They are also typically smaller than eukaryotic cells. For example, an average, rapidly growing *E. coli* cell is a cylinder about 1 × 0.5 μm, whereas a typical *S. cerevisiae* cell is round to ovoid and about 3 to 5 μm in diameter. The approximately 64-fold difference in cell volume is reflected in their internal cytoplasmic complexity, with *S. cerevisiae* containing the usual intracellular organelles such as mitochondria and endoplasmic reticulum, and prokaryotic cells having no real compartmentalization of function. In all the comparisons that follow, only the nuclear activity in the eukaryotic cells is considered, as modern evolutionary theory assumes that mitochondria and chloroplasts are descended from prokaryotic ancestors.

Another difference between the cell types is found in the way in which the cells carry out the processes of cell division and segregation of DNA. In both cases the cell volume increases during metabolism until a particular size is attained. At that point a complex series of events begins that culminates in the production of two daughter cells, each containing an exact copy of the DNA found in the parent cell.

Eukaryotic cells in general divide by a simple fission process coupled to mitosis that yields two equal-sized cells. However, in the case of many of the yeasts and *Saccharomyces* in particular, the process of cell division is called **budding,** because the new cell is produced as a small protrusion from the surface of the parent cell that rapidly enlarges and eventually pinches off (Fig. 1-1). During formation of the bud, the process of mitosis occurs. Spindle fibers form and attach to the centromeres on the already duplicated yeast chromosomes. The pairs of chromosomes line up and then are separated, moving along the spindle toward the poles of the elongating nucleus. The nuclear membrane persists at all times, unlike the situation in animals and plants. The nucleus continues to elongate, eventually entering the already large bud. When the nucleus splits in two, cytokinesis can occur to form the new cells.

The eubacteria reproduce by binary fission (Fig. 1-2), in which cell mass and volume enlarge linearly until the cell undergoes cytokinesis to yield two equal-sized daughters. The cell density remains roughly constant throughout the cycle. This mechanism is grossly similar to that in eukaryotic cells, but the process of mitosis is unknown in prokaryotic organisms. Moreover, there is no prokaryotic structure that is physically analogous to a centromere, and no microtubules have been seen. Therefore it is obvious that the cells must employ some other means to ensure proper segregation of their DNA molecules. The generally accepted theory, formulated by Jacob and co-workers, is that the replicating DNA molecules are attached to the plasma membrane, presumably at a site near the origin of replication. As each new round of replication begins, a new attachment site is formed on the membrane. The plasma membrane of a bacterial cell appears to grow primarily at the region along which the new septum will form. The insertion of new membrane material into this preexisting structure implies that two points lying astride the center line of the cell membrane and that are therefore initially close together gradually separate as the membrane grows (Fig. 1-3). Electron micrographic evidence indicates that the points of attachment of the replicating DNA molecules do lie on the plane of the future cell cleavage, and this mechanism apparently does

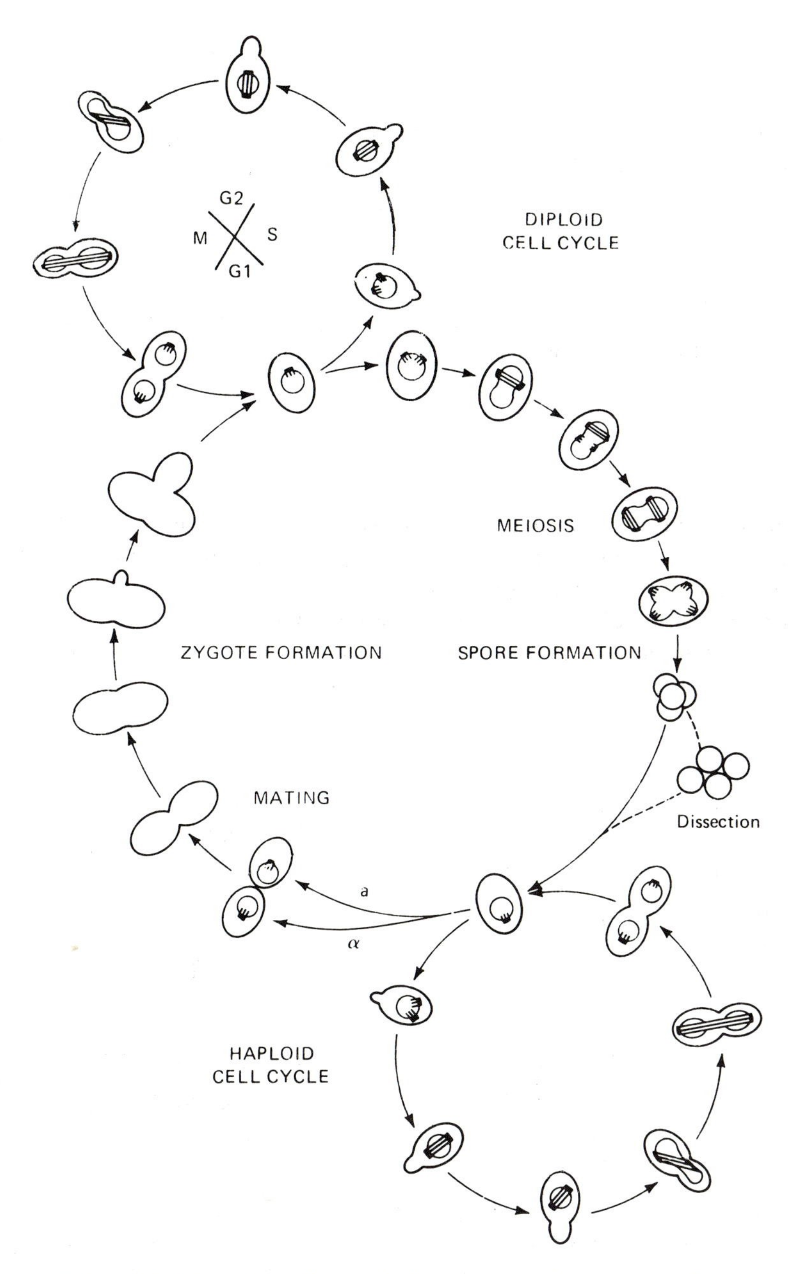

Figure 1-1. Possible life cycles of heterothallic strains of *S. cerevisiae*. In the diploid cycle there are two growth phases separated by DNA synthesis (S) and mitosis (M). The nuclear configuration is indicated for all dividing cells. From Dawes, I.W. (1983). Genetic control and gene expression during meiosis and sporulation in *Saccharomyces cerevisiae*, pp. 29–64. In: Spencer, J.F.T., Spencer, D.M., Smith, A.R.W. (eds.) Yeast Genetics. New York: Springer-Verlag.

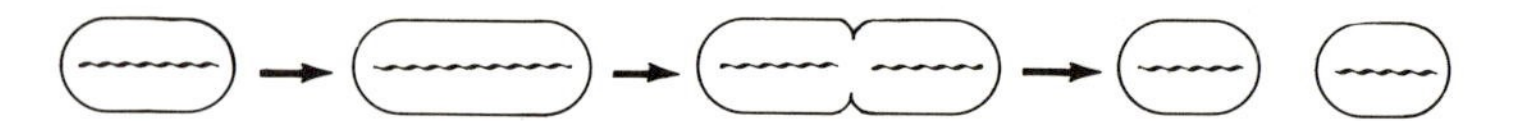

Figure 1-2. Simple binary fission as cell division. A cell elongates but changes little if at all in width. When the mass of the cell reaches a critical value, fission occurs.

shift the DNA molecules sufficiently to ensure proper segregation at the time of the binary fission.

Ploidy

In eukaryotic cells the process of mitosis serves to ensure that, after cell division, each daughter cell has the appropriate chromosome complement. Yeast cells may be either **haploid** (one copy of each chromosome) or **diploid** (two copies of each type of chromosome, one from each parent). The diploid and haploid forms are interconvertible via meiosis (diploid to hap-

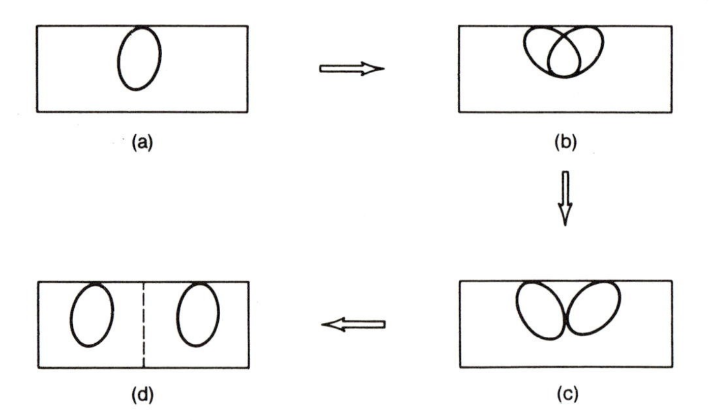

Figure 1-3. Segregation of replicating DNA. **(a)** A single DNA molecule is attached to the cell membrane. No DNA replication is occurring. **(b)** Replication has begun, and the origin region of the DNA (including the membrane attachment site) has been duplicated. The insertion of new membrane material has caused the two origin regions to become physically separated on the surface of the cell. **(c)** Replication of the DNA is almost complete. The membrane attachment sites have continued to separate until the physical connection between the DNA duplexes lies along the presumptive plane of cell division. **(d)** DNA replication has stopped, and a new round of replication has not yet been initiated. The cell will soon divide along the plane indicated by the dotted line.

loid) or mating (haploid to diploid), as shown in Fig. 1-1. In order for two haploid cells to mate, they must be of different mating types, one designated **a** and the other designated α. The necessity of two mating types makes *S. cerevisiae* a **heterothallic** yeast.

Because bacterial cells lack the ability to undergo mitosis or meiosis, they are fundamentally haploid. Otherwise there would be no way for a cell to guarantee the DNA content of its progeny. In the strict genetic sense, the term haploid means that there is only one copy of each piece of genetic information per cell. Obviously, shortly before a cell division the bacterial chromosome will have already replicated even though cytokinesis has not yet begun. This sequence of events does not violate the principle of haploidy, as the two chromosomes are identical, whereas in a true diploid they are permitted to be from different parents.

This description of the haploid state is complicated, however, by the fact that many bacteria are capable of growing at a rate such that the **generation time**, the average time interval between cell divisions, is shorter than the length of time required to replicate the entire DNA molecule in the cell (one round of DNA replication). The cell obviates this problem by beginning a second round of DNA replication prior to completion of the first. As the generation time decreases, the time interval between the initiation of new rounds of replication also decreases. The net result of these processes is that a rapidly growing bacterial cell actually has multiple copies of most genetic information. Moreover, the genetic information located near the origin of replication is present in proportionately greater amounts than that located near the termination site (Fig. 1-4).

Strictly speaking, it is therefore not possible to talk about the number of sets of information per cell (**genomes**) because most of them are incompletely replicated. The term **genome equivalent**, which refers to the number of nucleotide base pairs contained in one complete bacterial genome, is generally used instead. For example, a cell in which one genome has replicated halfway along the molecule contains 1.5 genome equivalents of DNA. Observe that a cell containing several genome equivalents of DNA is nonetheless haploid, as all the DNA is produced by replicating one original molecule. An example is shown in Fig. 1-4, in which there are four copies of *A* but only one of *J*. Fig. 1-5 demonstrates the relation between the number of genome equivalents and the growth rate for *E. coli*.

In the case of a bacterial cell that has received a new piece of DNA via some type of genetic process, it is possible to have two distinctly different sets of genetic information in the same cytoplasm. Such a cell is effectively diploid for that information. However, because most DNA transfer processes move only a fraction of the total genome (see Chapters 2 and 7 to 10), the resulting cell is only a **merodiploid**, or partial diploid. If the new piece of DNA is a **plasmid**, an independent DNA molecule that is capable of self-replication, the merodiploid state may persist indefinitely. If it is not a plasmid, only one of the daughter cells at each cell division

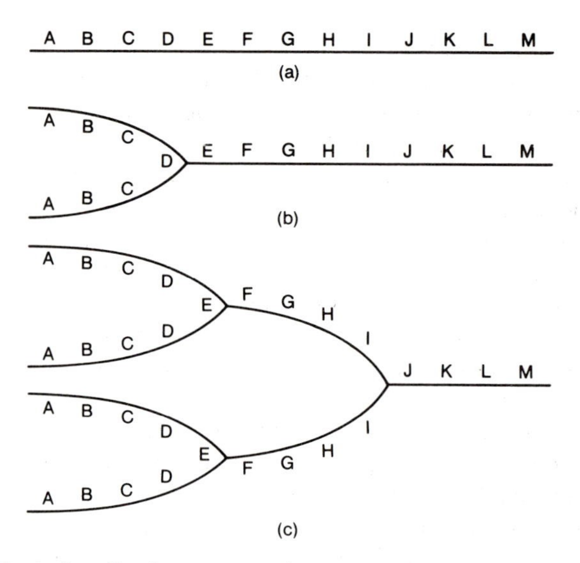

Figure 1-4. Effect of replication on gene dosage. (**a**) A nonreplicating DNA duplex. (**b**) The first round of replication has begun, initiated at the left-hand end of the duplex. (**c**) The second round of replication has begun before the first round of replication has finished. Once again the initiation occurred at the left-hand end, giving rise to two new replication forks. The same effect would be seen in a cell with a circular chromosome except that the DNA duplex would be longer and would be looped back on itself.

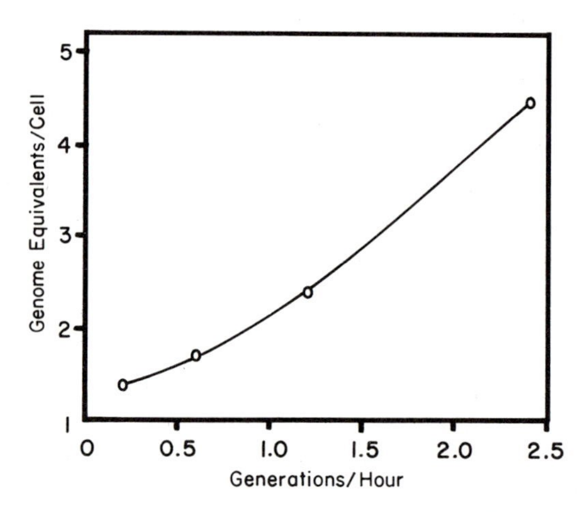

Figure 1-5. Relation between the amount of DNA per cell and the growth rate of the cell. Data are from Maaløe, O., and Kjeldgaard, N.O. (1966). The Control of Macromolecular Synthesis. Reading, Massachusetts: Addison-Wesley.

is a merodiploid, and the lone merodiploid cell is soon lost among the large number of haploid cells unless the extra nonreplicating DNA confers some selective advantage to the cell possessing it. Eventually, of course, the nonreplicating DNA is degraded and the bases reused, although it may not happen for several generations.

Chromosome Structure

Eukaryotes

The genetic material in eukaryotic cells is organized into **chromosomes**, precisely organized structures of protein and nucleic acid. The proteins incorporated into the chromosome are known as histones and are designated H1, H2A, H2B, H3, and H4. Two molecules of each protein except H1 join to form an octameric cylinder around which approximately 146 base pairs (bp[1]) of the DNA molecule are wrapped (Fig. 1-6). The individual cylinders of protein with their associated DNA are known as **nucleosomes**, and in the "higher" eukaryotes the individual nucleosomes are linked together by a spacer of about 40 to 60 bp of DNA coated by histone H1. The role of histone H1 in *Saccharomyces* is less clear, and there is some doubt as to its actual presence. The internucleosome distance is also reduced to only about 20 bp.

There are 17 chromosomes in *S. cerevisiae*, ranging in size from 150,000 to 2,500,000 bp. Physically the DNA molecules underlying these chromosomes are linear structures, although the ends of the DNA strands within a single helix seem to be joined by material that is neither protein nor DNA. The presumption is that the two strands of the DNA helix are actually one continuous thread that pairs with itself throughout most of its length.

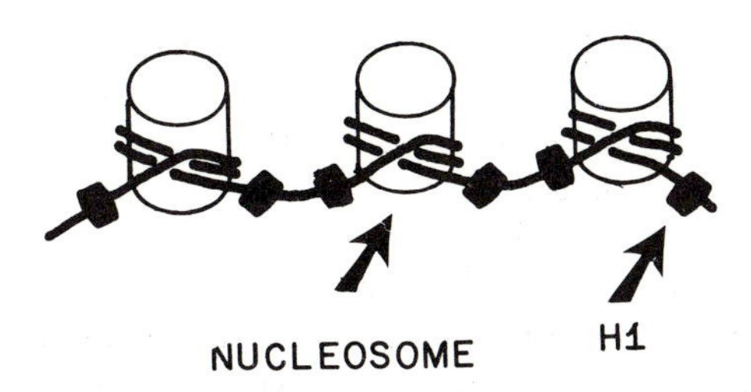

Figure 1-6. Chromosome structure of eukaryotic DNA. Each cylinder of protein is composed of two copies each of histones H2A, H2B, H3, and H4. There are approximately 146 bp of DNA wrapped around the histone core in a left-handed helix. In most organisms the DNA between each nucleosome is coated with histone H1.

[1]A DNA fragment 1000 bp (1 kb) in size has a molecular weight of 0.65×10^6 daltons or 0.65 megadaltons.

Prokaryotes

It was originally thought that bacteria did not have a structure that was truly comparable to that of the chromosome in eukaryotic cells. Indeed the special term "genophore" was coined to describe the physical DNA structure of a typical bacterium. More recently, however, it has become apparent that there are in fact some striking parallels between a eukaryotic chromosome and the DNA molecule as it is found in a bacterial cell.

The first clear visualization of a bacterial DNA molecule was achieved by Cairns, who used **autoradiography** (exposure of a film emulsion by the decay of tritium atoms incorporated into the DNA molecule) to demonstrate the circularity of an intact *E. coli* DNA molecule (Fig. 1-7). The DNA molecule has a length of about 1 mm and consists of about 4,500,000 bp (4,500 kilobase pairs [kb]) that are **covalently closed** (all joined by covalent chemical bonds). The contour length of the molecule when compared to the average size of an *E. coli* cell (~ 1 μm) indicated that the DNA must be highly folded within the cytoplasm. Stonington and Pettijohn subsequently showed that it was possible to extract intact DNA molecules from bacterial cells and preserve at least some of their folded configuration. These structures have been designated **nucleoids** or **folded chromosomes**; a typical nucleoid can be seen in Fig. 1-8. Immunolabeling studies of sections viewed in the electron microscope have shown that the core of the nucleoid is basically double-stranded DNA with single-stranded DNA at the periphery.

The predominant impression of the isolated nucleoid is one of coiled DNA. Pettijohn and co-workers have demonstrated that this complex structure consists of a series of about 43 **superhelical coils** that result from twisting the entire DNA helix over and above the normal helical turns. It is roughly equivalent to taking a stranded rope, coiling it on the ground, and then gluing the ends together so that the extra turns become an integral part of the structure. On a molecular scale, supercoils are added or removed by a group of enzymes known as **topoisomerases,** which act to change only the topology of the DNA molecule and not its base sequence. The superhelicity can be added in a positive sense, in which case the normal helix is coiled tighter, or in a negative sense, in which case the normal helix is coiled more loosely (Fig. 1-9). In the nucleoids studied thus far, all of the supercoiling has been negative supercoiling.

The superhelical density of the nucleoid has been estimated to be about one supertwist for each 15 normal helical turns. Each of the 43 superhelical loops is stabilized by protein interactions: If a **nick** (a broken phosphodiester bond) is introduced into such a loop, only that particular loop relaxes (unwinds). Maintaining the proper superhelical density is important to the normal functioning of the DNA, and there are several enzymes involved in the process. They are discussed later in connection with DNA replication.

The supercoiled DNA molecule of a bacterial cell is associated with

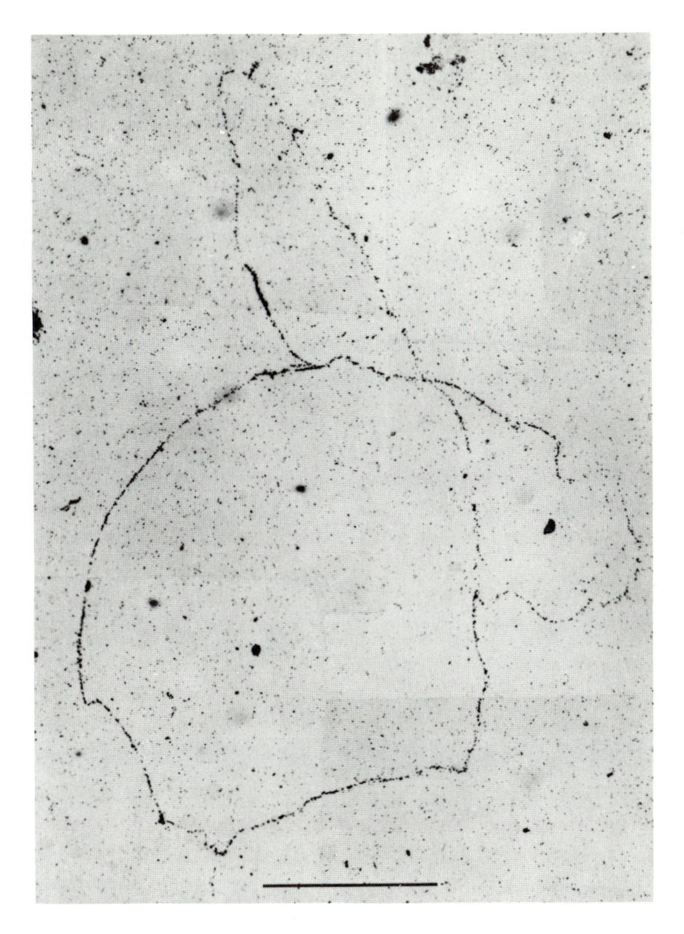

Figure 1-7. Autoradiograph of a DNA molecule extracted from *E. coli* Hfr3000. The DNA was labeled with tritiated thymidine for two generations and then extracted from the cell using the enzyme lysozyme, which attacks the cell wall. A photographic emulsion was overlaid on the DNA and exposed to the radioactive atoms for 2 months. As the tritium decayed, the beta particles that were emitted activated the emulsion in the same fashion as exposure to light. Upon development of the emulsion, silver grains were formed whose positions indicated the location of the original DNA molecule. The scale at the bottom represents a length of 100 μm; the length of the DNA, discounting the replicated portion, is estimated at about 1.1 mm. It should be remembered that the cell from which this DNA molecule was extracted was probably only a few micrometers in length. From Cairns, J. (1963). The chromosome of *E. coli*. Cold Spring Harbor Symposia on Quantitative Biology 28:43–45.

Figure 1-8. Membrane-attached *E. coli* nucleoid. Cells were gently lysed with lysozyme and detergent. The DNA was separated from cellular debris by sedimentation through increasing concentrations of sucrose and then mounted for electron microscopy using a monolayer of cytochrome C molecules on the surface of a formamide solution. The DNA was stained with uranyl acetate and coated with platinum to increase contrast. The remains of the cell envelope can be seen near the center of the photograph. The fine particles surrounding it are probably parts of the cell membrane. Continuous fibers of DNA in various states of supercoiling radiate out from the cell envelope. Short, kinky fibers of single-stranded RNA, representing transcription in progress, can be seen along the DNA. From Delius, H., Worcel, A. (1974). Electron microscopic visualization of the folded chromosome of *E. coli*. Journal of Molecular Biology 82:107–109.

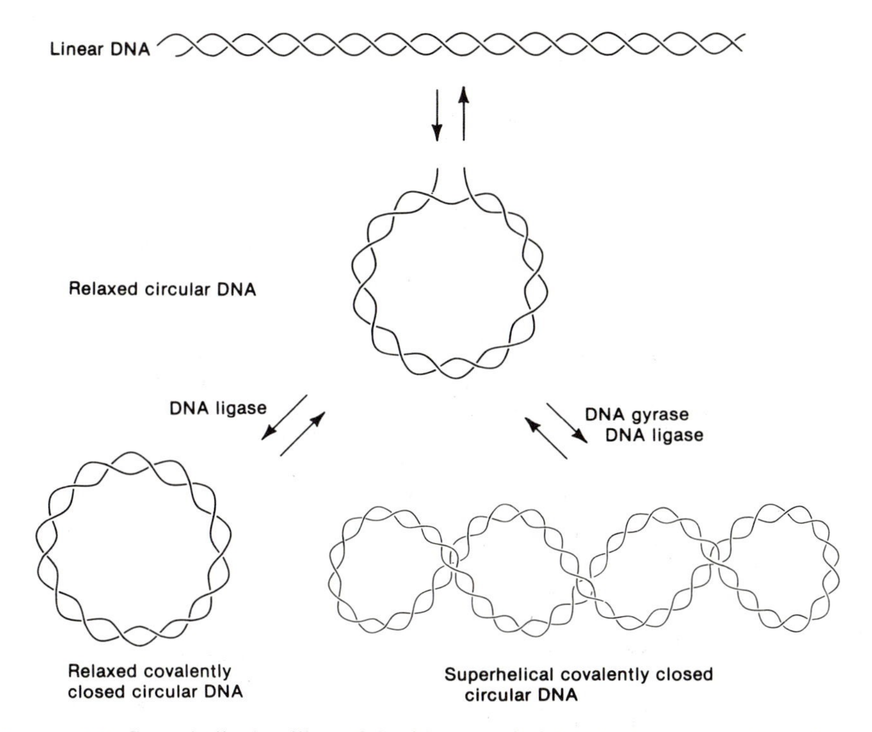

Figure 1-9. Superhelical coiling of double-stranded DNA. At the top of the figure is a linear DNA duplex. The process of forming a circle from this molecule can be envisioned as two independent ligation steps, one for each strand. The relaxed circular DNA represents an intermediate in this process. An additional ligation step can yield a molecule that is still relaxed but is covalently closed and circular (CCC). Alternatively, the enzyme DNA ligase can act in concert with DNA gyrase to yield a CCC molecule that also contains superhelical turns. These turns are inserted by rotating the free end of the DNA about the axis of the helix several times prior to the final ligation step. Depending on the direction of rotation, the superhelical turns may be considered to be either positive or negative, and the helix is slightly overwound or underwound.

certain "histone-like" proteins, basic proteins that lend regular structure to bacterial chromosomes. Predominant among them is a protein designated HU. HU is an abundant protein (2×10^4 to 2×10^5 copies per cell), and closely related proteins are found among the cyanobacteria and archaeobacteria. Broyles and Pettijohn have found that when a nuclease (an enzyme that cleaves nucleic acids) is added to DNA, it normally makes cuts every 10 bp, corresponding to one turn of the helix (Fig. 1-10). However, in the presence of purified protein HU, the cuts occur every 8.5 bp, indicating that the DNA helix has become overwound (more tightly coiled). They have also reported that the HU–DNA complex also contains negative supercoils induced by the interaction of the two components. The structure described is similar but not identical to that of a eukaryotic nucleosome; but electron microscopic observations indicate that the nucleosome-like

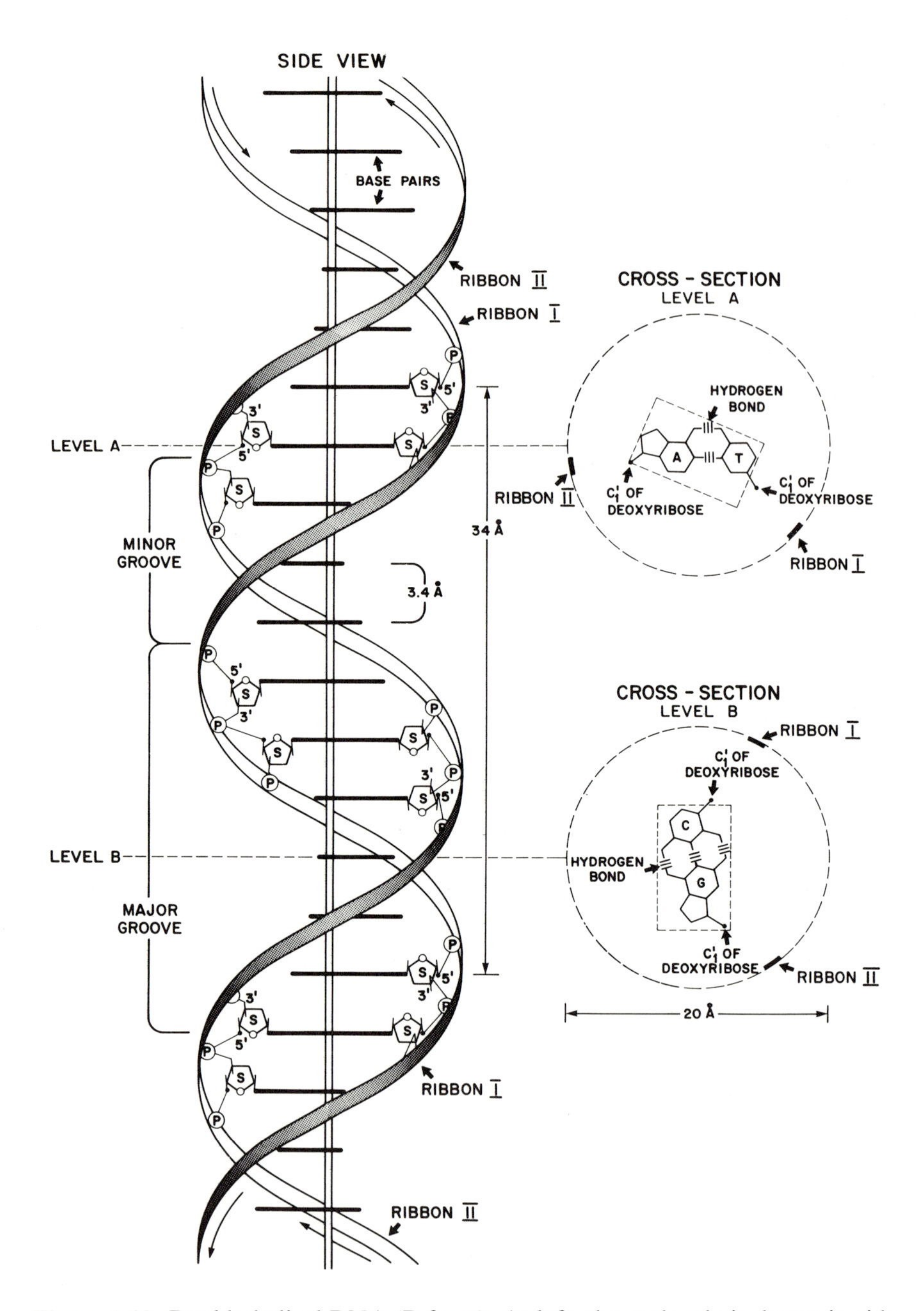

Figure 1-10. Double-helical DNA (B form). At left, the molecule is drawn in side view with the fiber axis indicated by the vertical rod. The backbone of the molecule consists of two polynucleotide chains that form right-handed helices. These chains are coiled together in a plectonemic (i.e., intertwined and not freely separable) manner to form a double helix having two grooves, one shallow (minor) and one deep (major), and an overall diameter of 2 nm.

structure, if it exists, is extremely sensitive to disruption of the cell and, unlike true nucleosomes, does not persist outside the cell for extended periods. Moreover, eukaryotic DNA has a nucleosome associated with each supercoiled loop. There are significant numbers of unrestrained supercoils in the bacterial nucleoid.

In summary, it seems clear that there is substantial structure to the prokaryotic chromosome. Although not explicitly considered in many models for genetic processes, it is certain that such structure affects genetic processes in subtle ways that remain to be determined.

Nucleic Acids

Prokaryotic cells, like their eukaryotic counterparts, contain both DNA and RNA. The general outline of the structure and function of the nucleic acids is identical for the two types of cell, but there are some specific differences.

Each chain is composed of D-2′-deoxyribose sugar moieties (S) linked by phosphate groups (P), thus forming 3′,5′-phosphodiester bridges and producing a long unbranched polymer. The individual bases are attached to the sugar molecules through β-*N*-glucosyl linkages. The two chains are antiparallel with the 5′ to 3′ direction proceeding upward for one chain but downward for the other. This 5′ to 3′ direction is illustrated by the arrows at the top and bottom of the diagram. For the sake of clarity, the molecular structure of the sugar–phosphate backbone is shown only over small regions. The two ribbons represent the continuity of the two chains, the shaded regions being closest to the viewer.

The hydrogen-bonded base pairs, represented by horizontal heavy lines, are planar molecules occupying the central core of the helix (the region indicated in cross section within the dotted rectangles at the right of the diagram). Only the bases lie in the plane of the cross sections, and thus only the base pairs are drawn, with attachment to the sugar merely being indicated. The position of each ribbon at either of the two cross-sectional levels is indicated.

The broken line forming a circle indicates the outer edge of the double helix that would be observed when the molecule is viewed end on. An adenine (A)–thymine (T) pair is shown as the pair representative of level A, whereas a guanine (G)–cytosine (C) pair is shown to represent level B.

The surface planes of the bases are perpendicular to the vertical axis and are separated from each neighboring base pair by a vertical distance of 0.34 nm. There are 10 bp per complete turn of the helix so that each turn of the helix has a vertical length of 3.4 nm, and each base pair is rotated 30° relative to its nearest neighbors. As a result of this rotation, the successive side views of the base pairs appear as lines of varying lengths depending on the viewing angle. The hydrogen bonding between the bases and the hydrophobic interactions resulting from the parallel "stacking" of the bases stabilizes the helical structure. From Kelln, R.A., Gear J.R. (1980). A diagrammatic illustration of the structure of duplex DNA. Bioscience 30:110–111.

DNA Structure

All known cells contain DNA, and the physical structure of the DNA molecules is generally assumed to be some variant of the double helical structure proposed by Watson and Crick. This structure has a built-in chemical polarity due to the position of various substituents on the deoxyribose moiety (Fig. 1-10). It is customary to refer to the 5′- or 3′-end of a linear nucleic acid, depending on the point of attachment of the last substituent (phosphate or hydroxyl group) to the pentose ring of the last nucleotide in the chain. In Fig. 1-10, for example, the arrowheads are always at the 3′-hydroxyl end of the chain. Each chain is referred to as a single **strand** of DNA.

In DNA the bases are paired such that a pyrimidine (thymine, cytosine) from one strand is always opposite purine (adenine, guanine) in another strand. The pairing of the bases is stabilized by hydrogen bonds (three for G-C pairs, two for A-T pairs) and generates a helical structure containing a major (wide) and a minor (narrow) groove (Fig. 1-10). In B-form DNA the helix is right-handed and has a constant width of about 2 nm. One complete turn of the helix requires about 10 bp. The helix is not always straight. Certain base sequences tend to cause a curve or bend in the DNA molecule, and the effect can be accentuated by the binding of specific proteins. A positive (+) curve means that the minor groove is opened.

It has been reported by Rich and Wells and their respective collaborators that an alternative physical state of DNA can be shown to exist. In this molecule the same base-pairing rules apply, but the stereochemical configuration of the bases is a strictly alternating *anti* and *syn*. The result is a double-stranded DNA molecule that is in a left-handed configuration and has a single deep groove (Fig. 1-11). This structure has been designated Z-DNA. Transitions between the B and Z forms of DNA have been shown to be all or none. The Z-DNA is a poor substrate for DNA polymerase I.

Unlike B-DNA, Z-DNA has sufficient tertiary structure to be a reasonable antigen, and Z-DNA-specific antibodies can be prepared. Such antibodies can be shown to specifically react with DNA isolated from whole cells, which seems to indicate that there are Z-DNA regions present within native DNA molecules (or else that the antibody can force the DNA into the Z configuration). The suggestion has been made that Z-DNA regions may serve a regulatory function. In *E. coli* it has been estimated from antibody binding studies that there is an average of one Z-DNA segment for every 18 kb. Certain bands of *Drosophila* chromosomes have also been shown to contain Z-DNA.

In the laboratory it can be shown that purines most easily assume the *syn* configuration necessary for Z-DNA structure. Therefore most work has been done with synthetic **oligonucleotides** (short chains of bases) that have a strictly alternating sequence of purines and pyrimidines, usually cytosine and guanine. Similar tracts have been observed in natural DNA

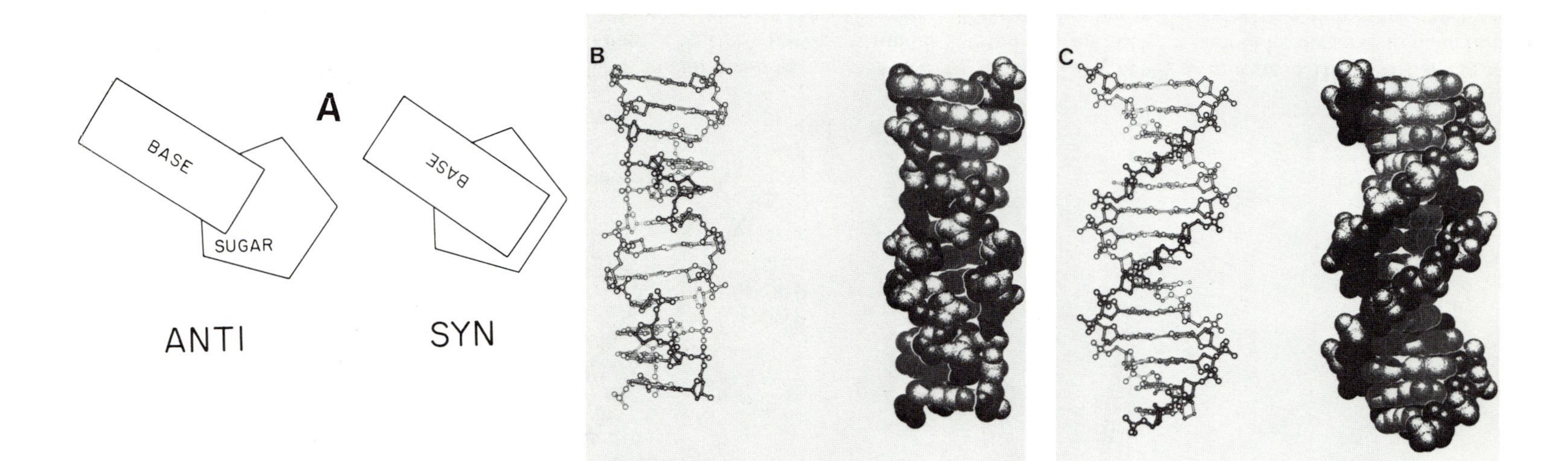

Figure 1-11. **(A)** Possible stereochemical configurations of a nucleic acid base. If the base is oriented primarily away from the plane of its attached deoxyribose, it is in the *anti* configuration. If the base lies primarily over the plane of the sugar, it is in the *syn* configuration. **(B)** Side views of a stick-and-ball model and a space-filling model of Z-DNA. The bases are in a strictly alternating *syn* and *anti* pattern, which gives the DNA a zig-zag arrangement. The sugar–phosphate backbone is the darker region of the diagram. Only the minor groove is visible, as the major groove is filled with cytosine C_5 and guanine N_7 and C_8 atoms. **(C)** Side views of a stick-and-ball model and of a space-filling model of B-DNA for comparison. Both major and minor grooves are visible. **B, C:** From Saenger, W. (1984). Principles of Nucleic Acid Structure. New York: Springer-Verlag.

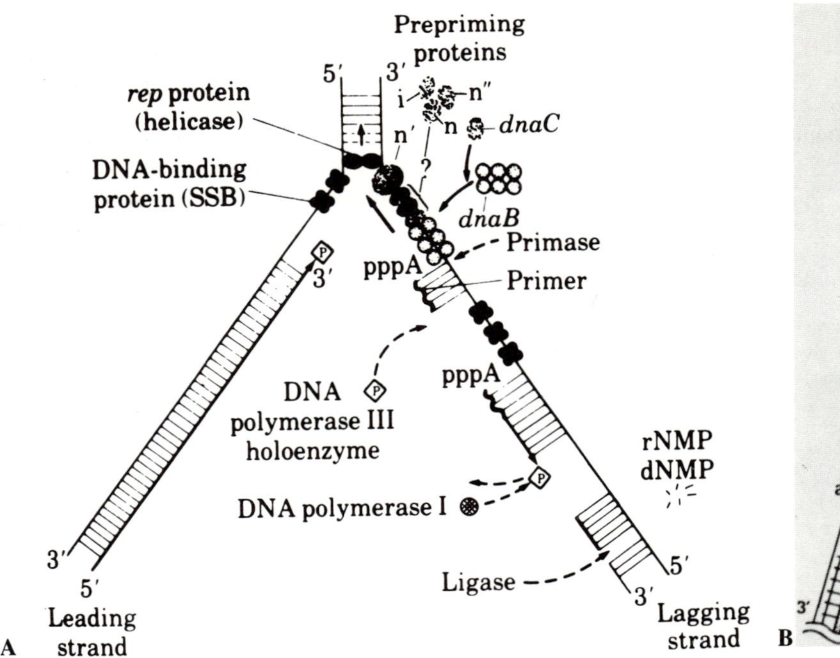

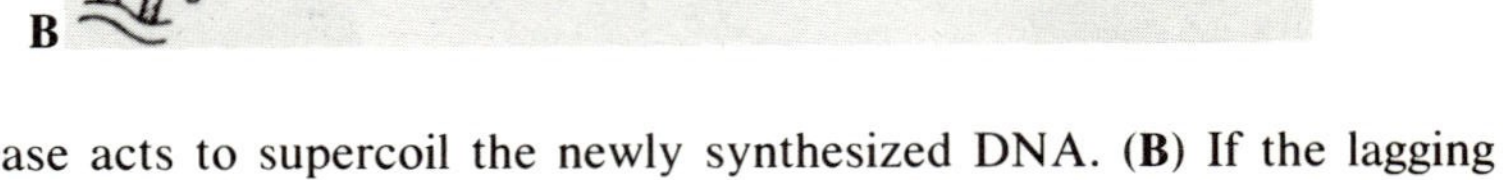

Figure 1-12. DNA replication. (**A**) The primer RNA is laid down by a primosome complex of the *dnaG* and *dnaB* proteins. The DNA polymerase III holoenzyme complex carries out the actual synthesis of the bulk of the DNA. The individual RNA primers are removed by DNA polymerase I, and the resulting nicks are sealed by DNA ligase. Topoisomerase I acts to relieve extra superhelical turns introduced into the DNA molecule by movement of the replication fork, and DNA gyrase acts to supercoil the newly synthesized DNA. (**B**) If the lagging strand is twisted appropriately, it should be possible to have both replication complexes moving in the same direction along the molecule. From Kornberg, A. (1984). Enzyme studies of replication of the *Escherichia coli* chromosome, pp. 3–16. In: Hübscher, U., Spadari, S. (eds.) Proteins Involved in DNA Replication. New York: Plenum Press.

molecules. Work from Wells and his collaborators indicates that runs of A•T pairs can also form the Z-DNA configuration, provided they are flanked by three to five G•C pairs.

DNA Replication

Replication is the in vivo synthesis of new DNA molecules and occurs in essentially the same fashion in both eukaryotic and prokaryotic cells. It is particularly well studied in *E. coli*, and the subsequent discussion focuses on that organism.

As postulated by Watson and Crick and first demonstrated by Meselson and Stahl, DNA replication is a semiconservative process in which new bases are hydrogen-bonded with old bases and then joined covalently to create a new strand. The resulting DNA duplex is thus composed of one new strand and one old strand. The enzymes that catalyze this process are said to be DNA-dependent DNA polymerases in that they copy DNA into more DNA. In conventional usage they are referred to simply as DNA polymerases.

However, if a DNA polymerase is presented with a single strand of DNA, it is not capable of synthesizing the complementary strand. This property of the enzyme is taken to mean that a DNA polymerase cannot start a new strand de novo but, rather, must have some sort of primer. The primer can be a short oligonucleotide, either RNA or DNA, so long as its nucleotide sequence matches a portion of the existing strand. New bases are added to the primer following a 5′ to 3′ polarity. In other words, the first base added is joined to the primer via the hydroxyl group on its 5′-carbon and has a free hydroxyl group on its 3′-carbon to which the next base will be attached.

If RNA synthesis is inhibited in a cell, DNA replication also stops, yet completed DNA does not contain any RNA sequences. This observation seems to indicate that RNA synthesis is necessary for DNA synthesis, possibly in the role of a primer that is later removed. Experimentally this suggestion can be verified by isolating newly synthesized DNA and demonstrating that transiently each newly synthesized piece of DNA carries a few RNA bases (three or four in the case of *Bacillus subtilis*) at its 5′-end. The enzymes involved in primer synthesis are discussed below. Several animal viruses and phage ϕ29 (see Chapter 5) are the only DNA replication systems known to prime without RNA.

Because DNA synthesis always is in a 5′ to 3′ direction, it is not possible for both strands of a DNA helix to begin replicating simultaneously. Instead, one strand (the leading strand) must be replicated for some distance and open up the helix so that the second or lagging strand can then begin its replication (Fig. 1-12). Note that the lagging strand must be synthesized by the production of a series of discrete fragments (called **Okazaki frag-**

ments after their discoverer). Each fragment of the lagging strand must have its primer removed and be joined together in order to complete the replicative process.

A variety of DNA polymerase enzymes has been identified in most organisms. In *Escherichia coli* there are two primary enzymes, designated DNA polymerase I and DNA polymerase III. DNA polymerase III is responsible for the bulk production of DNA. Its action is rapid, and it can join together some 15,000 to 60,000 nucleotides per minute. It has proofreading capability, a 3′→5′ exonuclease activity that allows the enzyme to "backspace" if an incorrect base has been inserted. A core enzyme comprised of the α, ε, and θ subunits can have other proteins added to it to form a holoenzyme complex (Table 1-1).

DNA polymerase I, on the other hand, functions more slowly and seems to be involved primarily in repair activities. It has 5′→3′ polymerase, 3′→5′ exonuclease, and 5′→3′ exonuclease activities, which means that the enzyme can add new bases to the 5′-end of an RNA primer, proofread, or if it attaches to a piece of DNA on the 5′-side of a primer degrade the primer and replace the RNA bases with DNA bases. Once all of the RNA primer has been removed, the final **nick** (missing phosphodiester bond) can be closed by the enzyme DNA ligase.

In practice, DNA replication in *E. coli* proceeds bidirectionally from a single origin of replication (*oriC*), a sequence of 245 bp that is both necessary and sufficient to initiate replication of a supercoiled plasmid. This single site is a major contrast to *S. cerevisiae,* which has some 400 distinct bidirectional initiation sites on its chromosomes. *E. coli* primer synthesis in vitro can be carried out by either of two DNA-dependent RNA polymerases: RNA polymerase itself, which handles the major transcription activities of the cell (see below), or DNA primase, encoded by the *dnaG* gene (Table 1-1), whose sole function is the synthesis of primers. The *dnaA, B,* and *C* proteins bind to *oriC*, and the *dnaB* helicase activity unwinds the DNA in both directions. The complex of DNA primase with the *dnaB* and *dnaC* proteins is referred to as a primosome and serves to prime all of the Okazaki fragments as well as the *oriC* region. Complete initiation requires RNA polymerase, DNA polymerase III, and the HU protein as well as the primosome.

Analysis of the RNA molecules transcribed from *oriC* in vitro shows that transcripts may begin at several sites, yet replication of the bacterial chromosome always begins at essentially the same spot. This specificity is enforced by the RNase H enzyme, which acts to degrade the RNA found in RNA•DNA hybrids. Presumably there is an inherent stability in the "true" primer that makes it relatively resistant to the action of RNase H, whereas the abnormal primers are easily degraded. It now appears that the leading strands of both the clockwise- and counterclockwise-moving replication forks are primed by separate RNA transcripts.

The DNA polymerase III holoenzyme complex attaches to the primer

Table 1-1. Some genetic loci of *E. coli* involved in DNA replication of that bacterium and its phages

			Necessary for replication[a]				
Locus	Synonym	Enzymatic function	*E. coli*	λ	T7	T4	P22
cou		DNA gyrase: coumermycin-sensitive subunit	+	nt	+	−	nt
dnaA		Initiation at origin, replication complex	+	−	−	−	−
dnaB		ATPase, mobile promoter	+	+	−	−	−
dnaC	*dnaD*	Chain initiation and elongation	+	−	−	−	−
dnaE	*polC*	α-Subunit of DNA polymerase III	+	+	−	−	+
dnaG		Rifampin-resistant RNA polymerase (primase)	+	+	−	−	nt
dnaI		Initiation at origin	+	−	−	−	nt
dnaN		β-Subunit of DNA polymerase III	+	+	nt	nt	nt
dnaP		Membrane defect at initiation	+	−	−	−	nt
dnaQ	*nutD*	ε-Subunit of DNA polymerase III	+	nt	−	−	nt
dnaX		δ-Subunit of DNA polymerase III	+	nt	−	−	nt
dnaY		$tRNA^{arg}$	+	nt	nt	nt	nt
dnaZ		τ-Subunit of DNA polymerase III	+	+	−	−	nt
gyrA	*nalA*	DNA gyrase nalidixic acid-sensitive subunit	+	nt	nt	−	nt
lig		DNA ligase	+	+	−	−	nt
nrdA	*dnaF*	Ribonucleotide phosphate reductase subunit B2	+	nt	nt	nt	nt
polA		DNA polymerase I (polymerase activity only)	−	−	−	−	−
rpoB	*rif*	RNA polymerase β-subunit	+	+	−	−	nt

[a] +, the protein is required for DNA replication in vivo; −, the protein is not required; *nt*, not tested. Note that independence of a phage from a particular host function may mean either that the phage does not require that function or that the phage can produce its own protein to carry out the function.

Adapted from Wickner, W. (1978). DNA replication proteins of *E. coli*. Annual Review of Biochemistry 47:1163–1191.

and begins to lay down the new strand of DNA (Fig. 1-12A). The helix ahead of the replication fork is unwound by a DNA helicase enzyme (*rep*), and the resulting single strands of DNA are stabilized by the *ssb* or single-stranded DNA binding protein. Kornberg proposed that appropriate folding of the single strands of DNA would permit a pair of identical DNA polymerase III oligomers to synthesize both leading and lagging strands in tandem (Fig. 1-12B).

The extra twists introduced into the unreplicated DNA structure by the act of unwinding the DNA duplex are removed by topoisomerase I (formerly known as the ω protein). The normal supercoiled state of the newly replicated DNA helices is restored by DNA gyrase acting after the replication fork has passed. The amount of superhelicity in the *E. coli* chromosome is therefore the result of the relative balance between these two enzyme activities. Experimentally it is observed that any change in the level of either topoisomerase is detrimental to the cell unless offset by a corresponding change in the other enzyme.

Termination of replication occurs at two loci positioned approximately 180° around the circular genetic map from *oriC*. The two terminator sites are separated by about five map units and are located roughly at positions 28.5 and 34.0 on the *E. coli* genetic map shown on the front endpapers of this book. Each region can be shown to inhibit the passage of a replication fork across it in one direction but not in the other. Thus any given replication fork passes by one terminator site and stops at the next one, generating a short region of overlap. The mechanism by which it occurs is not known. Another unknown quantity is precisely how the ends of the DNA from the various replication forks are joined together. One potential problem that could arise would be if the incorrect ends were joined (Fig. 1-13). If this situation were to occur, a linear **concatemer,** or molecule of greater than unit length, would be formed. Another possibility is the formation of a **catenane,** or interlocking circles. Concatemeric molecules can be reduced to unit length by homologous **recombination** (rearrangement of the phosphodiester bonds linking bases together so that altered DNA molecules are formed), and catenanes can be separated through the action of topoisomerase I.

RNA Structure

Cellular RNA molecules are predominantly single-stranded, although there are some examples among the animal viruses of double-stranded RNA molecules. Single-stranded nucleic acids are always available to form hydrogen bonds with other nucleic acids, and therefore it is easy to produce artificial double-stranded RNA molecules or RNA•DNA hybrid molecules.

Because the formation of hydrogen bonds gives a lower energy configuration to the molecular system, single-stranded RNA has a strong tendency to form internal hydrogen bonds that hold it into a definite three-dimensional configuration. One of the most common of these structures is the stem-and-loop (Fig. 1-14). Complex RNA structures can be held together by the usual hydrogen bonding rules plus the possibility of a single G•U hydrogen bond. For long RNA molecules, the number of

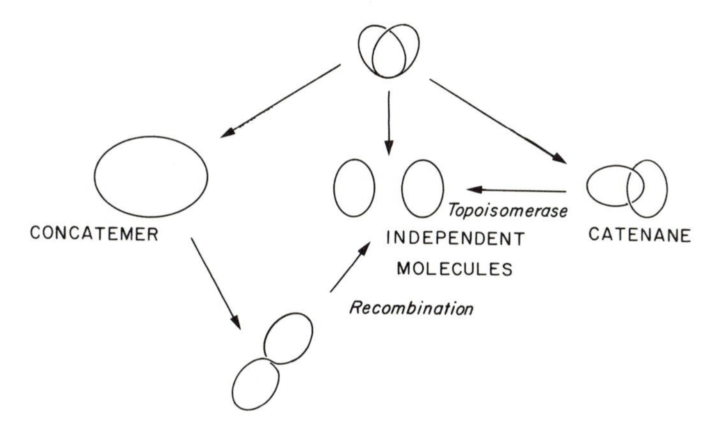

Figure 1-13. Termination of DNA replication. Depending on the way in which the ends of the newly replicated DNA helices are joined together, the product of replication might be one circular concatemer consisting of two complete genomes (left), two separate circular molecules (center), or two interlocked circular molecules (a catenane). Catenanes can be separated by a topoisomerase, and concatemers can be turned into monomers by recombination between homologous regions on the two component genomes.

•••ATGAAAAACGGACAUCACUCCAUUGAAACGGAGUGAUGUCCGUUUUAC•••

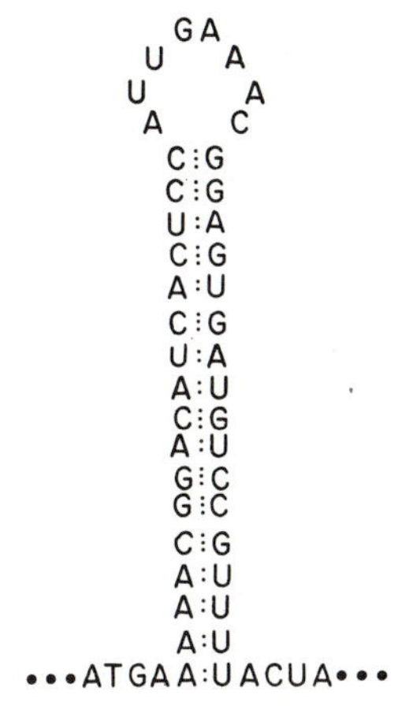

Figure 1-14. Formation of a stem-and-loop structure in an RNA molecule. A•T pairs can form two hydrogen bonds, G•C pairs can form three hydrogen bonds, and G•U pairs can form one hydrogen bond. The more hydrogen bonds that can form, the lower is the overall energy state of the molecule and the more stable is that particular configuration.

slightly different folded configurations that can be drawn is large. Presumably there is a favored structure that has a maximum number of hydrogen bonds and a minimum energy configuration. Unfortunately, there is no consensus on rules that might be used to calculate the minimum energy configuration for a large RNA molecule, and workers may disagree as to the exact structure that might be obtained in a given situation.

Transcription

Transcription is the synthesis of single-stranded RNA from double-stranded DNA using a DNA-dependent RNA polymerase. Eukaryotic cells in general, and yeast cells in particular, have three RNA polymerase activities designated A, B, and C. In the eubacteria there is only one RNA polymerase core or apoenzyme consisting of four protein chains, two α and two β. The holoenzyme is assembled by adding the appropriate protein(s).

The RNA polymerase holoenzyme binds to a specific **promoter** site on the double-stranded DNA molecule. A promoter has an intrinsic directionality to it and is located upstream (5′) of the first nucleotide of the transcript. The yeast promoters have three elements: an upstream regulatory element, a TATA box, and the initiator site for transcription. A conventional prokaryotic promoter has only two of these elements (Fig. 1-15). There are two regions of contact between the RNA polymerase and the holoenzyme, one centered at base −10 (the so-called Pribnow-Schaller box, which is roughly equivalent to the yeast TATA box), whose consensus DNA sequence is 5′TATAAT3′, and one centered at −35, whose consensus DNA sequence is 5′TGTTGACA. Studies measuring the amount of holoenzyme binding to variant promoter sequences have shown that the spacing between these sites is important, but the actual bases in the interval from −17 to −26 are not critical. For optimal promoter activity it is also important that the template DNA molecule be in the negatively supercoiled configuration, which has the effect of slightly loosening the DNA helix. Transcription begins at base +1, as is the case in yeast, but the spacing in prokaryotes (10 bp) is much less than in yeast (40 to 120 bp).

Figure 1-15. Promoter site on a DNA molecule. The bases comprising the promoter site are given negative numbers to indicate that they are not a part of the final transcript. Base +1 is the start of the transcript, and therefore there is no base numbered zero. The wavy lines indicate the points of contact between the promoter and the RNA polymerase holoenzyme.

Recognition of the promoter site is controlled primarily by proteins called sigma factors, which show notable structural similarities among organisms. After a sigma factor is bound, the holoenzyme then searches for a promoter whose -35 sequence is appropriate for that sigma factor. The initial binding to the DNA results in the formation of a **closed complex**, the normal DNA duplex with the RNA polymerase attached. Hydrogen bonds between about 10 bp in the region of base -5 are then broken, and the RNA polymerase settles into the major groove of the DNA to form an **open complex**. The broken hydrogen bonds in the open complex allow nucleoside triphosphates to pair with the template strand of the DNA helix to start the transcript (Fig. 1-16). Like DNA synthesis, the new strand of RNA is synthesized in a $5' \rightarrow 3'$ direction. The act of initiation "triggers" the RNA polymerase holoenzyme and results in the release of the sigma factor some 5 seconds later.

The RNA polymerase continues along the DNA helix until it reaches a termination signal. The type of terminator recognized can be controlled by proteins bound to the core enzyme. An example of such a protein is NusA which is discussed in connection with phage lambda (see Chapter 6). If the transcript should happen to extend into the *oriC* region of a plasmid, it can trigger replication of that DNA molecule. As the enzyme travels along the DNA, hydrogen bonds are broken in front and re-formed behind so that the size of the open complex remains essentially the same. The RNA transcript begins to fold into an internally hydrogen bonded structure as soon as it is of reasonable length. At DNA sequences containing a run of thymines, the RNA polymerase may "pause" briefly before

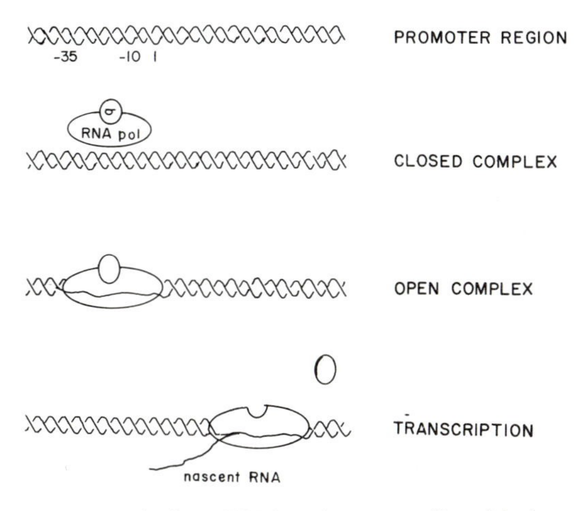

Figure 1-16. RNA transcription. RNA polymerase first binds to the promoter to form a closed complex. Melting of hydrogen bonds yields an open complex, and transcription begins at base $+1$. The sigma factor is released shortly after transcription begins.

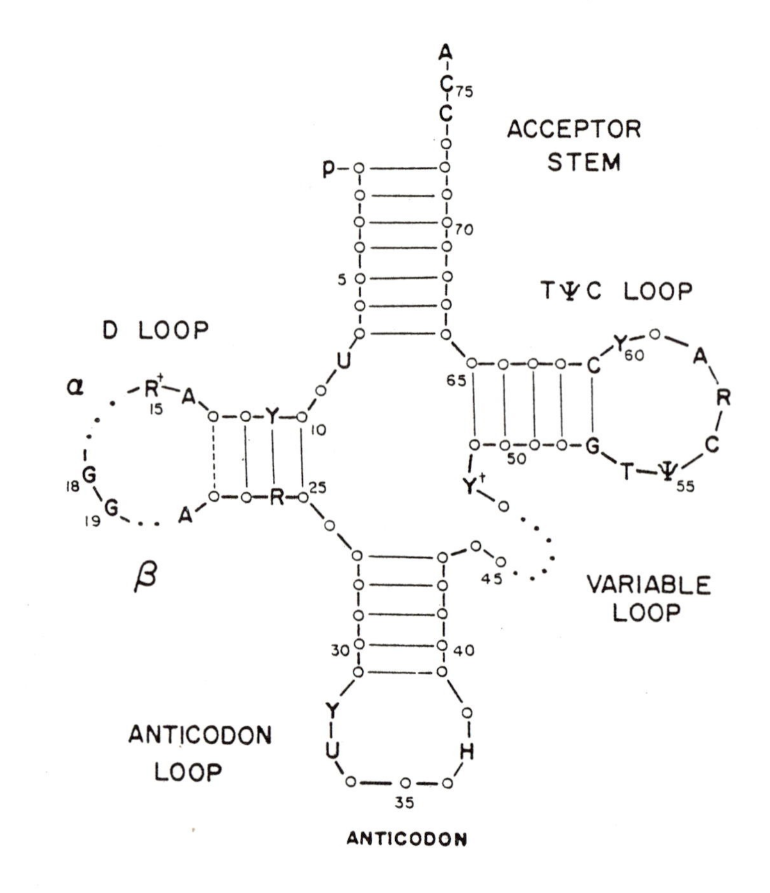
ACCEPTOR STEM
TΨC LOOP
D LOOP
VARIABLE LOOP
ANTICODON LOOP
ANTICODON
α
β
p
A
C
C
75
70
5
65
60
55
50
45
40
35
30
25
18
19
15
10

A

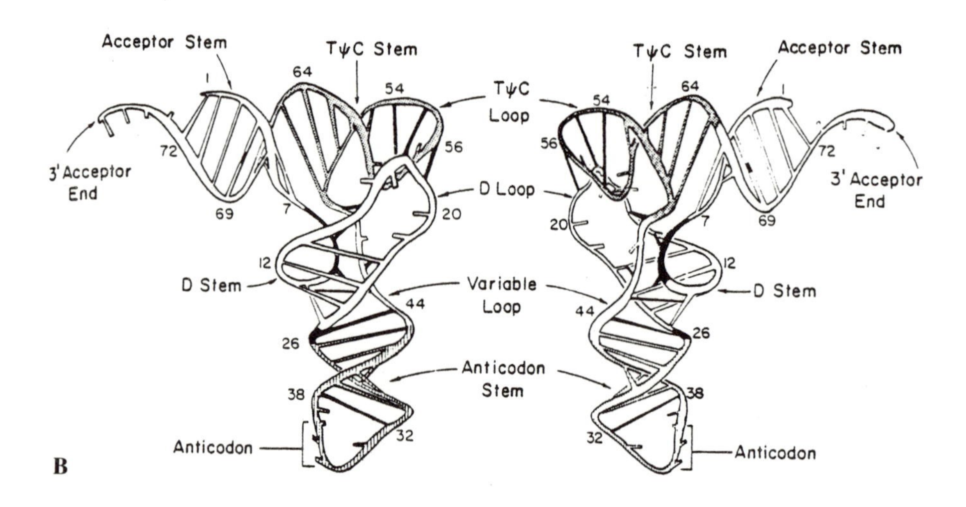
Acceptor Stem
TψC Stem
TψC Loop
3' Acceptor End
D Loop
D Stem
Variable Loop
Anticodon Stem
Anticodon
1
64
54
56
72
69
7
20
12
44
26
38
32

B

continuing. The three-dimensional configuration of the recently synthesized RNA has an important impact during a pause. If a stem-and-loop structure of the appropriate type has formed, transcription is terminated.

There are basically two kinds of stem-and-loop structures that can cause termination. One is an absolute terminator, and the other depends for its action on the presence of a protein factor called rho within the RNA polymerase. Rho is a helicase that can unwind RNA–DNA duplexes in a $5' \rightarrow 3'$ direction using the energy from nucleoside triphosphates, an effect that should contribute to the release of the transcript. It binds to an untranslated region relatively rich in cytosine residues that is upstream from the terminator. The rho-independent terminator has the approximate DNA sequence $5'CGGG(G/C)(T)_{4-8}T^{*}CTG3'$, where the asterisk indicates that the RNA transcript ends to the left of that base. Both types of terminator can function in the absence of supercoiling but can be blocked by closely coupled translation of the RNA. The presence of the ribosome is assumed to prevent formation of the stem-and-loop.

RNA Processing

In eukaryotic cells there is a great deal of RNA processing that must occur before transcripts are ready to be exported from the nucleus into the cytoplasm. In prokaryotic cells most of this activity is not needed because there is no physical boundary between the site of RNA synthesis and the cytoplasm. The type of RNA processing that does occur in prokaryotes depends on the type of the transcript.

Transfer RNA (tRNA) (Fig. 1-17) is used to carry amino acids for protein synthesis (see below). In *S. cerevisiae* there are about 360 genes coding for 28 types of tRNA molecule. These genes are distributed throughout all of the chromosomes and generally are not adjacent to one another.

Figure 1-17. (**A**) All tRNA sequences except initiator tRNAs. The position of invariant and semi-invariant bases is shown. The numbering system is that of yeast $tRNA^{Phe}$. Y stands for pyrimidine, R for purine, H for a hypermodified purine. R^{+}_{15} and Y^{+}_{48} are usually complementary. Positions 9 and 26 are usually purines, and position 10 is usually G or a modified G. The dotted regions α and β in the D loop and the variable loop contain different numbers of nucleotides in various tRNA sequences. (**B**) Two side views of yeast $tRNA^{Phe}$. The ribose–phosphate backbone is depicted as a coiled tube, and the numbers refer to nucleotide residues in the sequence. Shading is different in different parts of the molecule, with residues 8 and 9 in black. Hydrogen-bonding interactions between bases are shown as cross-rungs. Tertiary interactions between bases are shown as solid black rungs, which indicate either one, two, or three hydrogen bonds between them. Those bases that are not involved in hydrogen bonding to other bases are shown as shortened rods attached to the coiled backbone. From Rich, A., RajBhandry, U.L. (1976). Transfer RNA: molecular structure, sequence, and properties. Annual Review of Biochemistry 45:805–860. © by Annual Reviews Inc.

They do contain **intervening sequences**, segments of RNA located on the 3′-side of the anticodon that are not found in the mature product. These intervening sequences are removed posttranscriptionally by the process of **RNA splicing**. The 3′-end of each tRNA must have the bases CCA added to it, and each tRNA must have certain bases chemically modified (e.g., a uracil may be converted to a pseudouracil).

Bacteria do not usually carry more than one copy of a gene, but in *E. coli* that is specifically not true for the case of the tRNAs. Bacteria in general differ from yeast in that there are no intervening sequences (except in the case of the archaeobacteria). The tRNA genes in the eubacteria tend to be clustered. The most extreme case is that of *Bacillus subtilis* in which 21 tRNA genes are contiguous in one instance and 16 genes in another. In *E. coli* the tRNA genes are often found as spacers within ribosomal RNA (rRNA) transcripts (see below). There is one report of an *E. coli* transcript that contains tRNA, rRNA, and messenger RNA (mRNA). In *B. subtilis* the two large sets of tRNA genes occur at the ends of rRNA transcripts.

In any of these cases in which tRNA is part of a larger molecule, it is neccessary to cut out the tRNA precursor. This step is accomplished at the 5′-end by the endonuclease RNase P. The endonucleolytic cut at the 3′-end may be made by one of several endonucleases. Often the released tRNA is still too long and must be trimmed to size from the 3′-end. In *E. coli* all of the tRNA genes include the CCA 3′-end and do not require any additional synthesis. They do, however, have the same requirement for base modification as is seen in yeast.

Processing of rRNA is less complicated than that of tRNA. Each rRNA transcript contains the information for three rRNA molecules and may carry one or more tRNAs that act as spacers. In *E. coli* endonuclease processing releases one to three tRNA molecules and the various rRNAs (5S, 16S, and 23S). The large number of ribosomes found in a typical eubacterium necessitates multiple copies of the rRNA genes. In *E. coli* there are six sets of rRNA genes, three of which appear to be identical in terms of the single tRNA spacer used. *B. subtilis* has ten sets of rRNA genes, but only two contain spacers. The tRNA molecules included in the rRNA transcripts are, of course, processed as described above. After processing, the rRNA is assembled into the appropriate subunits. One 16S rRNA molecule associates with 21 proteins to give a 30S ribosomal subunit, and the 5S and 23S RNA molecules associate with some 26 different proteins to give a 50S ribosomal subunit.

In yeast, all cytoplasmic mRNA molecules are polyadenylated (carry a chain of adenyl residues on their 3′-end). The 5′-end of the mRNA is capped with a 7-methylguanosine triphosphate linked via its three phosphate groups in a 5′→5′ bond with the first base of the mRNA coding sequence. These modifications are involved in preparing the mRNA for export to the cytoplasm and facilitating its translation and are not found in prokaryotes. The original mRNA transcript in eukaryotic cells is longer than the final processed version and must be processed to remove intervening sequences, as in the case of the tRNA molecules. Once again the RNA splicing reaction occurs at specific sites characteristic of each type of mRNA. As a general rule RNA splicing reactions are not found in prokaryotic cells, although certain of their viruses use them (e.g., phage T4, discussed in Chapter 4).

Translation of the Genetic Message

The lack of a physical boundary between the bacterial nucleoid and the cytoplasm means that translation can proceed at the same time as transcription. The net effect is that mRNA is rarely if ever found as a separate entity in the bacterial cell. Instead, it is promptly located by ribosomes that attach and follow closely along behind the synthesizing RNA polymerase. As each ribosome moves along the growing mRNA molecule, a new ribosome takes its place. The complex of mRNA and translating ribosomes is designated a **polysome**. Amino acids are brought to the polysome by specific tRNA molecules.

A typical two-dimensional structure for tRNA is shown in Fig. 1-17. The anticodon pairs with the appropriate bases on the mRNA molecule during translation, and the translational efficiency of the tRNA molecule is determined primarily by the sequence of the anticodon arm. The characteristic amino acid is attached to the CCA or 3′-end of the molecule via its carboxyl group. The specificity for this attachment is presumed to lie in the three-dimensional structure of the tRNA molecule, but definitive experimental proof is lacking.

The actual mRNA binding site for the ribosome is not at the 5′-end of the mRNA molecule but, rather, somewhat internal to it. It is a region of some 9 bp known as the Shine-Dalgarno box and is complementary to the 3′-end of the 16S ribosomal RNA. The Shine-Dalgarno box is located within a larger region of nonrandom sequence that extends from base −20 to base +13 and is not the major nonrandom element within that region. The spacing between the Shine-Dalgarno box and the initiator (first) codon is important. The formation of the 30S initiation complex requires the following: a 30S ribosome; protein initiation factors IF1, IF2, and IF3; mRNA; and a special initiator tRNA molecule, usually carrying the amino acid methionine. The amino group of the initiating amino acid is blocked with a formyl group so that protein synthesis can proceed in only one direction. Thus the first tRNA bound usually carries the amino acid *N*-formylmethionine (fMet). Then the 50S ribosomal subunit is added to form the active 70S ribosome. Assuming that the first amino acid is to be methionine, the **codon** AUG (triplet of bases coding for the amino acid methionine; see Chapter 3) would be bound to the ribosome. By this point the IF proteins are released.

The fMet tRNA is located in the so-called peptidyl, or **P site,** on the ribosome (Fig. 1-18). Adjacent to it is a tRNA carrying the amino acid corresponding to the next codon of the mRNA. This binding site is the **A site,** or aminoacyl site. The fidelity of the codon–anticodon binding is

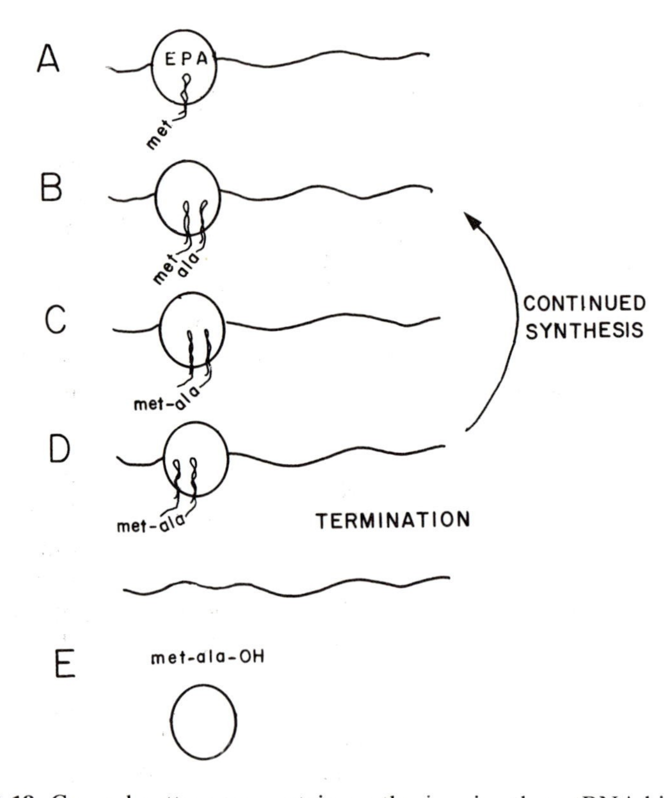

Figure 1-18. General pattern for protein synthesis using three tRNA binding sites as proposed by Nierhaus. The ribosome binds to a specific site on the mRNA molecule. *A:* The first codon, usually for methionine, aligns with the P site, and a correctly charged tRNA is bound. *B:* A second charged tRNA molecule corresponding to the next codon is bound to the A site. *C:* The chemical bond holding the amino acid to the tRNA bound to the P site is transferred to form a peptide bond with the amino acid held at the A site. *D:* As translocation occurs, the ribosome moves one codon down the mRNA molecule, and the two tRNA molecules move over one site. The growing peptide chain is now held at the P site. During continued synthesis, a new charged tRNA arrives at the A site and the uncharged tRNA is simultaneously released from the E site. The process now repeats from step *B*. *E:* At the end of the coding segment, the chemical bond holding the peptide chain to the tRNA is hydrolyzed, and the ribosome and uncharged tRNA molecules are released from the mRNA.

mediated not only by the usual base-pairing rules but also by the ribosomes themselves. In the **translocation** step, the ribosome together with elongation factor (EF) G forms a peptide bond between the two amino acids, perforce releasing the fMet from its tRNA molecule and moving the ribosome one codon down the message. The two tRNA molecules remain bound to the ribosome. The uncharged tRNA may now be located in the **E site**, or exit site, postulated by Nierhaus and co-workers; and the other tRNA, which is still attached to the growing peptide chain, is located in the P site.

The next round of amino acid addition commences when a newly charged tRNA molecule binds to the A site and, Nierhaus postulated, in the presence of elongation factor Tu causes release of the tRNA bound to the E site. The model of Nierhaus and his co-workers proposed that tRNA occupation of the A and E sites is mutually exclusive, so there are never more than two tRNA molecules bound to the ribosome at any given moment.

There are three specific codons for termination of protein synthesis (see Chapter 3). When one of these codons arrives at the A site, there is no corresponding tRNA to act as a translator, and the protein chain is released from the ribosome. The *N*-formyl group, or in some cases the entire *N*-formylmethionine, is cleaved from the end of the protein, and it is ready to function.

The fate of the ribosome after termination of translation depends on the nature of the mRNA molecule. Unlike eukaryotic cells, many prokaryotic mRNA molecules are polyinformational, i.e., contain the information to make more than one protein. In such a case, the 70S ribosome may remain attached to the mRNA and glide along until it arrives at another initiator codon. This action is a distance phenomenon; and the greater the separation of the terminator signal for one protein and the initiator codon for another, the greater the probability that the ribosome will release from the mRNA. When the ribosome does release, it dissociates into the subunits again before reinitiating.

Messenger RNA molecules do not last indefinitely. A typical *E. coli* mRNA molecule may remain functional for only 0.5 to 2.0 minutes. Degradation of the mRNA seems to occur by various means, as it may proceed from either the 3′- or 5′-end of the molecule. RNA turnover is important for regulatory processes (see Chapter 12).

Summary

DNA molecules are found as the genetic material in all cells. Those from bacteria are circular and supercoiled in order to allow such long molecules to fit into such small cells. They have associated with them certain histone-like proteins that may impart a nucleosome-like structure to the bacterial chromosome. If such a structure truly exists, it is much less stable than a eukaryotic chromosome because it cannot be isolated as an independent entity.

DNA molecules are replicated in the usual semiconservative fashion beginning at a single point of origin. Two replication forks are produced so the amount of time required to complete replication is reduced. The primer for DNA synthesis is a short RNA molecule. DNA polymerase III holoenzyme can work only in the 5′→3′ direction and therefore can lengthen one primer indefinitely to form the leading strand of newly replicated DNA. However, the opposite, or lagging, strand can be synthesized only as relatively short Okazaki fragments. The RNA primer segments are replaced with DNA by DNA polymerase I, and the resulting pure DNA fragments are joined together by the enzyme DNA ligase.

Transcription of the DNA is initiated at multiple promoter sites that are recognized by individual sigma factors located within the RNA polymerase holoenzyme complex. Only one strand of DNA, the sense strand, is transcribed in any one direction from a given promoter. Termination of transcription is by means of loop formation within the transcribed RNA. The RNA transcripts may contain information for tRNA, rRNA, mRNA, or any combination of the three. Substantial processing is required to prepare both tRNA and rRNA for their functions. In contrast to eukaryotic cells, bacteria do not have RNA splicing mechanisms, capping, or polyadenylating of the mRNA.

Translation converts the information of mRNA into a sequence of amino acids in a protein. The mediator of this activity is the ribosome, aided by the tRNA molecules. Each tRNA molecule occupies successively two sites on the ribosome, the A and P sites. In the A site the tRNA is bound to only one amino acid, and in the P site the tRNA is bound to the growing peptide chain. A third, or E, site has been suggested that would follow the P site and in which the tRNA would not be attached to any amino acid. Occupancy of the E and A sites would be mutually exclusive.

References

General

Adams, R.L.P., Burdon, R.H., Campbell, A.M., Leader, D.P., Smellie, R.M.S. (1986). The Biochemistry of the Nucleic Acids, 10th ed. New York: Chapman & Hall.

Brendel, V., Hamm, G.H., Trifonov, E.N. (1986). Terminators of transcription with RNA polymerase from *Escherichia coli*: what they look like and how to find them. Journal of Biomolecular Structure & Dynamics 3:689–723.

Deutscher, M.P. (1984). Processing of tRNA in prokaryotes and eukaryotes. CRC Critical Reviews in Biochemistry 17:45–71.

Drlica, K. (1984). Biology of bacterial DNA topoisomerases. Microbiological Reviews 48:273–289.

Drlica, K., Rouviere-Yaniv, J. (1987). Histonelike proteins of bacteria. Microbiological Reviews 51:301–319.

Hanes, S.D., Koren, R., Bostian, K.A. (1986). Control of cell growth and division in *Saccharomyces cerevisiae*. CRC Critical Reviews in Biochemistry 21:153–223.

Holland, I.B. (1987). Genetic analysis of the *E. coli* division clock. Cell 48:361–362.

Huberman, J.A. (1987). Eukaryotic DNA replication: a complex picture partially clarified. Cell 48:7–8.

Pettijohn, D.E., Sinden, R.R. (1985). Structure of the isolated nucleoid, pp. 199–227. In: Nanninga, N. (ed.) Molecular Cytology of *Escherichia coli*. London: Academic Press.

Spencer, J.F.T., Spencer, D.M., Smith, A.R.W. (eds.) (1983). Yeast Genetics. Fundamental and Applied Aspects. New York: Springer-Verlag.

Struhl, K. (1987). Promoters, activator proteins, and the mechanism of transcriptional initiation in yeast. Cell 49:295–297.

Vold, B.S. (1985). Structure and organization of genes for transfer RNA in *Bacillus subtilis*. Microbiological Reviews 49:71–80.

Zyskind, J.W., Smith, D.W. (1986). The bacterial origin of replication, *oriC*. Cell 46:489–490.

Specialized

Baker, T.A., Funnell, B.E., Kornberg, A. (1987). Helicase action of DnaB protein during replication from the *Escherichia coli* chromosomal origin in vitro. The Journal of Biological Chemistry 262:6877–6885.

Brennan, C.A., Dombroski, A.J., Platt, T. (1987). Transcription termination factor rho is an RNA-DNA helicase. Cell 48:945–952.

Brill, S.J., DiNardo, S., Noelkel-Meiman, K., Sternglanz, R. (1987). Need for DNA topoisomerase activity as a swivel for DNA replication for transcription of ribosomal RNA. Nature 326:414–416.

Broyles, S.S., Pettijohn, D.E. (1986). Interaction of the *Escherichia coli* HU protein with DNA: evidence for formation of nucleosome-like structures with altered DNA helical pitch. Journal of Molecular Biology 187:47–60.

Calladine, C.R., Drew, H.R. (1987). Principles of sequence-dependent flexure of DNA. Journal of Molecular Biology 192:907–918.

De Massy, B., Bejar, S., Louarn, J., Louarn, J-M., Bouché, J-P. (1987). Inhibition of replication forks exiting the terminus region of the *Escherichia coli* chromosome occurs at two loci separated by 5 min. Proceedings of the National Academy of Sciences of the United States of America 84:1759–1763.

Garcia, G.M., Mar, P.K., Mullin, D.A., Walker, J.R., Prather, N.E. (1986). The *E. coli dnaY* gene encodes an arginine transfer RNA. Cell 45:453–459.

Hoheisel, J.D., Pohl, F.M. (1987). Searching for potential Z-DNA in genomic *Escherichia coli* DNA. Journal of Molecular Biology 193:447–464.

Holck, A., Lossius, I., Aasland, R. (1987). DNA- and RNA-binding proteins of chromatin from *Escherichia coli*. Biochimica et Biophysica Acta 908:188–199.

Rheinberger, H-J., Nierhaus, K.H. (1986). Adjacent codon—anticodon interactions of both tRNAs present at the ribosomal A and P or P and E sites. FEBS Letters 204:97–99.

Rosenthal, E.R., Calvo, J.M. (1987). Effect of DNA superhelicity on transcription termination. Molecular and General Genetics 207:430–434.

Chapter 2

Genetic Processes and Procedures

The molecular and cellular processes in prokaryotes were reviewed in Chapter 1. In this chapter some of the special techniques used to study prokaryote genetics are discussed. After a review of the types of genetic exchange that can occur there is an introduction to some of the techniques fundamental to modern genetics: DNA sequencing, DNA hybridization, and DNA splicing.

Selection: An Essential Element of Microbial Genetics

The concept of **selection** is important to an understanding of microbial genetics. When dealing with microorganisms, the sheer number of individuals in any culture (generally on the order of 10^8 cells/ml) precludes any possibility of examining every cell involved in a particular set of genetic exchanges as can be done for macroscopic organisms such as pea plants or fruit flies. Instead, mixed cultures of cells differing in one or more biochemical pathways are used for genetic experiments. When placed under the appropriate set of nutritional or environmental conditions (the selection), growth of the original cell types in the culture is prevented so that only those cells that have acquired a new **phenotype** (one or more new observable characteristics) via DNA transfer are able to grow and divide. Examples of selective agents that have been used to prevent growth of the parental cells are antibiotics, bacteriophages, and required growth factors.

The transferred DNA may be in the form of a chromosomal fragment or an intact plasmid. In the case of fragmentary transfer, recombination must occur in order for the new genetic information to replicate. Each recombinant cell ultimately divides many times to produce a **colony**, a macroscopically visible mound of cells on the surface of an agar plate. Each colony is usually assumed to represent one genetic event, even though there is evidence that multiple rounds of recombination may occur.

In eukaryotic genetics, the distance between genetic loci is often expressed in terms of the **recombination frequency**, the total number of recombinants obtained divided by the total number of progeny from the cross. Under selective conditions, however, the geneticist is examining only a subset of recombinant progeny, those that survived the selective process. The total number of recombinant progeny of all types is in fact unknown. To the extent that recombination frequencies are calculated in microbial genetics, they represent the number of recombinant progeny obtained from the selection divided by the number of minority parent cells (assuming that the limiting factor in the exchange process is the number of each type of parental cell). Implicit in this definition is the assumption that the number of recombinants must always be less than or equal to that of the least abundant (minority) parent.

The ability to apply stringent selection to a large number of individuals has made microbial genetics a powerful tool for studying rare events. If a particular event is expected to occur one time in 10 million, the classic macroorganismal geneticist has a considerable problem. Yet 10 million *Escherichia coli* cells represent only about 0.1 ml of a typical growing culture. One billion *E. coli* cells, growing at their maximum rate, can easily be accommodated in 10 ml of culture. Moreover, their growth rate does not even begin to slow down until there are eight to ten times that many cells in the same culture volume.

It is important to bear in mind, however, that selection always introduces a bias into the sample population. Any cell that fails to promptly form a viable recombinant, for any reason, is lost to the sample. For example, if a cell does not display a recombinant phenotype for several generations after the genetic exchange, and if selection is applied too early, the cell is lost, and the genetic exchange is not detected even though it did occur. This problem is discussed further in Chapter 3. For now it is sufficient to remember that, typically, bacterial genetic experiments examine only those instances in which the entire process of genetic exchange and expression was successful.

Major Genetic Transfer Processes Observed in Microorganisms and Their Viruses

Although each genetic transfer process is discussed at length in subsequent chapters, it is advantageous to briefly introduce them now. The reason is that whereas textbooks can be divided into neat categories, the actual research cannot. As a result, people who are studying the nature of transduction may resort to some conjugation experiments and vice versa. Therefore to provide maximum flexibility in the forthcoming discussion, the major features of each process are reviewed.

Genetic Transformation

Genetic transformation was first observed in bacteria and was the first bacterial system for genetic exchange to be discovered. Although it occurs naturally in only certain bacteria, under laboratory conditions it seems to be possible to carry out genetic transformations with any cell type, prokaryotic or eukaryotic. In bacteria the process begins when a bacterial cell (living or dead) releases some DNA into the surrounding medium. This DNA is of course vulnerable to degradation but may encounter another bacterial cell before any significant change can occur. The second cell may take up the DNA, transport it across the cell wall and cell membrane, and allow it to recombine with the homologous portion of the resident bacterial chromosome. The resulting recombinant cell is called a **transformant**. In theory, any piece of genetic information may be transferred by this method, although the amount of DNA transferred per event is small, on the order of 10 kilobases (kb) in length. For a more complete discussion, see Chapter 8.

Transduction

In transduction a bacterial virus (**bacteriophage** or **phage**) is intimately involved in the genetic transfer process. Phage infections begin with adsorption of the virus particles to specific receptor sites on the surface of the cell. The nucleic acid contained inside the viral protein coat is then transferred to the cytoplasm of the bacterial cell, where it becomes metabolically active and undergoes replication and transcription.

Typically there are two possible outcomes to a phage infection. With the lytic response, the virus produces the structural components of new phage particles, packages its nucleic acid inside them, and then causes the cell to lyse and release the progeny phage. With the temperate response, the virus establishes a stable relationship with the cell in which some phage functions are expressed but not those that lead to uncontrolled DNA replication or the production and assembly of new particles. Instead, the viral DNA is replicated along with the host DNA, usually as an integral part of the same molecule, and is transmitted to all progeny cells. Occasionally cells carrying a temperate phage (**lysogens**) undergo a metabolic shift that reactivates the viral DNA. The result is the same as for an initial lytic response. Some phages may give only lytic responses and some only temperate; some, however, may give either response, depending on the growth conditions.

During the course of a phage infection of a bacterial cell, some or all of the viral DNA inside an individual **virion** (virus particle) may be replaced by bacterial DNA. This situation may occur only rarely or with great frequency. After such an altered phage particle is released into the medium, it may encounter another bacterial cell and attempt to initiate an infection. In so doing, however, it transfers the DNA fragment from the previous host's chromosome. If the newly infected cells are not killed and if the DNA fragment can either replicate or recombine, the result is the production of **transductants**.

The amount of DNA transferred by this means varies considerably but generally has as a maximum the amount of DNA normally present in a single bacteriophage particle. In some cases, it approaches 200 kb in length. The actual amount of DNA recombined is generally somewhat less and, in addition, depends on whether the transduction is generalized or specialized.

During generalized transduction the phage enzyme system that packages the viral DNA attaches to the bacterial chromosome and packages some of that DNA instead. The DNA that is packaged is chosen on a more or less random basis, and as a result it is possible for any piece of host genetic information to be transferred. Specialized transduction, on the other hand, specifically involves a temperate phage that has physically integrated its DNA into the bacterial chromosome at a specific site. As mentioned above, such an integrated phage may be stable for long periods of time. However, it may reactivate and replicate itself independently of the bacterial chromosome. During the reactivation process, it is possible for a mistake to occur so that some of the bacterial DNA located immediately adjacent to one end of the viral DNA is also excised from the chromosome instead of the appropriate DNA from the other end of the viral genome. Because the overall size of the excised DNA must be nearly constant, only certain pieces of genetic information can be transferred, and their size depends on the physical nature of the mistake that caused their production. For further discussion of both types of transduction, see Chapter 7.

Conjugation

The term conjugation can be used in several senses in biology. For example, in yeast the result of conjugation is fusion of haploid cells and formation of a diploid cell type (see Fig. 1-1). In a bacterium such as *E. coli*, instead of cell fusion there is a unidirectional transfer of DNA from the donor cell (which carries a **conjugative plasmid**) to the recipient cell beginning at a definite point on the DNA molecule and proceeding in a linear fashion. The transferred DNA may be all or part of the plasmid and may include a portion of the host DNA as well. By analogy to the other bacterial transfer processes, the recombinant bacteria are called **transconjugants**. The amount of bacterial DNA that can be transferred by conjugation ranges from a few kb to the entire chromosome. This process is discussed at length in Chapters 9, 10, and 11.

Protoplast Fusion

Protoplast fusion has been used successfully for many years with eukaryotic cells. Its use with prokaryotic cells is comparatively recent, but apparently the technique is applicable to most cells. In order for the process to work, **protoplasts** (cells that have been stripped of their walls) must be prepared by various enzymatic or antibiotic treatments. Fusion of the cell membranes is aided by a high concentration of polyethylene glycol. The resulting diploid cell usually segregates haploid offspring, many of which show extensive recombination of parental characters. There have also been some reports of the formation of stable diploid bacterial cells. In at least one case the diploid cells were reported to carry an inactivated gene, which may imply that genetic stability is not possible for a diploid bacterial cell in which both chromosomes are functional. Successful fusions have been reported with *Actinoplanes, Brevibacterium, Bacillus, Mycobacterium, Providencia, Staphylococcus,* and *Streptomyces.*

Bacteriophage Genetic Exchange

Viral genetics can be studied effectively by the straightforward process of arranging the virus/cell ratio so that the same cell is infected by more than one virus particle at approximately the same time. Assuming that the two viruses are genetically distinguishable, selection is applied to prevent parental-type phage particles from successfully completing an infection. Under these conditions, only cells in which phages carrying recombinant DNA have been produced yield progeny virus particles. The resulting virions are tested for phenotype, and recombination frequency is calculated in the same manner as for bacteria. For a more extensive discussion of this subject, see Chapter 4.

Nomenclature

Prior to the adoption of rules proposed by Demerec and co-workers in 1966, the nomenclature of bacterial genetics had developed in a somewhat haphazard fashion. The rules are intended to bring the nomenclature more in line with the conventions of eukaryotic genetics. The key provision is that each observable trait is given a three-letter, italicized symbol that is an abbreviation of a mnemonic. For example, a mutation that affects proline biosynthesis is designated *pro,* and a mutation that affects the utilization of proline as a carbon source is designated *put*. In many cases discrete genetic loci can be shown to affect the same phenotype. These loci are differentiated by assigning capital letters, e.g., *proA, proB, proC.* As each new mutation is isolated, it is assigned a unique allele number that identifies it in bacterial pedigrees. An example of a complete notation for one mutation is *proA52*. When it is not certain if more than one genetic

Table 2-1. Some of the better known genetic stock centers

E. coli Genetic Stock Center, Department of Human Genetics, Yale University School of Medicine, New Haven, Connecticut
S. typhimurium Genetic Stock Center, Department of Biology, University of Calgary, Alberta, Canada
Bacillus Genetic Stock Center, Department of Microbiology, The Ohio State University, Columbus, Ohio

locus can affect a particular trait or the precise locus is not yet known, the capital letter is replaced by a hyphen (e.g., *pro-106*). Some of the genetic stock centers that serve to coordinate naming and numbering of mutations are listed in Table 2-1. The commonly used symbols are given in Table 2-2.

Table 2-2. Frequently encountered genotype abbreviations for bacterial genetics[a]

Abbreviation	Phenotype
ace	Acetate utilization
ade	Adenine requirement
ala	Alanine requirement
ara	Arabinose utilization
arg	Arginine requirement
aro	Aromatic amino acid requirement
asn	Asparagine requirement
asp	Aspartic acid requirement
azi	Azide resistance
chl	Chlorate resistance
cys	Cysteine requirement
cyt	Cytosine requirement
fla	Flagella biosynthesis
gal	Galactose utilization
glt	Glutamic acid requirement
gln	Glutamine requirement
glp	Glycerophosphate utilization
gly	Glycine requirement
gua	Guanine requirement
his	Histidine requirement
hut	Histidine utilization
ile	Isoleucine requirement
ilv	Isoleucine + valine requirement
lac	Lactose utilization
leu	Leucine requirement
lys	Lysine requirement
mal	Maltose utilization
man	Mannose utilization

(*continued on following page*)

Table 2-2. *Continued*

Abbreviation	Phenotype
mel	Melibiose utilization
met	Methionine requirement
mtl	Mannitol utilization
nal	Nalidixic acid sensitivity
pan	Pantothenate requirement
phe	Phenylalanine requirement
pho	Alkaline phosphatase activity
pro	Proline requirement
pts	Phosphotransferase system
pur	Purine biosynthesis
put	Proline utilization
pyr	Pyrimidine biosynthesis
rec	Recombination proficiency
rha	Rhamnose utilization
rif	Rifampin resistance
rpo	RNA polymerase activity
rpsE	Spectinomycin resistance (ribosomal protein, small subunit)
rpsL	Streptomycin resistance
ser	Serine requirement
spo	Spore formation (*Bacillus*); magic spot production (*E. coli*)
str	Streptomycin resistance = *rpsL*
sup	Suppressor (usually a tRNA)
thi	Thiamine (vitamin B_1) requirement
thy	Thymine requirement
ton	Phage T1 resistance
trp	Tryptophan requirement
tsx	Phage T6 resistance
tyr	Tyrosine requirement
ura	Uracil requirement
uvr	Ultraviolet radiation sensitivity
val	Valine requirement
xyl	Xylose utilization

[a]Occasionally one or more of these abbreviations is printed with "Δ" as a prefix to indicate that the corresponding DNA is missing (deleted) from the chromosome.

The rules in the preceding paragraph concern the **genotype** of an organism (the total catalog of the genetic capabilities of an organism, regardless of whether they are presently in use). A convention has developed in the literature that when referring to the phenotype of an organism the same three-letter abbreviation is used, except that it is not italicized and the first letter is capitalized. It is possible, therefore, to talk about the

"Pro" phenotype of an organism, so that a strain carrying *proA52* would be phenotypically Pro$^-$ (i.e., unable to synthesize the amino acid proline).

There is one holdover from the old genetic nomenclature that may still be found in the contemporary literature and can confuse students. It has to do with suppressor mutations (see Chapter 3). Current nomenclature designates a strain not carrying such a mutation as *sup*$^+$ and a strain that does carry the mutations as *sup*$^-$. In the old system the terminology was exactly reversed: *sup*$^+$ = su$^-$ and *sup*$^-$ = su$^+$. Sometimes bacterial strains that are thought to be free of suppressor mutations are described as being su^0 or *sup*^{0}.

Analytic Techniques for DNA Molecules

One of the major changes in genetics over the past decade has been the implementation of methods that permit a geneticist to work directly with the DNA molecules under study. These techniques have rapidly become so fundamental that a basic knowledge of them is essential for understanding most contemporary work.

DNA Sequence Analysis

It is now comparatively easy to determine the nucleotide sequence of a DNA molecule. There are two primary methods for this determination, one developed by Sanger and the other by Maxam and Gilbert. Both are based on the principle of generating from a DNA molecule a set of labeled oligonucleotides of random length but with a specific base at the 3′-end of the fragment. The individual fragments are then separated by polyacrylamide gel electrophoresis (PAGE).

With this process a very thin gel of polyacrylamide is prepared, a mixture of oligonucleotides from either procedure is layered on the top, and a potential gradient of several thousand volts is applied across the gel (Fig. 2-1). The oligonucleotides migrate toward the anode at a rate inversely proportional to their size. By using a gel of appropriate composition, it is possible to separate olignucleotides that differ in length by only one base. Detection of the DNA bands within the gel is usually by means of autoradiography with either ^{32}P or ^{35}S as the label.

Obviously, for the DNA sequencing technique to be interpretable, only single-stranded DNA can be used as a source of DNA fragments. Otherwise it would not be possible to tell which fragments came from which strand. Where the Maxam-Gilbert and Sanger methods differ is in the method used to generate the single-stranded oligonucleotide fragments. Unless you understand how the fragments are generated, it is not possible to interpret the polyacrylamide gel patterns.

In the Sanger, or dideoxy, technique, the fragments are the result of

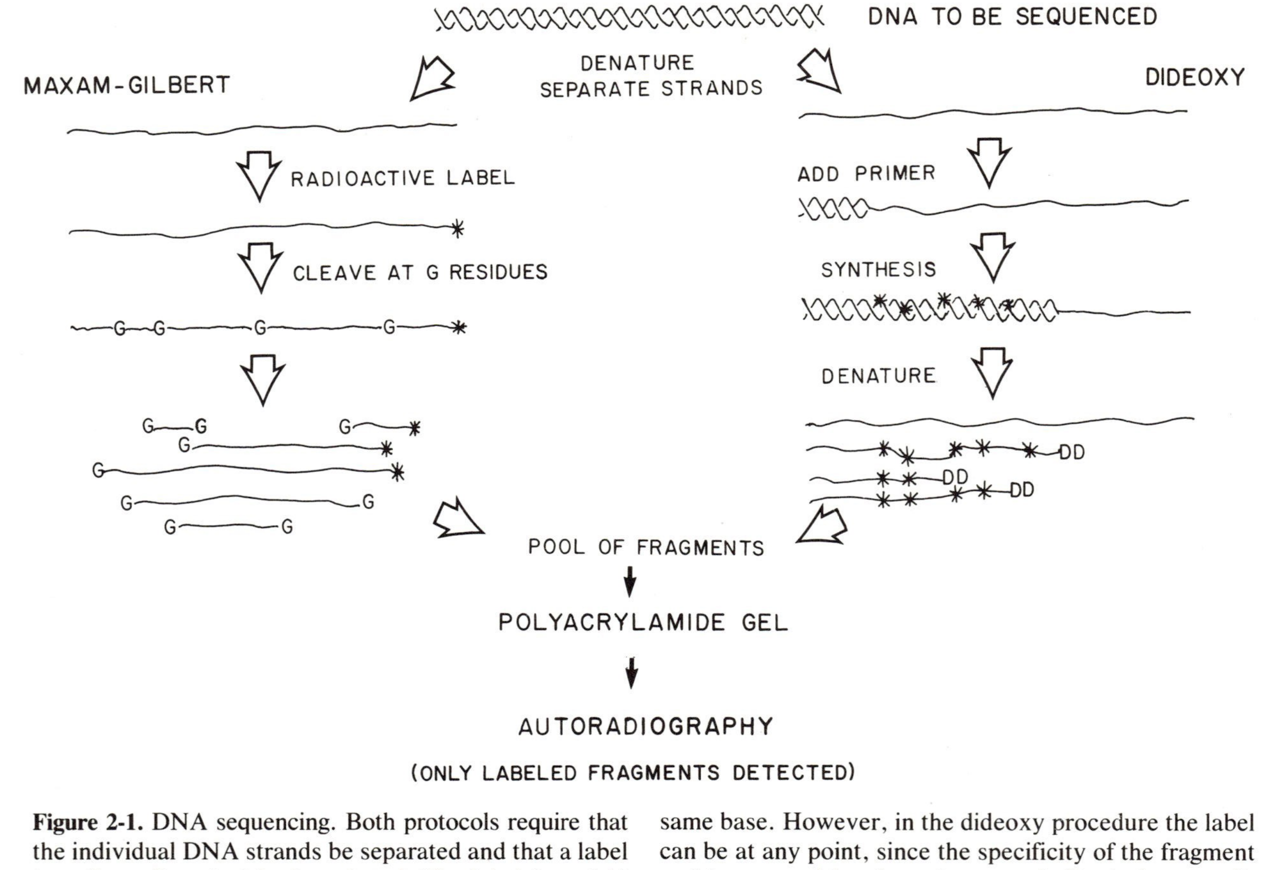

Figure 2-1. DNA sequencing. Both protocols require that the individual DNA strands be separated and that a label (usually radioactive) be introduced. The label (asterisk) must be at the end of the fragment for the Maxam-Gilbert procedure to guarantee that all fragments begin at the same base. However, in the dideoxy procedure the label can be at any point, since the specificity of the fragment end is ensured by the primer used. Typical autoradiograms produced by these procedures are shown in Figure 2-3.

the interruption of DNA synthesis by insertion of an inappropriate base. Referring back to Fig. 1-10, it can be seen that for a base to be accommodated within the DNA structure it must have hydroxyl groups on both its 3′- and 5′-carbons. A dideoxy base (Fig. 2-2) has an available 5′-hydroxyl group but not one on the 3′-carbon. Therefore whenever such a base is incorporated into a DNA molecule, it becomes the last base in the fragment. In practice, a shortened protein chain from DNA polymerase I known as the Klenow fragment is used to replicate purified single strands of DNA. The Klenow fragment has lost all error-correcting ability and therefore inserts an incorrect base but cannot remove it.

The point at which replication begins is controlled by the base sequence of the primer used. Four reaction mixtures are run, each containing trace amounts of a different dideoxy base. Whenever the dideoxy base is inserted into the growing DNA chain, all further synthesis must stop. For example, when the concentration of dideoxy adenosine is properly adjusted, a series of fragments of all possible lengths ending in adenine is generated. The various fragments can be made radioactive by adding one or more appropriately labeled nucleoside triphosphates to the reaction mixture so that they are incorporated into the oligonucleotides.

When the four reaction mixtures are subjected to electrophoresis, a ladder of individual bands is produced (Fig. 2-3). Because each lane on the gel corresponds to fragments ending in a particular base, the DNA sequence is determined by beginning with the smallest band and reading up the gel, noting which lane has the next smallest fragment.

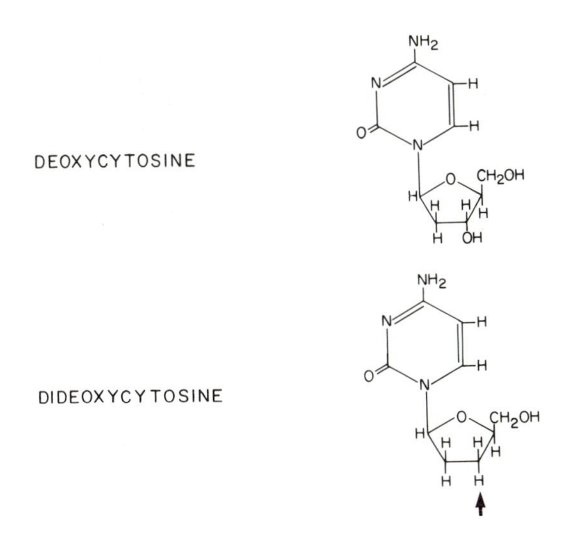

Figure 2-2. A dideoxy base compared with the normal deoxy base. The arrow indicates the position of the missing oxygen atom.

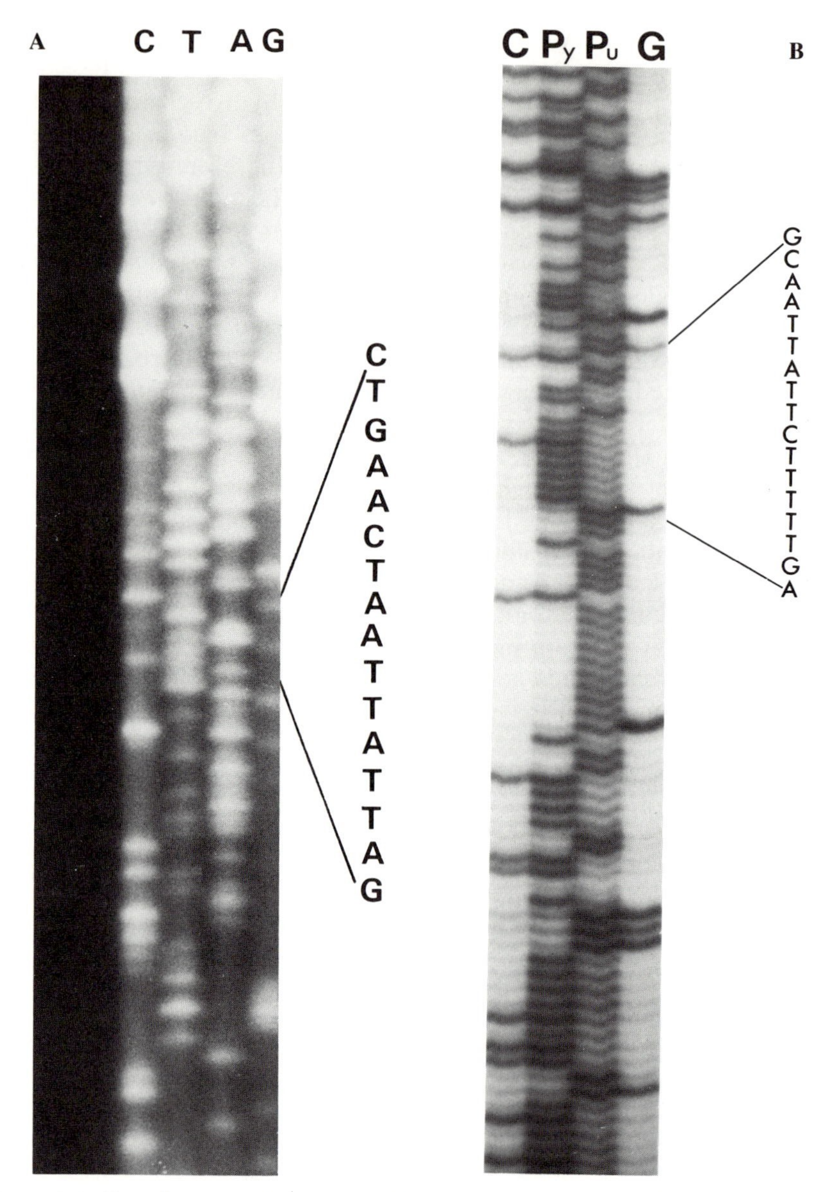

Figure 2-3. Sample autoradiograms of DNA sequencing reactions. Base abbreviations are as in Figure 1-10. (**A**) The pattern produced by the Sanger method. (**B**) Pattern produced by the Maxam-Gilbert method. **A:** Courtesy of L.G. Pearce, The Genetics Center, Scottsdale, Arizona. **B:** Courtesy of Dr. E. Goldstein, Department of Zoology, Arizona State University.

In the Maxam-Gilbert method, the double-stranded DNA is first made radioactive at the 5′- or 3′-end using an appropriate enzyme. The molecule is then denatured as outlined below to give single strands; and the strands are separated, usually by electrophoresis. The physical basis for the separation is unclear, but the technique works and yields two strands of identical length but complementary sequence. Choosing one of the two separated strands, chemical reactions are used to randomly cleave specific bonds within the DNA molecules. The four standard reactions cleave the DNA either at guanine residues, all purine residues, cytosine residues, or all pyrimidine residues.

Conditions of the reaction are controlled so that complete cleavage is not obtained, and once again a series of fragments of different lengths is obtained. Because the fragments are visualized by autoradiography, only those DNA fragments that include the labeled end are detectable. Separation of the fragments is achieved via the same electrophoretic technique used in the Sanger method.

The reading of the sequence is a trifle more complex than in the case of the dideoxy gels (Fig. 2-3) because some of the reactions cut at two bases. Thus a band present in both the (C) and (C + T) lanes indicates that a cytosine residue is present, but a band present only in the (C + T) lane indicates that a thymine residue is present. One advantage to the Maxam-Gilbert procedure is that the complementary strand can also be sequenced and used to confirm the sequence deduced for the original strand. The dideoxy method would require that the sequence be determined and a primer prepared that is the same as the 3′-end of the sequenced DNA. This primer could then be used to initiate DNA replication of the complementary strand.

Custom Synthesis of DNA

Microprocessor-controlled instruments are now available that can prepare short pieces of single-strand DNA of predetermined sequence. The chemistry of the system is somewhat complicated, but the basic outline is clear (Fig. 2-4). A small bead of material has attached to it the first base of the desired sequence. A solution containing a highly reactive form of the next base is then added and given time to react. The bead serves as an anchor to allow complete washing before the next base is added. One complete step can be carried out in about 15 minutes under optimal conditions.

One limitation to the process is the completeness of addition of each base. To the extent that the chemical reaction is not 100% complete, some DNA molecules have an incorrect base sequence. Obviously, the proportion of incorrect DNA molecules increases with the length of the DNA synthesized. Thus for practical reasons most custom DNA molecules have a length of about 100 bases or less. Another limitation to the technique is the price. As of this writing DNA synthesis instruments are retailing

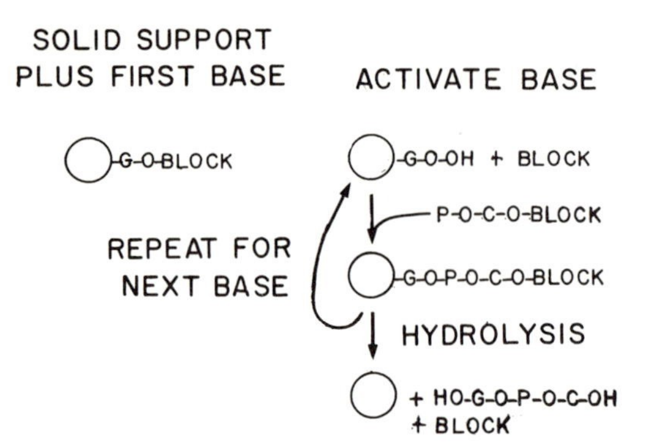

Figure 2-4. Custom DNA synthesis. The reactions are carried out in a small tube with the product held in place via a chemical linkage to an inert material. Reagents are passed sequentially through the tube. To prevent unwanted side reactions, each new base has its 3′-hydroxyl group chemically blocked. This block must be removed before the next base can be added. The final hydrolysis step removes the last blocking group and cleaves the chemical bond to the solid support so that the single strand of DNA can be isolated in pure form.

in the $50,000 price range, and one batch of chemicals may cost several hundred dollars.

Footprinting

The technology used for DNA sequencing can also be of value in experiments to determine precisely where a particular molecule binds to DNA. The general idea is that if a protein is bound to a specific DNA base sequence, it effectively covers that region and protects the DNA against degradative enzymes. If pure end-labeled DNA is randomly digested with DNase I, an endonuclease that makes double-strand cuts in the molecule, a family of DNA fragments of all possible lengths is produced. However, if the digestion is carried out on DNA to which protein is bound, certain fragment lengths are not produced because the endonuclease is prevented from reaching the DNA by the bound protein. When the fragments are later separated by electrophoresis on a polyacrylamide gel, certain bands are missing entirely. If sequencing samples are run on the same gel, the protected base sequence can be read directly (Fig. 2-5).

Detection of Specific DNA Sequences

There are many instances in genetics when it would be useful to be able to identify DNA sequences that are the same or similar to a particular reference sequence. Usually it is not feasible to completely sequence all of the candidate DNA molecules, and therefore some sort of rapid screening procedure is required. The methods for such screening take advantage

Figure 2-5. Footprinting. The gel shows the results of an experiment designed to demonstrate the binding of a protein to the regulatory region for *recA* (see Chapter 13). DNA fragments were made radioactive at their 5′-end prior to the experiment. *Lane 1:* Fragment after partial degradation by the restriction enzyme *Alu*I (see below). *Lane 2:* Maxam-Gilbert reaction which yields all possible fragments ending in a purine. *Lane 3:* Partial digestion with DNase I in the absence of added protein. *Lane 4:* A similar digestion in the presence of the *lexA* protein. The *lexA* protein was bound to the region of the *recA* DNA just upstream from the translation start site. From Little, J.W., Mount, D.W., Yanisch-Perron, C.R. (1981). Purified *lexA* protein is a repressor of the *recA* and *lexA* genes. Proceedings of the National Academy of Sciences of the United States of America 78:4199–4203.

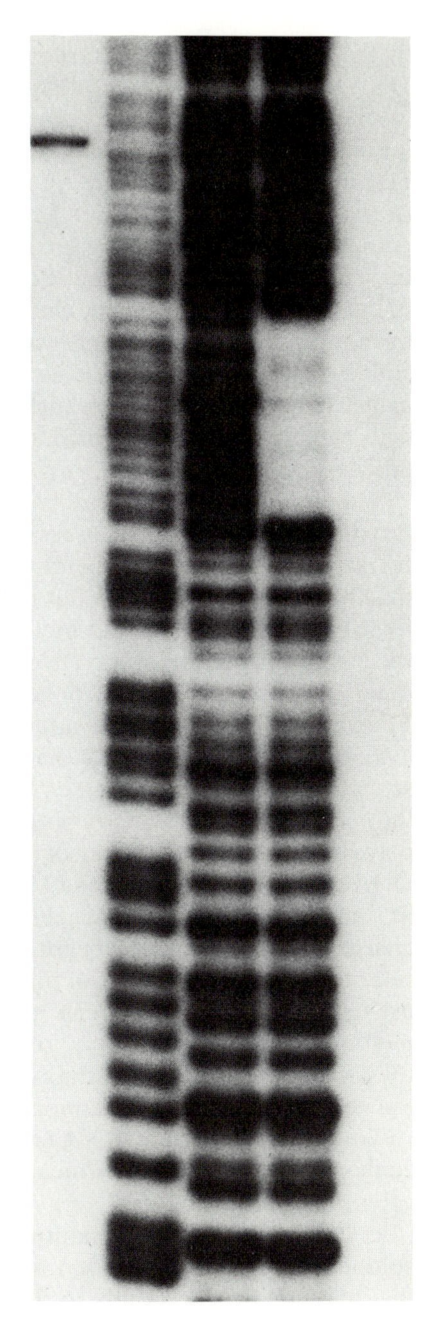

of the natural complementarity of nucleic acids to form **heteroduplexes**, DNA duplexes comprised of strands of DNA from different sources.

Natural DNA molecules contain two perfectly complementary DNA strands (each adenine is paired to thymine, and each guanine is paired to cytosine). If the temperature of a solution of DNA is raised to near 100°C or the pH is raised to about 12.3, the hydrogen bonds that link the two complementary strands are broken, and the individual strands **denature,** or separate, but the covalent bonds remain intact. If the temperature or pH is then rapidly reduced, the hydrogen bonds do not have time to re-form properly, and the single strands of DNA remain in the solution but coiled on themselves. This state is not stable, and the DNA strands gradually attempt to **rehybridize** (make as many hydrogen bonds as possible with neighboring molecules). However, at room temperature it is a slow process.

Maximum hydrogen bonding is obtained when the DNA is held at approximately 60°C or kept in a buffer with a mild denaturing agent, e.g., formamide. Either of these conditions prevents the formation of mismatches in the double helix but is not sufficient to break correctly formed hydrogen bonds. The exact incubation procedure to be used depends on the base composition of the DNA, as G:C base pairs have three hydrogen bonds and are therefore stronger than A:T pairs, which have only two. Thus an A:T-rich region denatures more easily than a G:C-rich one.

Large DNA molecules, either double- or single-stranded, can also be separated on the basis of size by electrophoresis. However, in this case the gel is not polyacrylamide but, rather, agarose, a high-molecular-weight polysaccharide that in dilute solution forms a porous matrix through which DNA fragments can pass. The DNA molecules to be examined are placed in wells at one end of the gel slab, and an electric current is applied. The charged DNA molecules begin to migrate through the agarose but are retarded by the latticework of polysaccharide. The result is that the smallest molecules (those that can most easily slip through the latticework) migrate the fastest. Typically, the bands of DNA in the gel are visualized by staining the gel with the fluorescent dye ethidium bromide and viewing the gel under ultraviolet illumination.

Testing for particular DNA sequences within the separated DNA molecules is generally done by means of **blotting**, by which the DNA from a gel is transferred (blotted) to a solid support such as a nylon membrane (Fig. 2-6) and the DNA denatured in place. The blot is then soaked in a solution containing a single-stranded nucleic acid probe under conditions that allow hydrogen bonds to form. If the corresponding sequences are present, the probe hybridizes to the blotted DNA. The probe is usually made radioactive or linked to biotin. Tests for the presence of the probe are easily performed. These procedures include autoradiography or addition of avidin, a chemical that binds very strongly to biotin. Usually the avidin is conjugated to an enzyme molecule that can be localized by addition of a substrate that yields a colored product. A typical blot prepared

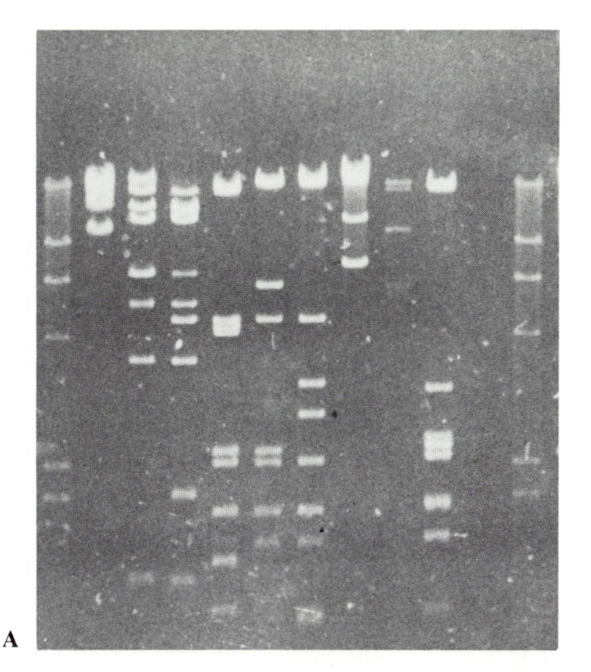

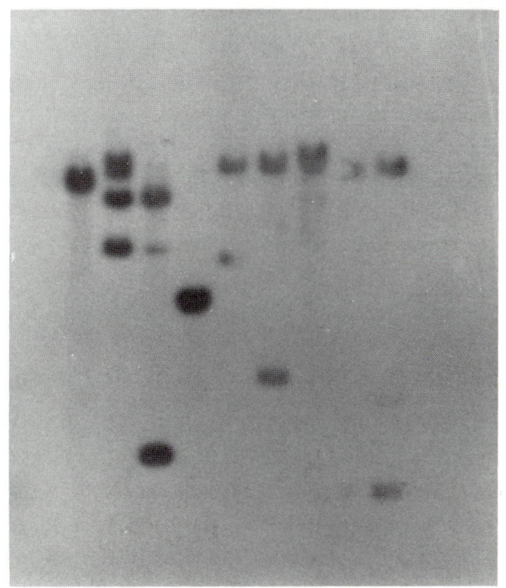

Figure 2-6. Southern blotting. (**A**) Agarose gel stained with ethidium bromide and viewed under ultraviolet radiation. The two outermost lanes are pure phage λ DNA digested with *Hin*DIII and used as size standards. The other lanes are a human DNA clone in a phage λ vector that has been digested with various restriction enzymes singly and in pairs. (**B**) An autoradiogram of a blot of the gel from **A** probed with radioactive human DNA. The bands from the original gel that show up darkly on the blot are the human DNA, and the bands that do not appear on the blot are the vector DNA. Courtesy of L.G. Pearce, The Genetics Center, Scottsdale, Arizona.

using a radioactive probe is shown in Fig. 2-6. This technique was invented by Earl Southern using a DNA gel and is therefore usually referred to as a Southern blot. If an RNA gel is used, the procedure is known as Northern blotting, although no one by that name was involved in developing the procedure.

Molecules of any size can be made to form heteroduplexes. Whenever large molecules such as entire plasmids or viral genomes are used, the resulting heteroduplexes can be visualized with a transmission electron microscope. A basic protein such as cytochrome C is employed to neutralize the charge on the DNA molecule and allow it to lay out on the surface of the support film. Formamide is used to stabilize any single-stranded regions that may occur. An electron-dense material such as uranyl acetate is applied to stain the DNA so that it can be easily seen. For a set of electron micrographs showing some examples of heteroduplex molecules and giving their interpretation, see Fig. 5-1.

DNA Splicing

One of the basic tenets of genetics is that indiscriminate exchange of genetic information is disadvantageous to the species. In eukaryotic cells problems with chromosome pairing during mitosis and meiosis often prevent cells that have acquired foreign chromosomes from surviving. However, because segregation of the nucleoid in prokaryotic cells requires no such elaborate mechanism, other strategies must come into play. In particular, many bacterial cells and their viruses use a system of restriction and modification to tag their own DNA and disrupt any foreign DNA that may be present.

Enzymology of DNA Restriction and Modification

As implied by their name, DNA restriction and modification systems actually consist of two enzymatic activities that may be on the same or different protein molecules. Each enzymatic activity is centered on a **recognition sequence**, a series of more or less specific bases to which the protein molecule(s) binds, but there the similarity ends. A **modification enzyme** makes a chemical alteration, usually adding a methyl group, to a specific site or sites within the recognition sequence. It acts immediately after the replication fork has passed, meaning that its normal substrate is a DNA duplex that has one strand methylated and the other not. A **restriction enzyme,** on the other hand, is stimulated by the presence of an unmodified recognition sequence to act as an **endonuclease** (an enzyme that cuts a phosphodiester bond within a DNA molecule) and cut the DNA one or more times. The effect of this cutting is to prevent or reduce the ability of foreign bacterial DNA to incorporate itself into the bacterial chromosome and to prevent or reduce the establishment of an infection by an invading viral DNA.

Restriction endonucleases are designated by italicized letters representing the genus and species name of the producing organism. A fourth letter may be added to indicate the specific strain that produces the enzyme. In cases where an organism produces more than one enzyme, the enzymes are differentiated by Roman numerals. Examples of this nomenclature and the types of base modifications that have been identified are given in Table 2-3.

Restriction enzymes actually fall into three broad classes. Type I enzymes are exemplified by the *E. coli* B and K restriction and modification systems. The individual proteins are combination restriction and modification enzymes (called *Eco*B and *Eco*K, respectively) that are composed of nonidentical subunits and have a high molecular weight (about 300,000 daltons). They require ATP, magnesium ions, and *S*-adenosylmethionine (SAM) in order to function properly. The enzyme binds to both the recognition sequence and some neighboring bases in the presence of SAM. If both strands of DNA are methylated, nothing happens. If only one strand is methylated (a recently replicated DNA duplex), the methylase activity is stimulated. If the DNA is unmethylated (foreign DNA), the enzyme remains bound to the recognition sequence, and one end of the enzyme uses energy from ATP to form a gradually enlarging loop of DNA by moving along the DNA to which it is bound (Fig. 2-7). When the loop reaches a size between 1000 and 6000 bp, the enzyme produces a gap of about 75 nucleotides in one strand of the duplex, releasing the excised bases as single nucleotides. While the first enzyme continues to hold the gapped DNA in a looped configuration, a second enzyme molecule attaches itself and introduces a second cut, but in the other (intact) DNA strand. The result of all this activity is a double-strand break in the DNA at a distance of about 1000 nucleotides from the original recognition sequence. Because the double-strand break is seemingly randomly located at any suitably distant point, the process is considered nonspecific.

In contrast to the mode of action just described, the type II restriction endonucleases always make their cuts at a definite point relative to the recognition sequence and always have separate methylase and restriction proteins. Because of their specificity of action, it has been easier to obtain information about the mode of action of these enzymes, some of which is summarized in Table 2-3. (There are more than 600 known restriction endonucleases, so only a small selection can be presented.) As might be expected from the fact that the recognition sequences are generally only four to seven bases in length, enzymes recognizing the same sequences have been isolated from different organisms. These duplicate enzymes are given the name **isoschizomers**, and the first three-letter abbreviation assigned is considered to have priority and is to be used in preference to any other.

Table 2-3. Properties of some commonly encountered restriction endonucleases[a]

Enzyme	Source	DNA sequence recognized	No. of cleavage sites	
			Phage λ	Simian virus 40
Type I				
*Eco*B	*E. coli* B	TGA*(N_8)TGCT	9	1
Type II				
Symmetric				
AluI	*Arthrobacter luteus*	AG/C*T	143	34
*Hae*III	*Hemophilus aegyptius*	GG/C*C	149	18
MboI	*Moraxella bovis*	/GA*TC	116	8
*Bsm*I	*Bacillus stearothermophilus* NUB36	5′GAATGCN/N3′ 3′CTTAC/GNN5′	46	4
*Bam*HI	*Bacillus amyloliquefaciens* H	G/GATC*C	5	1

*Eco*RI	*Escherichia coli* RY13	* * G/AATTC	5	1
*Pst*I	*Providencia stuartii*	* * CTGCA/G	28	2
*Bgl*I	*Bacillus globigii*	* GCC(N_4)/NGGC	29	1
*Not*I	*Nocardia otidis-caviarum*	* GC/GGCCGC	0	0
Degenerate symmetric				
*Hae*II	*Hemophilus aegyptius*	* PuGCGC/Py	48	1
Asymmetric				
*Hga*I	*Hemophilus gallinarum*	* 5′GACGC(N_5)/3′ 3′CTGCG(N_{10})/5′	102	0
Type III				
*Eco*P1	*Escherichia coli* (P1 lysogen)	* * * 5′AGACC3′ 3′TCTGG5′	49	4

[a]The recognition sequences are given as a series of letters beginning at the 5′ side in which A, T, C, and G represent the first letters of the corresponding bases, Py represents any pyrimidine, Pu represents any purine, and N represents any nucleotide. The site of methylation (if known) is indicated by an asterisk, and the point of cleavage is indicated by a slash. If only one strand is indicated, the position of the cut in the complementary strand is symmetric. Type II enzymes always cut the DNA at a precise site relative to the recognition sequence, whereas types I and III enzymes do not (see the text). Additional information on the properties of restriction enzymes may be obtained from Kessler and Holtke (1986).

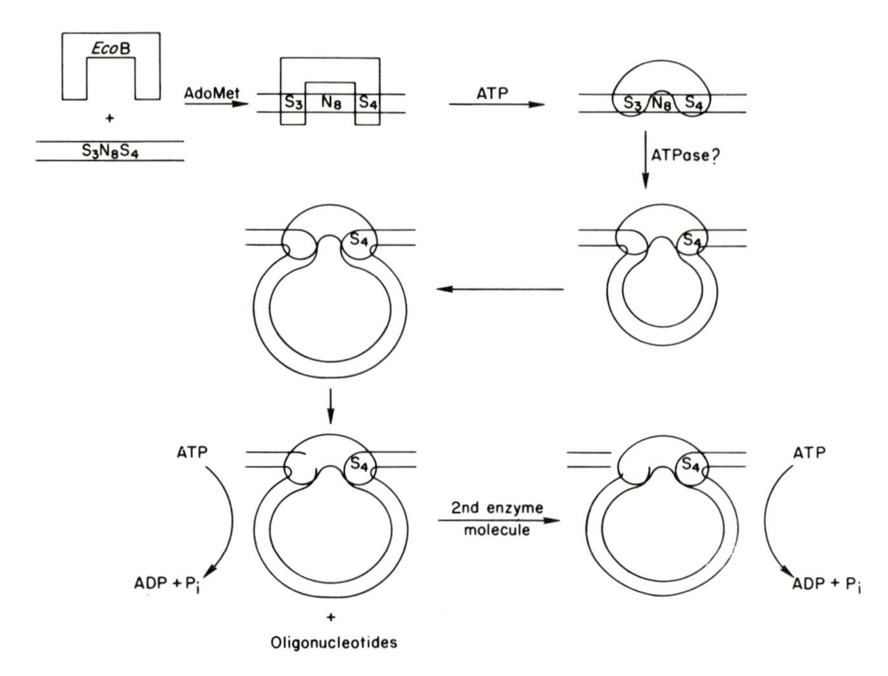

Figure 2-7. Hypothetical sequence of events by which *Eco*B cleaves unmodified DNA. The horizontal lines represent a portion of unmodified DNA containing a composite binding site for *Eco*B. This site consists of a trimeric segment and a tetrameric segment separated by eight nonspecific base pairs. In the presence of *S*-adenosyl-L-methionine, *Eco*B binds specifically to the trimer and tetramer. The addition of ATP causes a change in the conformation of the enzyme and a weakening of the attachment to the trimeric binding site. The energy of ATP is then used to create a loop of DNA at least 1000 bp long. When the loop is of suitable size, a random 75-bp gap is produced in one strand. The DNA and enzyme remain attached in this configuration until a second enzyme molecule binds to the complex and cleaves the other strand. At this time both enzyme molecules are released. From Rosamund, J., Endlich, B., Linn, S. (1979). Electron microscopic studies of the mechanism of action of the restriction endonuclease of *E. coli* B. Journal of Molecular Biology 129:619–635. © Academic Press Inc. (London) Ltd.

The specific restriction endonuclease activities have been divided into three subgroups based on the nature of the base sequence that each enzyme recognizes. In the case of *Eco*RI (Table 2-3), the sequence is considered symmetric because it has a twofold rotation symmetry. If the base sequence of both strands is written out, the symmetry can be more readily seen:

5′ GAATTC 3′
3′ CTTAAG 5′

Note that the sequence is a **palindrome**, one that reads the same regardless of the direction in which it is read. If the recognition sequence contains an odd number of bases, the symmetry is maintained if the center base is nonspecific. Symmetric enzymes are the most prevalent class, but asymmetric enzymes are known.

Type III enzymes again have both enzymatic activities in one protein. They make their cuts at a definite position some distance from the recognition site. An example is *Eco*P1, the restriction enzyme produced by phage P1. In this case the recognition site is 5 bp, but the base sequence on the two strands of DNA is not the same when read from the two respective 5′-ends. The enzyme makes its cuts some 24 to 26 bases from the 3′-sides of the recognition site.

The various endonucleases can produce three types of fragment when they cut the DNA. If they cut in the middle of the recognition site (e.g., *Alu*I), all fragments have blunt ends. If they cut to one side or the other, each fragment has a single-strand tail. If the cut is left of center (e.g., *Bam*HI) the tails have 5′-ends, whereas if the cut is right of center (e.g., *Pst*I) the tails have 3′-ends. Note that if a symmetric enzyme produces single-strand tails, all of the DNA fragments have identical tails. These single-strand tails can act as "sticky ends" to join various fragments by hydrogen bonding. The same is not true for the asymmetric enzymes, which seem to cut at some distance from the recognition site. The mechanism of this asymmetric cutting is not known, but its effect is to generate unique ends for each fragment (e.g., all *Hga*I fragments have five-base single-strand tails of nonspecific sequence).

Although methylation of bases is protective in the typical restriction system discussed above, that is not always the case. For example, *Hpa*II and *Msp*I are isoschizomers in the sense that they recognize the same base sequence. However, *Hpa*II makes its cuts when the sequence is not methylated, and *Msp*I makes its cuts when the sequence is methylated. An additional example of this activity is seen in the case of *E. coli* K-12, whose *mcrA* and *mcrB* functions restrict 5-methylcytosine-containing DNA.

The Use of Restriction Fragments to Make New DNA Molecules

The terms **DNA splicing** and **gene splicing** refer to the construction of new types of DNA molecule from DNA fragments such as those produced by the symmetric restriction endonucleases. The process is also frequently described as the production of recombinant DNA. However, this type of DNA rearrangement is not the same as the recombination of classic genetics, and it seems clearer in a text to reserve the term "recombination" for the naturally occurring in vivo process and to use the new term "gene splicing" for the in vitro process.

Gene splicing is a new technology, and a complete discussion requires information that has yet to be presented in this book. At the same time, the general outline of the procedure is readily understood and can be comprehended by the beginning student. Therefore a brief discussion is presented here, and a more thoroughgoing treatment can be found in Chapter 14.

The purpose of splicing fragments of DNA together is to produce a DNA molecule coding for a combination of traits that would be difficult or impossible to obtain by more conventional genetic techniques. If the new molecule is to be useful, it should be able to replicate itself (i.e., it must be a plasmid). Therefore the first step in any DNA splicing procedure is selection of a suitable plasmid **vector** (carrier) into which one or more DNA fragments can be inserted. Typical vectors are lambda phage (see Chapter 6), ColE1 (see Chapter 10), and various derivatives of R plasmids (see Chapter 10).

With the simplest form of DNA splicing, the vector DNA and the DNA to be spliced are both cut with the same restriction endonuclease, one that generates single-strand tails or "overhangs." Upon mixing the two sets of DNA fragments together, the single-strand regions can pair, reforming the original molecules or creating new combinations (Fig. 2-8). The new molecules can be stably joined by the enzyme DNA ligase in the presence of ATP. This step completes the actual splicing process. When the spliced DNA is introduced into a cell in which the vector can replicate, it reproduces and transmits itself to all of the daughter cells. Because a group of cells that are descended from one original cell is said to be a clone, the process of obtaining a set of cells that carries spliced DNA of a particular sequence is called **cloning** of the DNA.

A successful splicing reaction produces a new DNA molecule that is larger than either of the original DNA molecules. This procedure can be demonstrated by means of standard agarose gel electrophoresis as described above if the DNA molecules are not too large. Otherwise a more elaborate system such as pulsed field gradient gel electrophoresis must be employed. A sample of such a gel is shown in Fig. 2-9.

Summary

Natural genetic transfer in the prokaryotes is accomplished by genetic transformation, transduction, or conjugation. In the laboratory, protoplast fusion has also been shown to be possible. These techniques make possible a true study of genetics in bacteria. The nomenclature used to describe genetic traits is similar to that used for eukaryotes and is intended to apply to all bacteria.

Modern molecular biology has contributed some techniques that have become fundamental for the bacterial geneticist. Among them are the abil-

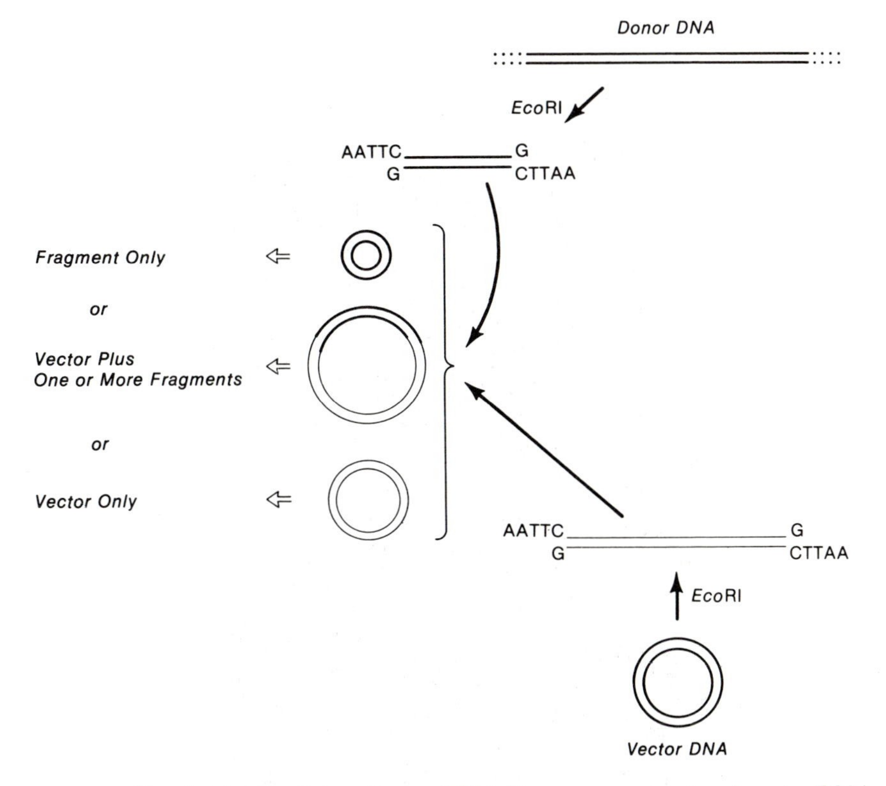

Figure 2-8. Simple DNA splicing. Donor DNA (from any source) and vector DNA (a plasmid) are cut into pieces with the same restriction endonuclease. The ends of each fragment have the same sequence of single-strand DNA and are therefore complementary. Double-strand circles of DNA can be formed by joining the complementary ends of one or more fragments. Only if the vector DNA is included will the new circular molecule be able to replicate.

ity to determine the sequence of a DNA molecule and to use the technique of heteroduplex formation to identify specific sequences within a DNA molecule. The most far-reaching discovery is that of restriction and modification enzymes within the bacteria. These enzymes normally maintain genetic exclusivity within a group of closely related organisms. However, because certain of the restriction enzymes make specific and reproducible cuts within DNA molecules, it is possible to create families of DNA fragments, all of which have identical single-strand ends. These fragments can then be reassorted in any desired order. If one or more of the fragments derives from a plasmid, the resulting DNA construct is capable of self-replication and can be transmitted along a cell line.

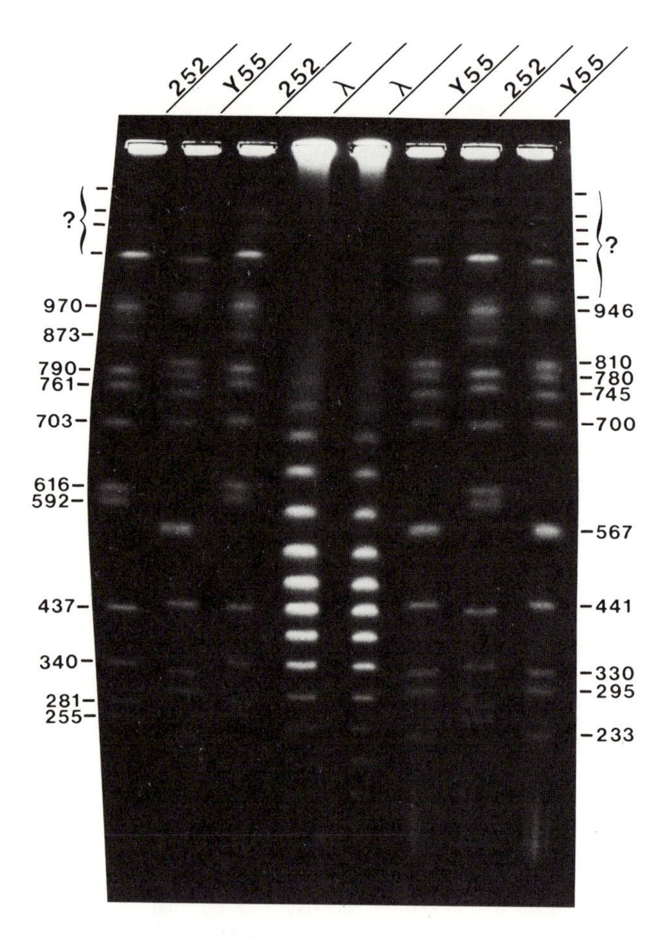

Figure 2-9. Pulsed field agarose gel. Conventional agarose gels such as that shown in Figure 2-6A have difficulty separating DNA fragments larger than about 20 kb. If the gel is run using a pulsed field in which the direction of the voltage gradient is switched by rotating the gel through a 90° angle, larger fragments have difficulty reorienting themselves in the agarose and move more slowly than smaller fragments. The gel shown here presents individual chromosomes from two strains of yeast, 252 and Y55, separated by pulsed field gel electrophoresis. The sizes of the chromosomes in kb are indicated where they are known. Note the size variation between strains. The lanes labeled λ contain DNA molecules extracted from phage λ and joined end to end to make a series of size standards of predictable lengths. Reprinted by permission of the publisher from Gemmill, R.M., Coyle-Morris, J.F., McPeek, F.D., Jr., Ware-Uribe, L.F., Hecht, F. (1987). Construction of long-range restriction maps in human DNA using pulsed field gel electrophoresis. Gene Analysis Techniques 4:119–132. © 1987 by Elsevier Science Publishing Co., Inc.

References

General

Adams, R.L.P., Burdon, R.H. (1985). Molecular Biology of DNA Methylation. New York: Springer-Verlag.

Demerec, M., Adelberg, E.A., Clark, A.J., Hartman, P.E. (1966). A proposal for a uniform nomenclature in bacterial genetics. Genetics 54:61–76.

Gilbert, W. (1981). DNA sequencing and gene structure. Science 214:1305–1312.

Kessler, C., Holtke, H-J. (1986). Specificity of restriction endonucleases and methylases—a review (edition 2). Gene 47:1–153.

Moores, J.C. (1987). Current approaches to DNA sequencing. Analytical Biochemistry 163:1–8.

Neumann, R., Rudloff, P., Eggers, H.J. (1986). Biotinylated DNA probes: sensitivity and applications. Naturwissenschaften 73:553–555.

Specialized

Akamatsu, T., Sekiguchi, J. (1987). Genetic mapping by means of protoplast fusion in *Bacillus subtilis*. Molecular and General Genetics 208:254–262.

Lévi-Meyrueis, C., Sanchez-Rivas, C. (1984). Complementation and genetic inactivation: two alternative mechanisms leading to prototrophy in diploid bacterial clones. Molecular and General Genetics 196:488–493.

Raleigh, E.A., Wilson, G. (1986). *Escherichia coli* K-12 restricts DNA containing 5-methylcytosine. Proceedings of the National Academy of Sciences of the United States of America 83:9070–9074.

Sanchez-Rivas, C., Lévi-Meyrueis, C., Lazard-Monier, F., Schaeffer, P. (1982). Diploid state of phenotypically recombinant progeny arising after protoplast fusion in *Bacillus subtilis*. Molecular and General Genetics 188:272–278.

Chapter 3
Mutations and Mutagenesis

The first problem facing the early bacterial geneticist was to prove that bacteria did have inherited traits. The earliest presumption was that bacteria and other microorganisms were too small to have any phenotypic traits that could be studied. That concept was disabused by the work of Beadle and Tatum, who demonstrated that biochemical reactions could be used as phenotypic traits and developed the famous "one gene—one enzyme" hypothesis. There was, however, still one remaining area of uncertainty about the existence of bacterial genetics. Many workers thought that the hypothesis of Lamarck regarding the inheritance of acquired traits was true for bacteria even though it had already been disproved for the higher eukaryotes. The first task of the fledgling science of bacterial genetics was to prove that the same processes of mutation that had already been shown to occur in eukaryotes also occurred in prokaryotes.

Bacterial Variation

Fluctuation Test

The controversy focused on experiments concerning the acquisition of resistance to some sort of selective agent, usually a bacteriophage or an antibiotic such as streptomycin. Everyone agreed that if bacteria growing on the surface of an agar plate were treated with the selective agent most of the cells died, but a few resistant colonies were observed to grow. The descendents of these resistant colonies were also resistant, meaning that the change was stable and inherited (i.e., genetic).

The theoretical interpretation of these results was in considerable dispute. The Lamarckian theorists maintained that any cell in the culture had a small but finite probability of surviving the selective treatment. If a cell did survive, the acquired trait of resistance was passed onto all of the subsequent members of the clone. The opposing theory was one of preexisting mutations. Certain bacteria in a culture were naturally resistant to the selective agent because of a preexisting change in their genetic material (a mutation) that had occurred in the absence of the selective agent. Treatment with the selective agent did not cause the change but, rather, demonstrated its existence by eliminating sensitive cells.

An elegant theoretical and experimental analysis of this problem was prepared by Luria and Delbrück. They developed an experimental system, known as the **fluctuation test,** that not only greatly contributed to the resolution of the mutation question but still finds occasional use. The basic protocol was simple. A small number of cells taken from the same culture were inoculated into 10 to 100 tubes, each of which contained sterile medium. The tubes were incubated until the cells had reached the stationary phase of growth, and then either aliquots or the entire contents of each tube were placed under selective conditions by spreading them on agar that had been saturated with T1 phage; the number of resistant colonies was then counted.

Interpretation of the data is reasonably clear. If the selective treatment induces a change in the cells (acquired immunity), there should be no difference between applying the selective treatment to ten samples of the same culture or one sample from each of ten cultures. In both cases the number of treated cells is approximately the same, and therefore approximately the same number of resistant colonies would be expected. On the other hand, if the resistant colonies reflect the number of preexisting mutations in the culture, ten samples from the same culture should be similar, whereas samples from different cultures should be less similar because some cultures have mutations that occurred early and hence gave rise to a larger number of resistant progeny than others. The way to quantitate the difference is to measure the variance of the population, which is the square of the standard deviation (see Appendix).

Table 3-1 shows the type of data obtained by Luria and Delbrück. It can easily be seen that there is a dramatic difference in the variances of the two sample populations, which offers strong support for the theory of preexisting mutations. Note also the existence of certain **jackpot tubes,** tubes whose samples contained substantially more resistant cells than the average.

Plate Spreading

A more directly visible method of demonstrating the same type of result as that obtained by Luria and Delbrück was developed by Newcombe. The basic experiment consisted in taking several agar plates and spreading

Table 3-1. Results of a typical fluctuation test: number of resistant bacteria[a]

	Same tube	Different tubes
	46	30
	56	10
	52	40
	48	45
	65	183
	44	12
	49	173
	51	23
	56	57
	47	51
Mean	51.4	62
Variance	27	3498

[a]The values in each column indicate the number of T1-resistant bacteria observed per 0.05 ml sample. The total volume of the individual cultures was 10 ml.

Data are from experiment 11 of Luria, S.E., Delbrück, M. (1943). Mutations of bacteria from virus sensitivity to virus resistance. Genetics 28:491–511.

a uniform lawn of bacteria on each of them. After various periods of incubation, the plates were sprayed with a virulent phage, and resistant colonies were allowed to develop. However, before spraying, some of the plates had the bacteria on one-half of their surfaces spread around using a glass rod. If phage resistance were due to acquired immunity, the spreading should have no effect. On the other hand, if there were preexisting mutations, each of the preexisting mutants would give rise to a clone of cells that would be redistributed by the spreading. Therefore the spread side of the plate would be expected to have more resistant colonies than the unspread side. The actual results were fully in accord with this expectation. One set of six plates, for example, had a total of 28 colonies on the unspread sides but 353 on the spread sides.

Replica Plating

An entirely different type of experiment from the Lederbergs contributed evidence in favor of preexisting mutations and a valuable new technique for bacterial genetics. The technique is called **replica plating** and is a type of printing process (Fig. 3-1). A piece of sterile velvet or velveteen is laid across the top of a cylinder with the pile surface of the fabric facing up. A band holds the velvet tightly to the cylinder. A master plate containing

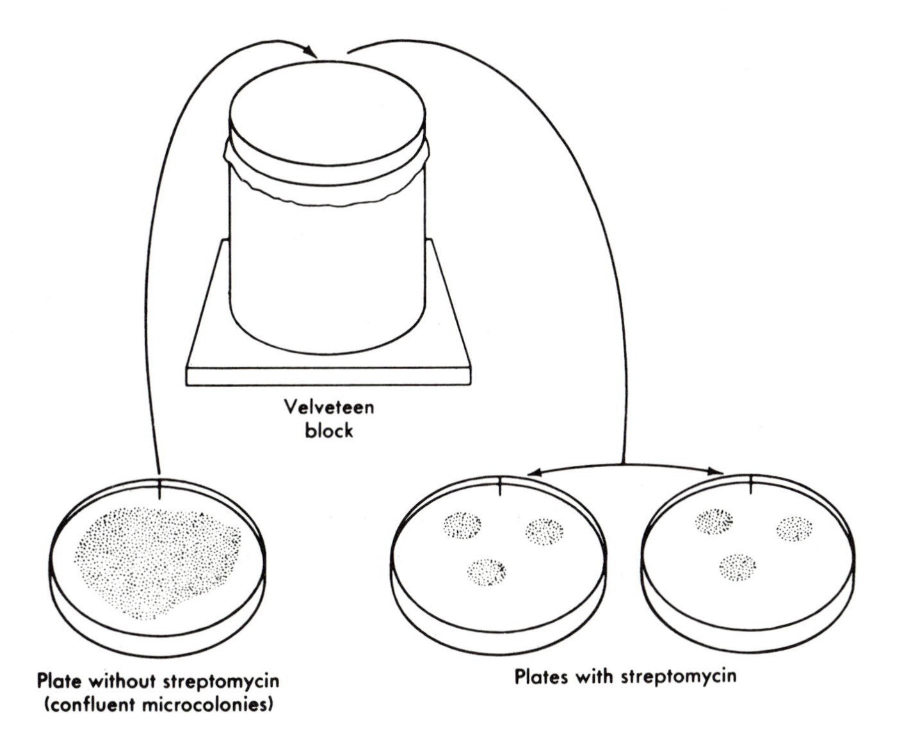

Figure 3-1. Use of replica plating to demonstrate undirected, spontaneous appearance of streptomycin-resistant mutants. About 10^5 sensitive cells were spread on a plate of drug-free solid medium and allowed to reach full growth (10^{10} to 10^{11} cells). Sterile velvet covering the ends of the cylindrical block was pressed lightly on this continuous heavy lawn ("master plate") and was then pressed successively on two plates of medium containing streptomycin at a concentration that killed sensitive cells. A few colonies of resistant cells appeared on each plate, usually in coincident positions. From Davis, B.D., Dulbecco, R., Eisen, H.N., Ginsberg, H.S. Microbiology, 3rd ed., 1980, p. 31. Philadelphia: Lippincott, Harper & Row.

bacteria and marked as to orientation is inverted and pressed gently against the surface of the velvet. As the fibers of the fabric poke into the agar, bacteria stick to their sides. After the master plate is removed, as many as 10 to 12 uninoculated plates, also marked as to orientation, may be pressed against the fabric. Each plate produces a faithful replica of the master plate after incubation because some of the bacteria that adhered to the fabric are transferred to the agar at each replication.

The Lederbergs' experiment consisted in replicating master plates of sensitive cells to two or more plates containing either streptomycin or T1 phage. When the replicas had grown they were compared, and any resistant colonies that appeared at the same position on all of the replica plates were marked. The corresponding area of the master plate was cut out and the bacteria on it resuspended in liquid medium. If the hypothesis of preexisting mutations were correct, the culture derived from these cells would be enriched for resistant mutants by virtue of the fact that only a small piece of agar (and therefore a limited number of cells) was removed from the master plate. The enriched culture could then be used to prepare a new master plate and the whole process repeated.

The final result was a master plate that contained nothing but resistant bacteria, even though the cells and their progenitors had never been directly subjected to selection. This process of indirect selection is entirely in accord with the hypothesis of preexisting mutation but not that of acquired immunity.

Sib Selection

One question left unanswered by the three experiments discussed above was whether preexisting mutations could account for all of the observed variation within a culture. Cavalli-Sforza and Lederberg devised a modification of the fluctuation test that directly addressed this question; it was called **sib(ling) selection.**

The experiment began with the results from a standard fluctuation test such as that shown in Table 3-1. Bacteria from the tube with the largest number of resistant cells (usually a jackpot tube) were used to inoculate a second set of tubes for a new fluctuation test, and the process was repeated several times. However, at each inoculation the sample size was chosen so that the probability of any single tube receiving a resistant mutant was less than 1. Because it is not possible to receive less than one mutant bacterium, any tube that did receive such a cell was relatively enriched for resistant mutants. For example, if the size of the inoculum were 10^5 cells, and the frequency of mutant cells were 10^{-6}, a tube that received a mutant cell would have a culture enriched so that the frequency of mutant cells was now 10^{-5}, a tenfold increase.

If mutation were the only factor producing bacterial variation, it should be possible to predict the exact degree of enrichment for mutant cells at each step in the process. If acquired immunity contributed to the population of resistant cells, the total degree of enrichment at each step would be the sum of the enrichment due to inoculum size plus the amount of acquired resistance, and the total would be greater than that predicted by the simple mathematic calculation. In fact, the observed degree of enrichment was less than predicted. In order to account for this strange result, Cavalli-Sforza and Lederberg postulated that mutation was the only operative factor in bacterial variation and that the resistant mutants grew slightly slower than did sensitive bacteria. Thus the frequency of mutant cells in the population after growth was always less than before growth had taken place. When the resistant bacteria were isolated in pure culture and tested, it was indeed the case.

Measurement of Mutation Rate

The measurement of mutation rate in bacteria is more complicated than for other organisms owing to their large numbers and rapid rate of binary fission. The units by which the mutation rate for a particular trait is expressed are mutations/bacterium/cell division, and the observable quantities are the total number of bacteria at the beginning and end of the experiment as well as the total number of mutant cells.

In order to calculate the number of cell divisions that have occurred in the culture since an arbitrary zero time, it is possible to make use of the following relation:

Cells (n)	1		2		4		8		16		32		64	etc.
Cell divisions (n)		1		3		7		15		31		63	etc.	

Note that the total number of binary cell divisions required to obtain a particular number of cells is always one less than the number of cells. Therefore because the number of bacterial cells is generally large, the total number of cell divisions that have occurred since time zero is closely approximated by the increase in the total number of cells during the experiment. After many generations this approximation can use just the final number of cells, as the initial number of cells is negligible.

The calculation of the average number of cells present in the culture during any particular time interval is more complex. The cells are dividing in an asynchronous manner, and therefore the graph describing their growth is a smooth exponential curve rather than a step function. The method used for the calculation is differential calculus (however, an understanding of calculus is not necessary, as there is only one other example of it in this book). The rate of change in the number of bacteria in a culture per unit time is

$$dN/dt = \mu N \qquad [3\text{-}1]$$

where N is the number of cells in the culture at the beginning of the time interval and μ is a proportionality constant designated as the growth rate constant. Integration of Eq. 3-1 yields

$$N_t = N_0 e^{\mu(t - t_0)} \qquad [3\text{-}2]$$

where N_t is the number of cells in the culture at time t and N_0 is the number of cells in the culture at time zero. The length of time required for the number of cells in the culture to double (i.e., $N_t = 2N_0$) can be represented by g, the average doubling time, or generation time. Appropriate substitutions in Eq. 3-2 yield the relation

$$g = \frac{\ln 2}{\mu}. \qquad [3\text{-}3]$$

The above equations can be converted into a form suitable for the calculation of $\overline{N}$, the average number of cells in a culture during a particular time interval. The basic equation is

$$\overline{N} = \int_{t_1}^{t_2} N_t dt \,/ \int_{t_1}^{t_2} dt \qquad [3\text{-}4]$$

which upon substitution from Eq. 3-2 becomes

$$\overline{N} = \int_{t_1}^{t_2} N_0 e^{\mu(t - t_0)} dt \,/ \int_{t_1}^{t_2} dt. \qquad [3\text{-}5]$$

Equation 3-5 can be integrated by assuming that t_0 is zero and taking advantage of the following mathematic relation

$$\int e^{ax} dx = e^{ax}/a. \qquad [3\text{-}6]$$

Therefore

$$\overline{N} = N_0 \int_{t_1}^{t_2} e^{\mu t} dt / \int_{t_1}^{t_2} dt = N_0(e^{\mu t_2} - e^{\mu t_1})/(t_2 - t_1)\mu \qquad [3\text{-}7]$$

and substituting for μ from Eq. 3-3 yields

$$\overline{N} = N_0(e^{\ln 2 \cdot t_2/g} - e^{\ln 2 \cdot t_1/g})/\ln 2(t_2 - t_1)/g. \qquad [3\text{-}8]$$

However, the expression t/g is G, the number of division cycles (generations) that have occurred between time zero and t. Therefore Eq. 3-8 becomes

$$\begin{aligned} \overline{N} &= N_0(e^{G_2 \ln 2} - e^{G_1 \ln 2})/\ln 2(G_2 - G_1) \\ &= (N_0 2^{G_2} - N_0 2^{G_1})/\ln 2(G_2 - G_1) \\ &= (N_2 - N_1)/\ln 2(G_2 - G_1). \end{aligned} \qquad [3\text{-}9]$$

For one generation (one doubling), Eq. 3-9 becomes

$$\overline{N} = \frac{2N_g - N_g}{\ln 2(1)} = \frac{N_g}{\ln 2}.$$

The point is that the average number of bacteria in a subsequent generation can be obtained by dividing the number of bacteria at the beginning of the time period by the natural logarithm of 2, provided the cells are dividing by binary fission and randomly with respect to time. Using this information, it is now possible to demonstrate how to calculate a mutation rate.

Luria and Delbrück developed two methods for calculating mutation rates based on the fluctuation test. The first of these methods utilizes the Poisson distribution. A fluctuation test can be prepared so that, unlike the experiment shown in Table 3-1, some of the tubes have no resistant mutants. It can be done using small inocula and sampling a smaller number of cells than usual. If it is assumed that the mutant cells are randomly distributed throughout the culture, this system is ideal for applying the Poisson distribution (see Appendix) because the sample size is large, the probability of success (a mutation) is small, and therefore the average number of mutant cells per sample is moderate.

In the specific case presented by Luria and Delbrück (their experiment 23), the samples from the fluctuation test contained 2.4×10^8 bacteria per tube, and 29 of 87 tubes had no resistant cells. Solving the Poisson distribution for the zero case gives

$$29/87 = 0.33 = \frac{e^{-m}m^0}{0!} \qquad \text{and } m = 1.10 \text{ mutants per tube.}$$

This number must be divided by the number of bacteria per cell division, which can be calculated by using the above approximations as the average number of bacteria during the last cell division (which actually occurs on the selective plate) and is equal to $(2.4 \times 10^8)/\ln 2 = 3.4 \times 10^8$. Then the actual mutation rate is $1.10/(3.4 \times 10^8) = 3.2 \times 10^{-9}$ mutations/bacterium/cell division.

The principal disadvantage to this calculation is that it wastes much of the information available from the fluctuation test because it does not take into account the frequency distribution of the tubes that do contain resistant cells. By making some additional assumptions, it was possible for Luria and Delbrück to develop a graphic method to measure the mutation rate (now computerized by Koch). To utilize this method, it is necessary to assume that there is a certain density of cells in a culture such that the probability of at least one mutation occurring somewhere in the culture is high. If this assumption is correct, whenever this density is surpassed the sum of the mutational events at each cell division results in the same number of resistant cells in the final population (Fig. 3-2). Therefore if one designates the time at which the proper density is reached as time zero and ignores all mutations that occur earlier, it is possible to mathematically describe the behavior of the population of mutant cells.

Specifically, the equation that was developed is

$$M = d \cdot N_t \cdot \ln(N_t C d) \qquad [3\text{-}10]$$

where M is the average number of mutants per culture, d is the mutation rate, N_t is the total number of cells in each culture, and C is the number of cultures used in the fluctuation test. This method gives a value for the mutation rate of 3.5×10^{-8}, which is considerably higher than the value calculated by the Poisson method. The reason is that the graphic method assumes that no mutation occurs prior to the time that a mutation becomes a likely event for the entire population. If a jackpot tube occurs, it has a disproportionately large number of mutant cells in the final population. The result is an overestimate of the number of mutant cells arising during a particular period.

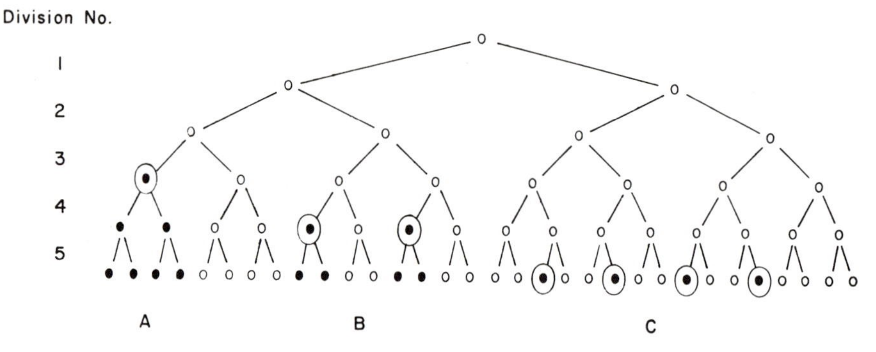

Figure 3-2. Constancy of the number of mutant progeny in a large population. Once the number of cells in a culture reaches a certain value, one or more mutations are expected to occur during each doubling of the cell number. Nonmutant cells are represented by open circles and mutant cells by filled circles. Each newly arisen mutant cell is circled to emphasize its initial appearance in the population. No mutations arose during divisions 1 or 2, but one mutation occurred during division 3, two during division 4, and four during division 5. Regardless of when the mutations were first observed, the effect at division 5 was the same, i.e., four mutant cells. Note that the overall proportion of mutant cells in the population is increasing, as the possibility of back-mutation has been neglected.

The final method of calculating mutation rate discussed here is based on the plate spreading experiment. Using Newcombe's experimental protocol, a number of plates are prepared. It is possible to spray them at different times and to take some control plates and wash off all the bacteria for counting and testing for the presence of mutant cells. It is then possible to know the change in total cell number and the change in the number of mutant colonies. The mutation rate is then:

d = change in no. of resistant colonies/change in total no. of cells/ln2

In this case the mutation rate is approximately 5.8×10^{-8} mutations/bacterium/cell division.

For most of the mutable sites on the normal *Escherichia coli* genome, the mutation rate is found to vary between 10^{-6} and 10^{-10}. Mutation rates have been determined for a number of organisms (Table 3-2). Note that for the bacteria and viruses the total mutation rate per organism is relatively constant, which means that the relative mutation rate per base must vary inversely with genome size.

It is also important to remember that this discussion has dealt only with **forward mutations.** The term forward is used to imply a change from some arbitrarily designated original genetic state to a new mutated state. The reverse process, i.e., a change from the mutant to the original state, is called **backward mutation,** or **reversion.** In practice, cultures grown for long periods of time, e.g., in a chemostat, tend to drift toward **genetic**

Table 3-2. Comparative forward mutation rates for various organisms[a]

Organism	No. of base pairs in the genome	Mutation rate per base pair replicated	Total mutation rate
Bacteriophage λ	4.8×10^4	2.4×10^{-8}	1.2×10^{-3}
Bacteriophage T4	1.8×10^5	1.7×10^{-8}	3.0×10^{-3}
Salmonella typhimurium	4.5×10^6	2.0×10^{-10}	0.9×10^{-3}
Escherichia coli	4.5×10^6	2.0×10^{-10}	0.9×10^{-3}
Neurospora crassa	4.5×10^7	0.7×10^{-11}	2.9×10^{-4}

[a] The details of the calculations are given in the original paper.

From Drake, J.W. (1969). Comparative rates of spontaneous mutation. Nature 221:1132.

equilibrium. The proportion of mutant cells becomes constant as the number of forward mutational events equals the number of backward mutational events (allowing for possible differences in growth rates). In all of the foregoing discussion about mutation rates, it was tacitly assumed that the rate of reversion was negligible compared to the (forward) mutation rate. If it were not, all the calculated mutation rates would be too low.

Expression and Selection of Mutant Cell Phenotypes

Expression

In the previous section mutation was discussed as an all-or-none phenomenon. The cell was either mutant or it was not. However, strictly speaking, there is a period of transition during which the new phenotype is expressed (i.e., appropriate macromolecules are synthesized). The nature of the transition period depends in part on the nature of the final gene product. All cells have a gradual turnover of proteins and RNA molecules as existing molecules are degraded or diluted as a result of cell growth and the synthesis of new molecules produced according to the current needs of the cell. The rates of turnover of macromolecules vary widely, with mRNA being relatively unstable and proteins and the other RNA molecules being relatively stable (except during starvation conditions when considerable protein degradation occurs). The turnover rates may affect the timing, but not the nature, of the events described below. The types of events that occur following changes in regulatory regions of the DNA (i.e., DNA that does not have a macromolecular product) are discussed in Chapter 12.

One reason for the existence of a demonstrable transition period following a mutational event is the multiple genome copies found in an actively growing cell. As a general rule, a mutation occurs in only one copy of a particular piece of DNA, leaving the cell with several DNA copies coding for the unmutated product and only one coding for the mutated product. Once the normal process of transcription (and translation in the case of a protein) has occurred, the cytoplasm of the cell contains two kinds of macromolecules, mutant and nonmutant. Such a cell is said to be a transient merodiploid, and the question of the dominance of the mutation arises.

The dominance of a bacterial mutation is due to exactly the same biochemical processes as occur in eukaryotes. If the mutation confers the ability to carry out some biochemical process it has a dominant effect, and the phenotype of the cell changes as soon as enough of the new product has been synthesized to permit the reaction to take place at a significant rate. However, Figure 3-3 indicates that, even though the phenotype of the cell has changed, the number of cells with the mutant phenotype cannot increase for several generations because of the large number of genome equivalents and the fact that most of the DNA currently being replicated actually does not segregate until several cell divisions later (compare with Figure 1-4). This type of delay is called **segregation lag,** and its duration depends on the number of genome equivalents in the cell. An example of segregation lag is shown in Figure 3-4. The first recombinant cells appeared at about 10 minutes, but they did not begin to divide at the same rate as the rest of the cells until about 120 minutes. Therefore in this case the lag amounted to approximately 110 minutes, or 2.5 doublings, suggesting the presence of slightly more than four genome equivalents per original cell.

Recessive mutations experience a different sort of lag, called **phenotypic lag.** In this case the cytoplasm of the cell still contains nonmutant dominant-type product until after the segregation process shown in Figure 3-3 is complete and the genome is homogeneous. At that time, all the new product being produced is mutant, but there is still some nonmutant product remaining in the cytoplasm. As this nonmutant product decays away or is diluted, the phenotype of the cell changes to mutant. Note that whereas segregation lag consists of only one process phenotypic lag consists in two processes: segregation and replacement of macromolecules.

The lag phenomena can have important consequences in terms of applying a selective treatment to a bacterial culture. After a genetic exchange or a mutational event, sufficient time must be allowed for the expression of the new phenotype. If selection is applied too early, potentially recombinant or mutant cells may have their metabolism shut off before the DNA modification process has reached the point at which an altered product can be produced (see Chapter 13). In such a case, the cell is not scored as a recombinant or mutant because it fails to produce a colony under selective conditions.

Witkin demonstrated clearly the necessity for metabolic activity in the expression of a mutation. Working with a *trp* strain of *Salmonella typhimurium,* she set up parallel cultures. One culture was treated with ultra-

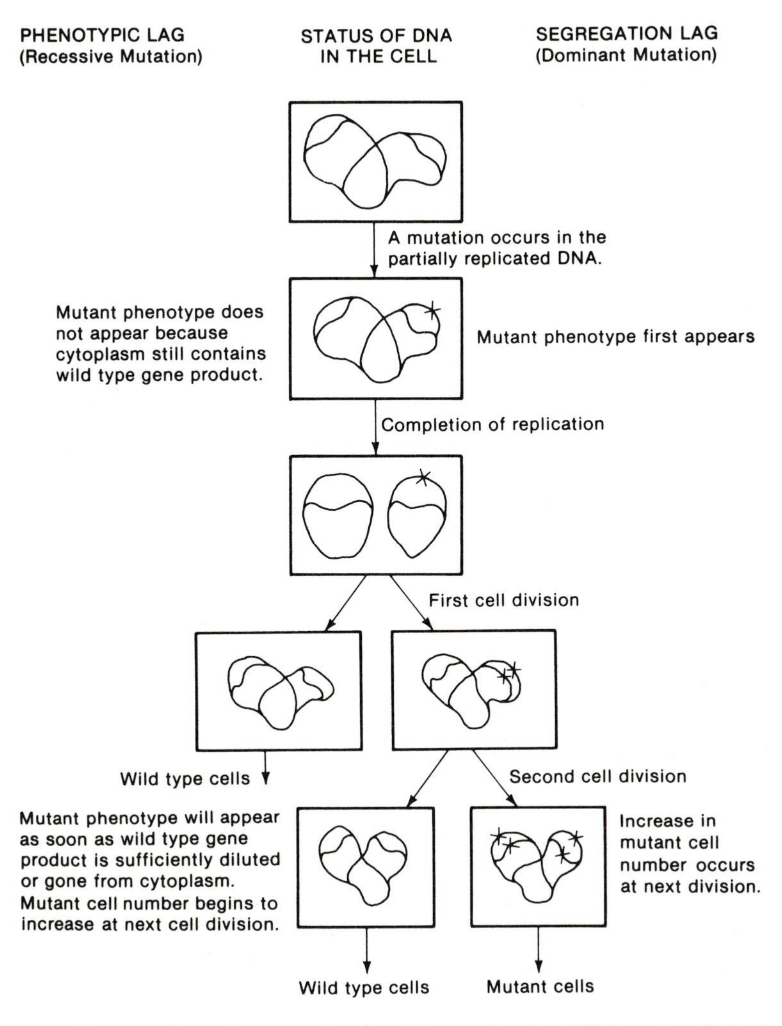

Figure 3-3. Phenotypic and segregation lag. The replicating DNA molecule is shown within the cell in much the same fashion as in Figure 1-4. A mutation is assumed to occur within one duplicated region at the point marked *X*. If it is a dominant mutation, the phenotypic effect of the mutation is observed immediately, but only one of the two daughter cells produced after division one is mutant (segregation lag). If the mutation is recessive, the appearance of the phenotype is delayed until the second cell division when all the DNA in the daughter cell is homogeneous (phenotypic lag). It is important to remember that the duration of either type of lag is a function of the number of genome equivalents of DNA in the cell.

violet radiation (UV) (see Mutagens, below), and the other was treated with both UV and generalized transducing phage grown on a UV-irradiated wild-type donor strain. The first culture would be expected to yield trp^+ mutants induced by the UV, and the second would yield both UV-induced trp^+ mutants and phage-induced trp^+ recombinants. After the appropriate

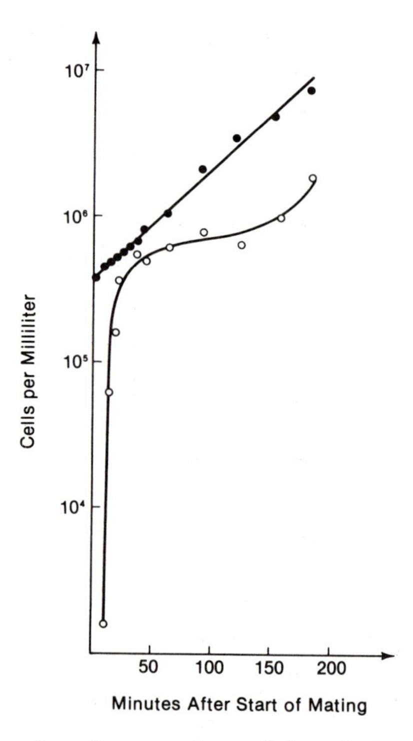

Figure 3-4. Segregation lag. A comparison of the relative growth of an *E. coli* culture (filled circles) and of Lac^+ cells newly arisen within that culture as a result of conjugation (open circles). The initial increase in the number of Lac^+ cells seen during the first 30 minutes of the experiment was due to the transfer of DNA into the cells. After 30 minutes no further transfer was possible because nalidixic acid was added to the culture (see Chapter 9). The increase in the number of Lac^+ cells after 125 minutes marks the end of the segregation lag. Note that the growth rates of the Lac^+ and Lac^- cells are essentially identical.

treatments the cells were plated on several agar plates containing sufficient tryptophan to allow one to six generations of residual growth, the plates were incubated, and the number of Trp^+ colonies was counted.

The maximum number of recombinant Trp^+ colonies was produced if sufficient tryptophan to allow one generation of residual growth was present, but the maximum number of mutants did not appear unless sufficient tryptophan to allow six generations of residual growth was present, even though the growth rates of mutant and recombinant Trp^+ cells were shown to be identical. Further experiments involving the shifting of cells from one type of medium to another indicated that the expression of both mutant and recombinant phenotypes was complete after one generation, even though the mutated cells needed enough tryptophan to allow six doublings. Apparently mutagenic processes require a higher level of metabolic activity, especially protein synthesis, than do recombinational processes.

Other workers have obtained similar results which, taken together, indicate that most but not all mutants or recombinants must be allowed time for expression of their new phenotype prior to selection. However, it is possible to have too much of a good thing, and excessive amounts of nutrients produce microscopic colonies from even nonmutant cells.

Selection

Until now most of the discussion has dealt only with the type of mutants that are dominant and easily selected. There are, however, many useful and important mutations that do not fall into this category, and some method is needed to find them. In cases of desperation, it is possible to find a particular mutant by the brute-force method of checking individual colonies until the appropriate phenotype is found. This method was utilized by DeLucia and Cairns when they were searching for the original *polA* mutant in a culture that had been treated with a chemical to enhance the rate of mutation. In that case, approximately 5000 individual colonies were tested in order to find one mutant. It is a rather ponderous method of searching, and a more efficient technique is generally used unless there is no alternative.

One such method was developed by Gorini and Kaufman and is based on the well known ability of penicillin or related antibiotics to preferentially attack growing cells. The penicillins do not alter existing peptidoglycan but do inhibit the new formation of the peptide cross-bridges, which tie the peptidoglycan structure of the cell wall into a cohesive unit. Without the cross-bridges, the peptidoglycan sacculus cannot be maintained, and the cell, having lost its wall, becomes osmotically fragile.

The **penicillin selection** method is of general applicability, requiring only that it be possible to stop the growth of the desired mutant by starvation, extreme temperature, etc. It can be used with either liquid cultures or agar plates. For the isolation of an auxotrophic mutant, a culture of cells in logarithmic growth phase is placed under starvation conditions with a defined medium in which prototrophs grow but auxotrophs cannot. After sufficient time has elapsed to allow the cells to exhaust any intracellular pools of nutrients; penicillin, ampicillin, cycloserine, or various combinations of them are added to the culture and incubation continued until cell lysis occurs. The penicillin is removed from the liquid culture either by pelleting the cells in a centrifuge or by filtration, followed by resuspension in fresh complete medium. Penicillin can be removed from agar plates by adding the enzyme penicillinase. The addition of fresh nutrients allows the mutant cells to grow again, giving rise to a new culture that is greatly enriched for auxotrophic mutants.

It should be borne in mind that this method has several limitations. Rossi and Berg have shown that different types of auxotroph are recovered with highly variable efficiency after penicillin selection. They theorized that it is due to incomplete shutdown of metabolism during starvation. Moreover, an inherent limitation of penicillin selection is that nonauxotrophs that are growing slowly for any reason tend to survive the treatment. Finally, if the growing cells lyse during the penicillin treatment, they may release enough nutrients to allow the auxotrophic cells to grow and be affected by the penicillin. Gorini and Kaufman controlled this last problem by raising the osmotic strength of their medium with sucrose until after the penicillin had been removed.

Another potential problem is that of cross-feeding (**syntrophy**), which can occur in either liquid or solid medium but which is most easily visualized on agar plates. Auxotrophic cells do not grow on a minimal medium lacking the required nutrient when plated as a pure culture. However, when plated as a mixed culture it is possible that the second type of cell may release a substance into the medium that permits the auxotroph to grow. The phenomenon may occur even if the second cell type is auxotrophic for the same nutrient, providing the block in the metabolic pathway of the second cell type is at a step that occurs later than the blockage of the first cell type. For example, for a simple biochemical pathway involving two enzymes

$$\text{Substrate} \xrightarrow{\text{enzyme 1}} \text{product 1} \xrightarrow{\text{enzyme 2}} \text{product 2}$$

a cell with a mutation in enzyme 2 tends to accumulate product 1. If product 1 can be released into the medium, it can act as a nutrient for a cell defective only in enzyme 1, permitting it to grow. In some cases, as product 2 is produced, some of the molecules diffuse through the medium to the other cell, also permitting it to grow. This type of analysis can be used to order the steps of a biochemical pathway by observing which cell types are cross-feeders and which are cross-fed. To the extent that cross-feeding occurs during penicillin selection, auxotrophic cells are also affected by penicillin and hence selected against.

Genetic Code

The universal form of the genetic code is presented in Table 3-3 in both its RNA and DNA forms. In the absence of any evidence to the contrary, it is assumed that all organisms conform to this code. There are, however, some notable exceptions, particularly among the eukaryotic organisms. In human mitochondria the DNA codon ATT codes for tryptophan instead of termination, and in both human and yeast mitochondria ATA is often used as an initiator codon instead of ATG. By contrast, the mitochondria of *Chlamydomonas reinhardii* seem to follow the universal code, so not all mitochondria are identical. Among the prokaryotes, *Mycoplasma capricolum* has been shown to use the codon ACT for tryptophan instead of ACC. Jukes theorized that the coding change is a consequence of the high (G + C) content (75%) for the DNA of this organism. The substitution of the tryptophan codon serves to lower the (G + C) content slightly.

The general tendency of the base in the third position of the codon to have little coding importance has been codified by Crick as part of his **wobble hypothesis** in which various tautomeric forms of the bases are assumed to "wobble" back and forth to generate hydrogen bonds between codon and anticodon. Wada and Suyama theorized that the wobble base normally counteracts the (G + C) content of the first two bases in the codon. They proposed that it provides a more uniform density to the DNA molecule, which in turn provides greater stability for the double helix. In a related vein it is clear that where more than one codon can code for a particular amino acid each organism exhibits a definite preference for certain codons, and codon preference tables characteristic of each organism can be prepared.

Simplified versions of the genetic code often present AUG as the only codon used to initiate protein synthesis. This statement is not strictly correct, as AUG is merely the most common and most efficient initiator codon. It is also possible to use GUG (next most common), UUG, AUA, or AUU to start a protein. The lack of an amino-terminal methionine on a given protein does not necessarily provide any information about the initiator codon, however, because there is an enzyme that specifically removes amino-terminal methionine residues.

An obvious problem with the use of methionine as the initiator amino acid is that there is only one possible codon for that amino acid. Therefore an AUG codon at the beginning of the mRNA must mean something different from an AUG codon located in the interior of the mRNA. The answer apparently lies in so-called **context effects,** alterations in the coding properties of the codon by neighboring bases. It has been shown that the beginning of an mRNA has a base sequence that is unlikely to self-pair to form a stem-and-loop structure, whereas the internal sequences have a strong tendency to do that. Therefore the initiator AUG codon is more likely to be exposed to the ribosome in a single-stranded region than is an internal codon.

Kinds of Mutation

The term **mutation** has been used throughout this chapter in the classic genetic sense of an abrupt, inherited change in an observable trait of an organism. It is now time to consider what is meant by the term mutation at the molecular level. However, before so doing it is necessary to redefine the term mutation. A mutation is henceforth considered to be any alteration in the base sequence of the nucleic acid comprising the genome of an organism, regardless of whether there is any phenotypic effect from the

Table 3-3. Genetic code[a]

	DNA							
	A		G		T		C	
A	AAA	Phe	AGA	Ser	ATA	Tyr	ACA	Cys
	AAG		AGG		ATG		ACG	
	AAT	Leu	AGT		ATT	Term	ACT	Term
	AAC		AGC		ATC		ACC	Trp
G	GAA		GGA	Pro	GTA	His	GCA	Arg
	GAG		GGG		GTG		GCG	
	GAT		GGT		GTT	Gln	GCT	
	GAC		GGC		GTC		GCC	
T	TAA	Ile	TGA	Thr	TTA	Asn	TCA	Ser
	TAG		TGG		TTG		TCG	
	TAT		TGT		TTT	Lys	TCT	Arg
	TAC	Met	TGC		TTC		TCC	
C	CAA	Val	CGA	Ala	CTA	Asp	CCA	Gly
	CAG		CGG		CTG		CCG	
	CAT		CGT		CTT	Glu	CCT	
	CAC		CGC		CTC		CCC	

alteration. This definition is deliberately broad in order to encompass the tremendous variety of mutational types. Except where specifically indicated, the following discussion concerns only mutations in DNA sequences coding for polypeptides.

Base Substitutions

The easiest type of mutation to visualize is that of **base substitution,** in which a single nucleotide base is replaced by another. If a purine is replaced

Table 3-3. *(Continued)*

	RNA							
	U		C		A		G	
U	UUU	Phe	UCU	Ser	UAU	Tyr	UGU	Cys
	UUC	Phe	UCC	Ser	UAC	Tyr	UGC	Cys
	UUA	Leu	UCA	Ser	UAA	Term	UGA	Term
	UUG	Leu	UCG	Ser	UAG	Term	UGG	Trp
C	CUU	Leu	CCU	Pro	CAU	His	CGU	Arg
	CUC	Leu	CCC	Pro	CAC	His	CGC	Arg
	CUA	Leu	CCA	Pro	CAA	Gln	CGA	Arg
	CUG	Leu	CCG	Pro	CAG	Gln	CGG	Arg
A	AUU	Ile	ACU	Thr	AAU	Asn	AGU	Ser
	AUC	Ile	ACC	Thr	AAC	Asn	AGC	Ser
	AUA	Ile	ACA	Thr	AAA	Lys	AGA	Arg
	AUG	Met	ACG	Thr	AAG	Lys	AGG	Arg
G	GUU	Val	GCU	Ala	GAU	Asp	GGU	Gly
	GUC	Val	GCC	Ala	GAC	Asp	GGC	Gly
	GUA	Val	GCA	Ala	GAA	Glu	GGA	Gly
	GUG	Val	GCG	Ala	GAG	Glu	GGG	Gly

[a]Each base is represented by a single letter: A = adenine; C = cytosine; G = guanine; T = thymine; U = uracil. Amino acid abbreviations are listed in Table 1-1. *Term* means translation termination. The leftmost base of the codon is the 3′-end of the DNA but the 5′-end of the RNA.

with a purine (e.g., adenine for guanine) or a pyrimidine with a pyrimidine (e.g., thymine for cytosine), the change is termed a **transition** (for an RNA virus the transition would be uracil for cytosine). If a purine is replaced by a pyrimidine or vice versa, the change is referred to as a **transversion.**

During the base substitution process a transient DNA heteroduplex is formed, and therefore these mutations can occur only if the natural repair systems in the cell (see Chapter 13) do not correct the error. Some errors are apparently more easily detected than others. For example x-ray diffraction studies on G•A base pairs indicate that they have the adenine in the *syn* configuration and guanine in the *trans* configuration. A G•A base pair thus turns out to be less symmetric than the normal pairing but much more symmetric than, for example, a G•T pair. It has been suggested that the added symmetry of a G•A base pair may make it less susceptible to standard repair mechanisms than some other combinations of bases, thereby producing transversion mutations.

Although a single base change is the simplest kind of mutation to visualize, it may also be one of the more difficult to detect owing to the redundancy in the genetic code, so that in many cases the same amino acid can be coded by many codons. For example, see the codons for leucine listed in Table 3-3. A DNA codon that was originally GAA could change to GAG with no change in the amino acid sequence of the coded polypeptide.

Another possible effect of a base substitution is what is known as a **missense mutation.** Instead of the original amino acid in the polypeptide chain, a different one is substituted. The phenotypic effect of the substitution can range from nonexistent to devastating. Certain types of amino acids, e.g., threonine and alanine, can often be substituted for one another with little effect on the secondary and tertiary structure of the protein. However, substitution of a proline into what is normally a helical region of the protein destroys the remainder of the helix and possibly the activity of the polypeptide.

A unique class of substitution mutations is the **nonsense mutations** (terminators), which lack any corresponding tRNA. Table 3-3 shows that there are three such DNA codons: ATC, ATT, and ACT. These codons normally act as punctuation marks within the genetic code, signaling the end of the polypeptide chain. When any one of them appears within the coding sequence for a polypeptide, it results in premature termination of the growing peptide chain and the formation of a truncated polypeptide consisting of the amino-terminus and a specific number of amino acids determined by the physical site of the mutation. The opposite class of mutations, those that result in a new TAC start signal, is also known but is much rarer because the start signal requires more than a single triplet codon.

The nonsense mutations have been given specific names in what began as a pun on the name of the person who identified the first mutant of this type. The codon ATC is an amber codon (a translation of the German word Bernstein). By analogy, ATT is the ochre codon, and ACT has been referred to as opal.

Because peptide chain termination is often associated with release of the ribosome from the mRNA molecule, nonsense mutations frequently exert **polar effects** (i.e., preventing translation of subsequent polypeptides using information located on the same mRNA molecule). The degree of polarity is apparently a function of the probability that the untranslated RNA will form a terminator loop and thereby prevent further transcription. Consequently, terminator mutations near the amino end of a polypeptide are much more polar than those close to the carboxy-terminus because of the greater length of untranslated mRNA in the former case.

Finally, there are base substitutions affecting DNA that do not code for proteins. Some of this DNA codes for structural RNA molecules (tRNA, rRNA). Changes in structural RNA can result in altered ribosome function or in tRNA molecules that have altered amino acid or anticodon specificity. Changes in nontranscribed DNA may result in altered regulatory functions such as those discussed in Chapter 12. Some nontranscribed DNA acts merely as a spacer between transcribed regions, and changes here would be expected to have little effect.

Insertion and Deletion Mutations

A **deletion mutation** is the removal of one or more base pairs from the DNA, whereas an **insertion mutation** is the addition of one or more base pairs. In practice, deletion and insertion mutations usually involve considerably more than one base pair. Insertion or deletion of base pairs in multiples of three results in the addition or elimination of amino acids in the polypeptide chain (Fig. 3-5). All other insertions or deletions result in frameshift mutations. The formation of at least some insertion and deletion mutations appears to be intimately connected to transposons, which are discussed in Chapters 11 and 13. In addition, certain types of DNA synthetic mutations such as *polA* appear to increase the probability of spontaneous deletion mutations. It has been proposed that many deletions arise from hairpin looping followed by DNA replication across the stem (Fig. 3-5).

Frameshift Mutations

When a ribosome translates an mRNA molecule, it must accurately determine the **reading frame,** as all possible triplet codons are meaningful. Proper framing is accomplished when the ribosome binds to the Shine-Dalgarno box adjacent to the initial AUG codon. Then it moves along the molecule in three base jumps. If an insertion or deletion of base pairs in other than multiples of three has occurred, the reading frame shifts, and gibberish is produced instead of the normal amino acid sequence. For example:

Normal RNA sequence:	AUG	AGU	UUU	AAA	GAC	etc.
Normal amino acids:	met	ser	phe	lys	asp	etc.
Deleted RNA:	AUG	A*UU	UUA	AAG	ACU	etc.
Gibberish sequence:	met	ile	phe	lys	thr	etc.

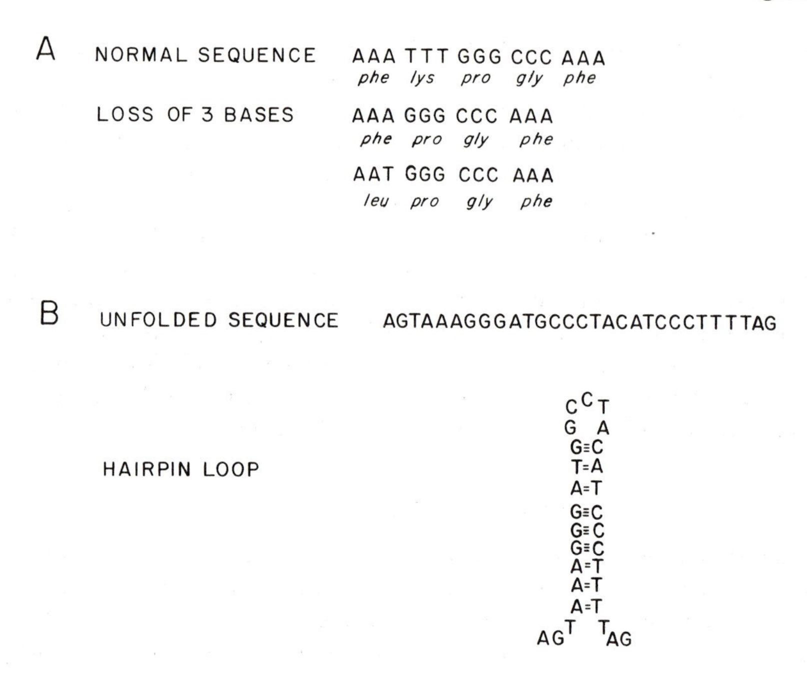

Figure 3-5. Production of deletions. **(A)** Removal of bases in multiples of three results in the loss of one or more amino acids (and the possible change of one of the amino acids flanking the deletion). **(B)** If a hairpin loop should form in the DNA duplex, it seems to be possible for DNA polymerase to replicate across the hairpin and create a deletion. Shown here is a single strand of DNA. A similar loop would form in the complementary strand.

All of the amino acids downstream from the frameshift are incorrect, although the amino-terminus is normal. Frequently frameshift mutations also result in the accidental production of chain-terminating codons, leading to the production of both gibberish and a truncated protein. It is also possible, of course, to have a frameshift such that the normal terminators are bypassed and the abnormal protein molecule is much longer than normal. Frameshift mutations are categorized according to the extent that the reading frame differs from the normal multiple of three: $+1$, $+2$, -1, or -2.

Glickman and his collaborators have completed a study of spontaneous mutations in the *lacI* gene (see Chapter 12). Of the 144 mutations they characterized, roughly two-thirds occurred in a frameshift hotspot with the sequence TGGCTGGCTGGC in which the usual mutation was the addition of 4 base pairs (bp). Other repeated sequences did not give frameshift mutations, so whatever causes spontaneous frameshifts requires more than simple repeats. When the remainder of the mutations were examined, 37% were deletions away from the hotspot, mainly associated with other repeated sequences. Another 34% were base substitutions with no obvious bias toward any particular base pair. The remaining mutations were mainly single base frameshifts, a few tandem duplications, and some 12% insertion mutations due to transposition (see below). Thus most of the mutations examined were either frameshifts or deletions.

Whether the same would be true for all coding regions is not yet known. However, in a similar study of mutations in the phage T4 *r*II region (see Chapter 4) it was found that about half of the frameshifts were consistent with a model in which misalignment of DNA with repeated sequences had occurred. In that particular instance, 25% of all frameshifts were found to be the addition or loss of a single base.

Suppressors of Mutations

A **suppressor** is something that eliminates the phenotypic effect of another mutation but not its genotypic effect; i.e., a cell that carries one mutation has a mutant phenotype, but a cell that carries the original mutation plus a suppressor mutation has a normal phenotype but a doubly mutant genotype. There are a variety of ways in which suppression can occur, but only a few of the more common types are discussed here.

Often suppression of a mutation is based on an alteration of the codon—anticodon interaction that occurs during translation. It may take the form of an error in the translation mechanism itself. In one case certain types of ribosomal mutation make it susceptible to perturbation by low concentrations of streptomycin. When the antibiotic is present, there is a strong tendency for misreading to occur in the wobble base. Given the appropriate initial mutation in a protein, it may be possible to make a streptomycin-induced error that restores functionality to the protein. Under other conditions ribosomal misreading may not occur at all (stringent translation), or even frameshift mutations may be suppressed. The use of a second-site mutation for suppression is called **intergenic suppression.**

The most common type of intergenic suppression involves changes in the tRNA molecule itself. If the original type of mutation was a nonsense mutation, any change in an anticodon that would allow pairing with the terminator codon would tend to alleviate the polar effects of the terminator mutation and possibly also restore function to the protein as well. For example, $tRNA_{gly}^{GAA}$ can become $tRNA_{gly}^{UAA}$ and thereby insert a glycine residue wherever the ochre codon was found. Other nonmutated tRNA molecules would be used by the cell to supply glycine for the normal codons.

There are apparently strong context effects on the ability of a tRNA anticodon to correctly translate a codon. Engelberg-Kulka has shown that it is possible to have a normal $tRNA_{trp}^{UGG}$ translate a UGA codon, provided the codon is followed by another codon beginning with A. Björk and co-workers have shown that the ability to ignore context effects in some measure depends on the correct modification of the base just before the anticodon (i.e., the 3'-base). When it is unmodified, suppressor tRNA molecules function much less efficiently.

Missense mutations are also susceptible to suppression by the same sort of mechanisms as those just described. In addition, it is known that in the case of $tRNA_{lys}$ misacylation (attaching an amino acid other than lysine) can occur. If the incorrect amino acid can compensate for the original mutation, protein function can be regained. Misacylation is not, however, common.

The efficiency of suppression naturally cannot be high. If it were, many normal proteins would be damaged by the same mechanisms that result in mistranslation of the mutated protein. The extent of suppression is rarely more than 10% and frequently less than 1%. Murgola estimated that in the case of a particular *trpA* mutation a suppression efficiency of 2% allows the mutated cell to grow. In a study of alterations in tRNA sequence created by genetic engineering, Yarus found that the anticodon loop could be modified to maximize suppressor function, but changes in the stem region usually minimized suppression. Moreover, changes in an anticodon that cause it to match another codon in the same row of Table 3-3 give a better efficiency of suppression than do changes that result in a match with another codon in the same column.

Another type of intergenic suppression is the development of an entirely new biochemical pathway to replace the one that has been blocked by mutation. Examples in *E. coli* are the *sbc* mutations that reverse the effect of *recBC* mutations (see Chapter 13) and the *ebg* mutation that codes for a new enzyme to replace the defective β-galactosidase in *lacZ* strains (see Chapter 14). This type of suppressor mutation could, of course, also be considered an example of evolution.

Examples of **intragenic** suppression are also known where the second mutation occurs in the same gene as the first. One of the simplest is the occurrence of a second frameshift mutation opposite in sign to that of the first. The combination of frameshift mutations results in a polypeptide with normal amino acid sequences at the amino- and carboxy-termini but with a region of gibberish somewhere in the interior of the molecule. It was by means of experiments involving this type of suppression that the triplet nature of the genetic code was verified.

Intragenic suppression may also occur for missense mutations. In this case, a compensating amino acid change elsewhere in the same polypeptide restores the normal secondary and tertiary structures of the molecule and hence its enzymatic activity. Yanofsky and co-workers provided a good example with their work on mutations affecting the enzyme tryptophan synthetase.

Suppressible mutations are important for work with bacteriophage because they are **conditional** (i.e., the mutant phenotype is expressed only under certain conditions). Mutant phage can be grown normally on a suppressor-carrying host strain and then transferred to a suppressor-free strain for genetic crosses. The phenotype accordingly shifts from normal to mutant, and the experimenter is able to prepare large quantities of phages for analysis even though the mutations they carry should be lethal. In phage work, suppressible mutations are frequently referred to as "sus" mutations.

In the case of bacteriophages, Paolozzi and Ghelardini have shown that it is possible to devise a selection that targets a defect in a specific function. It can be done by cloning a segment of wild-type DNA that includes the function in which mutants are needed into a plasmid. The phage particles are then treated as described below to induce mutations, and those carrying sus mutations are identified by growing them on a suppressor-carrying host strain and on a host strain that lacks the suppressor. Each suppressible mutant is then tested for its ability to grow on a suppressor-free host that carries the plasmid with the cloned DNA. Any phages that carry a suppressible mutation in the region of the DNA that has been cloned are able to grow using the normal gene product provided by the cloned DNA.

When talking of conditional mutations, it is also important to remember that some types of conditional mutations other than chain terminators are modified by temperature, either high or low. In effect, they are suppressed by environmental factors, and their phenotypes can be changed in the middle of an experiment. If the phenotype is mutant at high temperature, the polypeptide is thermolabile or temperature-sensitive (ts). If the phenotype is mutant at low temperature, the polypeptide is cold-sensitive (cs).

Mutagens

A **mutagen** is anything that increases the mutation rate of an organism. Mutagens are frequently used to increase the probability of finding a mutation by some selective process. In this section, a variety of mutagens are discussed and some indication of their modes of action given (summarized in Table 3-4). The list of mutagens is intended to be illustrative but by no means comprehensive. As might be expected, all of the mutagens inflict various types of damage on the DNA of the cell, damage that either cannot be repaired properly or that is so extensive as to overwhelm the repair mechanisms of the cell. A detailed discussion of repair processes can be found in Chapter 13, along with some discussion about mechanisms of mutagenesis.

Radiation

Two types of radiation are commonly used: UV and x-rays. They differ greatly in terms of the energy involved and therefore in their effects. X-rays are extremely energetic, and when they interact with the DNA the

Table 3-4. Some common mutagens and their properties

Mutagen	Structure	Mode of action
X-rays	5 nm wavelength	Single- and double-strand breaks
UV	254 nm wavelength	Pyrimidine dimers
Nitrous acid	HNO_2	Deamination?; intrastrand cross-links?
Hydroxylamine	NH_2OH	Hydroxylation of cytosine
N-methyl-*N*′-nitro-*N*-nitroso-guanidine	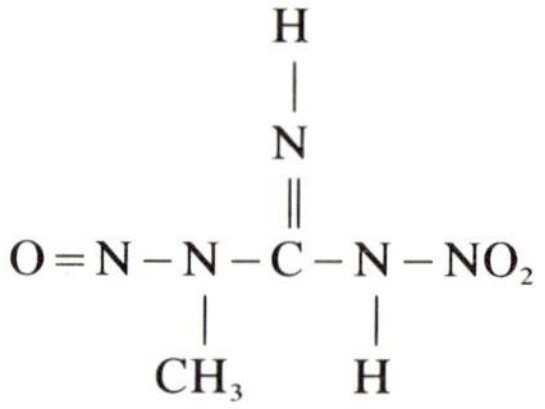 $O{=}N{-}N(CH_3){-}C({=}N{-}H){-}N(H){-}NO_2$	Production of 6-methyl guanine at replication fork; gives mainly transitions
Ethyl methane sulfonate	$CH_2SO_3CH_2CH_3$	Alkylation of purines, gives mainly transitions
Methyl methane sulfonate	$CH_3SO_3CH_3$	Alkylation of purines
2-Aminopurine	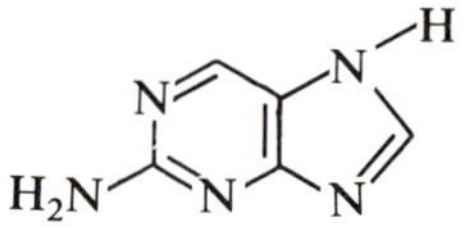 (ring structure; labels: H_2N, N, N, N, N–H)	May replace adenine; may hydrogen bond to cytosine
5-Bromouracil	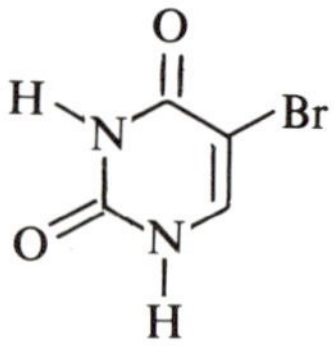 (ring structure; labels: O, H–N, Br, O, N–H)	May replace thymine; may hydrogen bond to guanine
Acridine orange	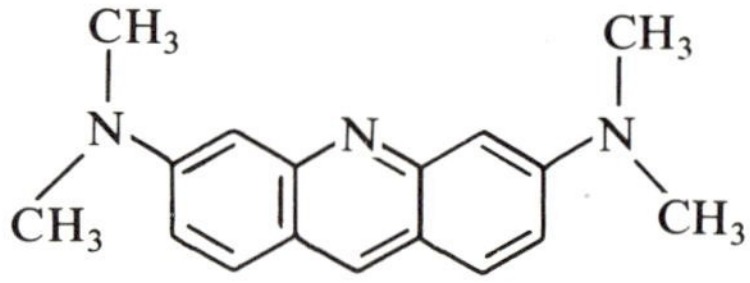 (ring structure; labels: CH_3–N–CH_3, N, N–CH_3, CH_3)	Production of frameshifts
ICR 191 (a nitrogen mustard)	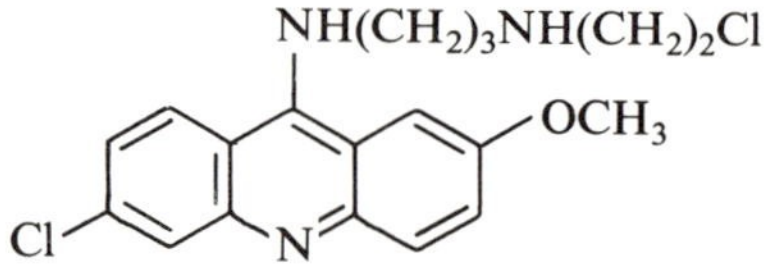 (ring structure; labels: $NH(CH_2)_3NH(CH_2)_2Cl$, OCH_3, Cl, N)	Production of frameshifts

result is usually a break in the phosphodiester backbone of the DNA. UV, on the other hand, catalyzes a reaction in which adjacent pyrimidine bases (on the same strand) form dimers. The presence of a dimer prevents the various polymerases from functioning until it is removed. Mutations may occur during the repair process.

Chemical Modifiers

One of the earliest mutagens used on bacteria was nitrous acid, whose primary effect has been assumed to be a deamination of cytosine and guanine. The deamination would result in a change in the hydrogen-bonding relations so that, at the next replication adenine or thymine, instead of guanine or cytosine, would be inserted. Although it has sometimes been stated that nitrous acid specifically induces transitions of the G•C to A•T type, it is probably an oversimplification, as even transversions are occasionally observed. It has been suggested that nitrous acid may introduce intrastrand cross-links that must be excised in the same manner as alkylated bases (see below).

Hydroxylamine is a moderately specific mutagen that reacts primarily with cytosine but may also attack uracil or adenine. A typical effect of the chemical is to replace a cytosine with a thymine residue.

Numerous **alkylating agents** have mutagenic activity. They attach ethyl or methyl groups at the 6-position on the purine ring, which results in mispairing of the base. Examples of alkylating agents are ethyl methane sulfonate (EMS), methyl methane sulfonate (MMS), and *N*-methyl-*N'*-nitro-*N*-nitrosoguanidine (MNNG). The last is an extremely potent mutagen that produces 6-methylguanine and has its greatest effect at the replication fork. In a culture treated with this mutagen, as many as 15% of the cells may be mutated for a specific trait such as maltose utilization. In fact, the greatest problem with MNNG mutagenesis is its tendency to produce multiple mutations. In a study parallel to that mentioned earlier, Glickman and co-workers examined 167 MNNG-induced mutations of *lacI*. All but three were G•C→A•T transitions.

Base Analogs

A **base analog** is a chemical that has a ring structure similar to one of the normal nucleic acid bases but that does not have the same chemical properties. Some base analogs, e.g., 5-bromouracil (5-BU) or 2-aminopurine (2-AP), are also structural analogs and are incorporated directly into the DNA in place of the normal bases (thymine and adenine, respectively). They tend to be more variable in their hydrogen-bonding properties (Fig. 3-6) and therefore may induce errors during replication, either by inserting themselves in the wrong position or by causing an incorrect pairing when acting as a template. In addition, base analogs may increase the sensitivity of the molecule to other mutagenic treatments (e.g., 5-BU makes DNA more sensitive to UV).

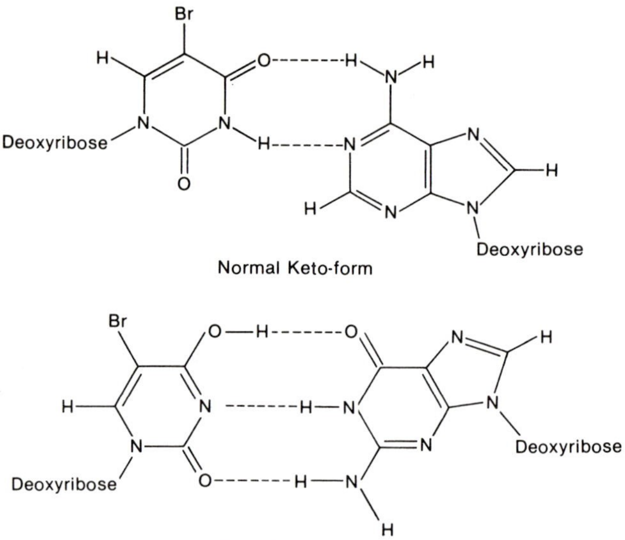

Figure 3-6. Possible hydrogen bonding relations of 5-bromouracil. In each case the base on the left is 5-bromouracil. In its keto-form it pairs with adenine (top), but in its enol-form it can pair with guanine (bottom). The dotted lines indicate hydrogen bonds, and the solid lines indicate covalent bonds.

Other types of base analogs act as **intercalating agents.** To intercalate is to slip between two things, and these chemicals have a ring structure similar to a base but no deoxyribose phosphate with which to be linked into the DNA. An intercalating agent interferes with proper base stocking during replication and may produce a gap in the newly synthesized strand or may result in the newly replicated strand having an extra base at a position corresponding to the point of intercalation. If the gap left by the mutagen is repaired, a base substitution may result. Examples of intercalating agents are acridine orange, proflavin, and nitrogen mustards.

Cross-Linking Agents

Certain chemicals result in the production of interstrand cross-links in the DNA that obviously prevent DNA replication until they are repaired. Examples of cross-linkers are mitomycin C and trimethyl psoralen. The latter compound has been widely used because it must be activated by exposure to 360 nm light. It gives the experimenter good control of the timing of the cross-linking events.

Transposons

Transposons are units of DNA that move from one DNA molecule to another, inserting themselves nearly at random. They are also capable of catalyzing DNA rearrangements such as deletions or inversions. An excellent example is bacteriophage Mu (see Chapter 6), which acts as a mutagen owing to its propensity for inserting itself randomly into the middle of a structural region of DNA during lysogenization, causing loss of the genetic function encoded by that stretch of DNA. The mutations produced are stable, as normal Mu, unlike many other temperate phages, is not inducible and therefore rarely leaves the DNA again. This mode of insertion is in marked contrast to that of a phage such as lambda, which has a specific site of integration for its DNA.

Mutator Mutations

Certain types of mutation that affect the DNA replication machinery have mutagenic effects. These mutations affect the fidelity of the replication process but do not appear to significantly impair the polymerization reactions. Mutations have been isolated in *E. coli* that tend to produce transitions, transversions, deletions, or frameshifts. For example, mutations in the *mutS*, *H*, and *U* functions affect mismatch repair and lead primarily to transition mutations. Mutations in *dnaQ* affect the epsilon subunit of DNA polymerase III, which is the 3′→5′ exonuclease activity. Some of these mutations have been designated *mutD*. In one study of *dnaQ*-induced mutations, it was found that 95% were transversions of the type G•C→T•A or A•T→T•A. An alternative class of mutations affecting the same components of the replication system has also been identified. These are antimutators that alter the replication machinery so as to reduce the error rate below the normal 10^{-10}/base incorporated.

Site-Directed Mutagenesis

The availability of cloned DNA has made possible **site-directed mutagenesis** (in vitro mutagenesis of a specific base or sequence of bases). The general approach is to separate a single strand of DNA that includes the region of interest. An artificial primer DNA sequence is then prepared whose base sequence differs from the original at one or more points (Fig. 3-7). These differences represent the specific mutations to be introduced. They can be base substitutions, insertions, or deletions. The primer is then extended using the Klenow fragment of DNA polymerase I as in the case of DNA sequencing. The resulting double-stranded DNA molecule is transformed into an appropriate host, and the culture is screened to identify mutants of the appropriate type.

One frequent problem with site-directed mutagenesis is that the mismatched bases are repaired by the host bacterium, thereby negating the effect of the primer mismatch. An interesting method to circumvent this problem was devised by Livak and Whitehorn, who used phage T4 DNA polymerase (see Chapter 4) to specifically degrade the original strand of DNA opposite the primer after the primer had been suitably extended (Fig. 3-7). When resynthesis of the degraded bases was permitted, both of the DNA strands were correctly paired and mutated.

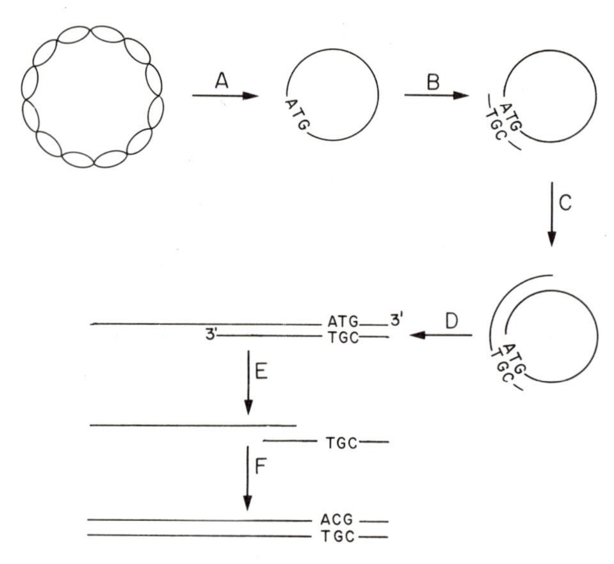

Figure 3-7. Site-directed mutagenesis as suggested by Livak and Whitehorn. The process begins with DNA cloned into a vector based on the phage M13. *A:* A single strand of DNA to be mutagenized is obtained from the phage particles. *B:* An appropriate primer is added that contains a mismatch at a single base, in this case a G•T mismatch. *C:* The primer is extended by the usual methodology. *D:* DNA synthesis is halted, and the circular DNA is linearized by digestion with a restriction enzyme in the double-stranded region near the beginning of the primer and 3′ to the mismatched base. *E:* A 3′-exonuclease is used to digest both old and new strands of DNA in a limited fashion. It removes the region of the old strand that includes the mismatch. *F:* DNA synthesis is allowed to resume, and all of the gaps are filled in. The result is a double-stranded DNA molecule that contains no mismatches but does have a mutation at a specific site. When transformed into a cell, the error correction enzymes do not affect this molecule.

Summary

Genetic variation in bacteria occurs by mutation in the same manner as it does in eukaryotic organisms. It can be shown by fluctuation tests, indirect selection, plate spreading, or sib selection experiments. The rates at which mutations occur in a bacterial culture can be calculated from either direct plate counts or the fluctuation test. In the latter case, assumptions must be made as to the nature of the distribution of the mutations within the population.

Mutant cells may be found by manual screening, but generally some enrichment procedure is used. Phenotypes are selected directly in some cases, but in others an indirect selection using penicillin to attack growing cells is the only feasible method. In any case, it is always necessary to allow sufficient time for segregation lag or phenotypic lag. These lag periods are due to the multiple-genome copies present in growing cells and involve dominance relations of the mutant and nonmutant products.

The genetic code is fundamentally a universal one, but exceptions are known within the mitochondria of certain organisms and within at least one bacterial species. Many types of mutation are known, including base substitutions, insertions, deletions, and frameshifts. An important subclass of base substitutions is the terminator class in which the mutation results in production of a shortened polypeptide. Modifications in tRNA molecules can lead to suppression of base substitutions or frameshifts, but large insertions and deletions can be suppressed only by the development of new metabolic pathways or additional insertions and deletions.

A plethora of mutagens has been identified. All produce some kind of damage to the DNA that fails to be repaired properly, thereby giving rise to a mutation. Radiation treatments, base analogs, cross-linking agents, chemical modifiers, mutations within the DNA synthetic apparatus, and transposable elements such as phage Mu have all been shown to have mutagenic effects. The efficiency and specificity of the mutagens vary widely. A modification of the procedure used for DNA sequencing allows the site-directed mutagenesis of cloned DNA.

References

General

Hatfield, D. (1985). Suppression of termination codons in higher eukaryotes. Trends in Biochemical Sciences 10:201–204.

Murgola, E.J. (1985). tRNA, suppression, and the code. Annual Review of Genetics 19:57–80.

Sirotkin, K. (1986). Advantages to mutagenesis techniques generating populations containing the complete spectrum of single codon changes. Journal of Theoretical Biology 123:261–279.

Wada, A., Suyama, A. (1986). Local stability of DNA and RNA secondary structure and its relation to biological functions. Progress in Biophysics and Molecular Biology 47:113–157.

Specialized

Bouadloun, F., Srichaiyo, T., Isaksson, L.A., Björk, G.R. (1986). Influence of modification next to the anticodon in tRNA on codon context sensitivity of translational suppression and accuracy. Journal of Bacteriology 166:1022–1027.

Brown, T., Hunter, W.N., Kneale, G., Kennard, O. (1986). Molecular structure of the G•A base pair in DNA and its implications for the mechanism of transversion mutations. Proceedings of the National Academy of Sciences of the United States of America 83:2402–2406.

Burns, P.A., Gordon, A.J.E., Glickman, B.W. (1987). Influence of neighboring base sequence on *N*-methyl-*N'*-nitro-*N*-nitrosoguanidine mutagenesis in the *lacI* gene of *Escherichia coli*. Journal of Molecular Biology 194:385–390.

Cavalli-Sforza, L.L., Lederberg, J. (1956). Isolation of pre-adaptive mutants in bacteria by sib selection. Genetics 41:367–381.

Cline, S.W., Yarus, M., Wier, P. (1986). Construction of a systematic set of tRNA mutants by ligation of synthetic oligonucleotides into defined single-stranded gaps. DNA 5:37–51.

Ganoza, M.C., Kofoid, E.C., Marliere, P., Louis, B.G. (1987). Potential secondary structure at translation-initiation sites. Nucleic Acids Research 15:345–360.

Gorini, L., Kaufman, H. (1960). Selecting bacterial mutants by the penicillin method. Science 131:604–605.

Jukes, T.H. (1985). A change in the genetic code in *Mycoplasma capricolum*. Journal of Molecular Evolution 22:361–362.

Kück, U., Neuhaus, H. (1986). Universal genetic code evidenced in mitochondria of *Chlamydomonas reinhardii*. Applied Microbiology and Biotechnology 23:462–469.

Koch, A.L. (1983). Mutations and growth rates from Luria-Delbrück fluctuation tests. Mutation Research 95:129–143.

Lederberg, J., Lederberg, E.M. (1952). Replica plating and indirect selection of bacterial mutants. Journal of Bacteriology 63:399–406.

Leong, P-M., Ksia, H.C., Miller, J.H. (1986). Analysis of spontaneous base substitutions generated in mismatch-repair-deficient strains of *Escherichia coli*. Journal of Bacteriology 168:412–416.

Livak, K.J., Whitehorn, E.A. (1986). Half-site editing; an in vitro mutagenesis procedure for truncating a DNA fragment and introducing a new restriction site. Analytical Biochemistry 152:66–73.

Paolozzi, L., Ghelardini, P. (1986). General method for the isolation of conditional lethal mutants in any required region of the virus genome: its application to the semi-essential region of phage Mu. The Journal of General Microbiology 132:79–82.

Richardson, K.K., Richardson, F.C., Crosby, R.M., Swenberg, J.A., Skopek, T.R. (1987). DNA base changes and alkylation following in vivo exposure of *Escherichia coli* to *N*-methyl-*N*-nitrosourea or *N*-ethyl-*N*-nitrosourea. Proceedings of the National Academy of Sciences of the United States of America 84:344–348.

Rossi, J.J., Berg, C.M. (1971). Differential recovery of auxotrophs after penicillin enrichment in *E. coli*. Journal of Bacteriology 106:297–300.

Trifonov, E.N. (1987). Translation framing code and frame-monitoring mechanism as suggested by the analysis of mRNA and 16S rRNA nucleotide sequences. Journal of Molecular Biology 194:643–652.

Witkin, E.M. (1956). Time, temperature and protein synthesis: a study of ultraviolet-induced mutation in bacteria. Cold Spring Harbor Symposia on Quantitative Biology 21:123–140.

Wood, R.D., Hutchison, F. (1987). Ultraviolet light-induced mutagenesis in the *Escherichia coli* chromosome: sequences of mutants in the *c*I gene of a lambda lysogen. Journal of Molecular Biology 193:637–641.

Wu, C-I., Maeda, N. (1987). Inequality in mutation rates of the two strands of DNA. Nature 327:169–170.

Chapter 4
T4 Bacteriophage as a Model Genetic System

This chapter, in combination with Chapters 5 and 6, present an overview of some of the most extensively studied bacteriophages. The intent of the three chapters is to illustrate the basic nature of genetic processes using some of the simplest genetic systems as examples. In Chapter 2 the distinction was made between phages that are always virulent and those that can temper their lytic response to form lysogens. Chapters 4 and 5 discuss the intemperate (i.e., virulent) phages, and Chapter 6 discusses the problems associated with lysogeny. The details concerning regulation of the metabolic activities discussed in these chapters are deferred until Chapter 12.

The early work on bacteriophages, as was true for many other facets of bacterial genetics, received its impetus and direction from Max Delbrück. It was at his insistence that most phage workers concentrated on just one or two phages during the formative years of phage genetics. Most of the early studies were devoted to bacteriophage T4, a member of the T series of phages that are numbered 1 through 7. The balance of this chapter is devoted entirely to this phage. The rest of the T-series phages, including T2 and T6, which are similar to T4, are discussed in Chapter 5.

Morphology and Composition

T4 is an extremely complex phage, familiar to most students via textbook pictures, that possesses an oblong head 80 by 120 nm in size, with a contractile tail measuring 95 by 20 nm (Fig. 4-1a). The various specific anatomic structures of which the **virion** (phage particle) is composed are indicated in Figure 4-1b. Most of the mature virion is composed of associations of protein subunits, but there are several important nonprotein

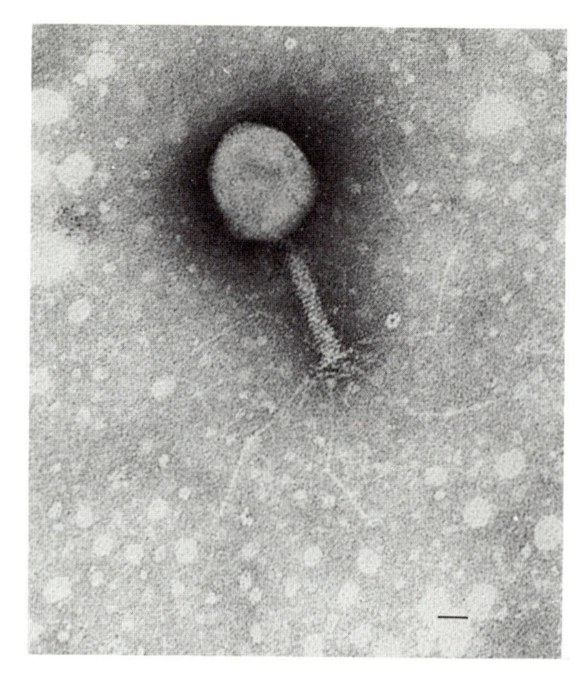

(a)

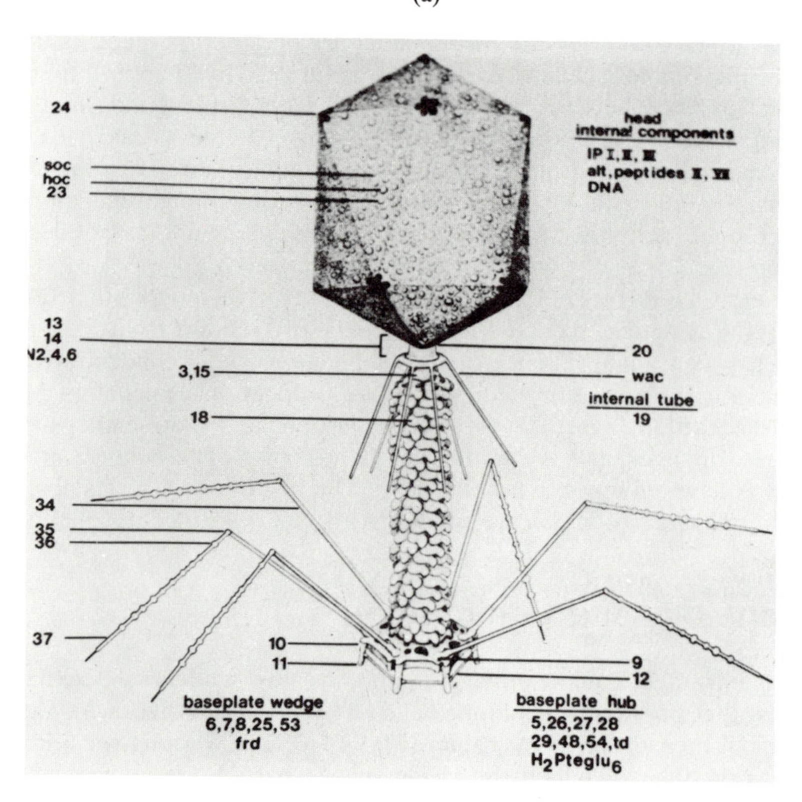
24
soc
hoc
23
head
internal components
IP I, II, III
alt, peptides II, VII
DNA
13
14
N2,4,6
20
3,15
wac
internal tube
19
18
34
35
36
37
10
11
9
12
baseplate wedge
6,7,8,25,53
frd
baseplate hub
5,26,27,28
29,48,54,td
H2Pteglu6

(b)

constituents that have been identified. Included among them are a large, linear, double-stranded DNA molecule with a molecular weight of 1.30 $\times$ 10^8 daltons (1.69 $\times$ 10^5 nucleotide pairs) contained within the head; certain polyamines (putrescine, spermidine, and cadaverine) associated with the DNA; ATP and calcium ions found associated with the tail sheath; and dihydropteroylhexaglutamate associated with the base plate.

One unique feature of T4 phage is its DNA composition. The DNA has had all of its cytosine residues replaced by hydroxymethylcytosine (Fig. 4-2), a base that does not normally occur in *E. coli*. This base has the same hydrogen-bonding characteristics as normal cytosine and therefore does not affect the genetic code, but it does offer a reactive site to which is attached one molecule of glucose (70% of the time in the α-linkage and 30% of the time in the β-linkage). These two differences permit the phage DNA to be unambiguously differentiated from host cell DNA in various experiments. They also have important physiologic functions. The hydroxymethylcytosine is necessary for the expression of certain late functions (see below), and the glucose moieties prevent restriction of the phage DNA by the host cell.

A virion with a genome the size of T4 can be expected to produce a large number of proteins, many of which are enzymatically active. Most of them are, of course, associated with the intracellular activity of the phage. However, certain enzymatic activities have been identified as being present within the free virion as well as in the cytoplasm of an infected cell. Some of these enzymes are listed in Table 4-1, along with their presumptive functions in the infectious process. In many cases a functional enzyme is not required for normal infectivity, suggesting that other means exist to provide the same function. Modifications in protein structure, however, produce physiologic changes. For example, strains T4B (Benzer) and T4D (Doermann) exhibit slight differences in their dihydrofolate reductase activity that appear to be reflected in a requirement by T4B, but

Figure 4-1. **(a)** Electron micrograph of a T4 virion negatively stained with potassium phosphotungstate. The bar indicates a length of 25 nm. **(b)** Structure of bacteriophage T4, based on electron microscopic structure analysis to a resolution of about 2 to 3 nm. Near the head and tail are the locations of the known major and minor proteins. The icosahedral vertices are made of cleaved gene 24 protein (gp24). The gene 20 protein is located at the connector vertex, bound to the upper collar of the neck structure. The six whiskers and the collar structure appear to be made of a single protein species, gp*wac*. The gp18 sheath subunits fit into holes in the baseplate, and the gp12 short tail fibers are shown in a stored position. The baseplate is assembled from a central plug and six wedges; and although the locations of several proteins are unknown, they are included here with the plug components. Diagram from Eiserling, F.A. (1983). Structure of the T4 virion, pp. 11–24. In: Mathews, C.K., Kutter, E.M., Mosig, G., Berget, P.B. (eds.) Bacteriophage T4. Washington, D.C.: American Society for Microbiology.

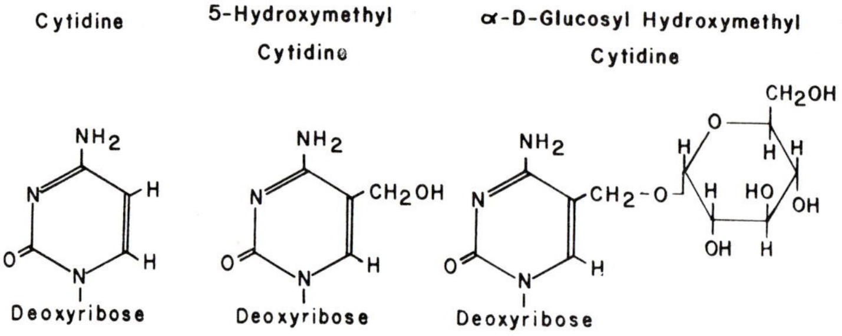

Cytidine | 5-Hydroxymethyl Cytidine | α-D-Glucosyl Hydroxymethyl Cytidine

Figure 4-2. Forms of cytidine found in T4-infected cells. The structural formula for the normal base is shown at the left, and the progressive modifications made by T4 are shown at the right. Although the glucosyl moiety is shown in an α-linkage, a β-linkage is also observed. Phages T2 and T6 may add a second glucosyl moiety attached to the first one in an α-linkage.

not T4D, for the presence of tryptophan in order to unfold the tail fibers from around the sheath.

Experimental Methods Used to Study Phage Infection

It is possible to study phage-infected cells in a variety of ways. The presence of a virion in a sample can be demonstrated by adding it to a culture of phage-sensitive bacteria and then plating the mixture in a soft agar (0.6%) overlay. Because T4 is a virulent phage, the infected bacterial cells lyse, releasing virions that can then infect other sensitive cells. As the bacteria increase in numbers, so do the virions by constantly infecting new bacterial cells. The multiple rounds of infection give rise to a hole,

Table 4-1. Virus-specific enzymatic activities associated with T4 virions

Enzyme	Location and/or function
Dihydrofolate reductase	Located in baseplate, may have role in unfolding tail fibers
Thymidylate synthetase	Found in baseplate; necessary for infectivity
Lysozyme	Found in baseplate; penetration through cell wall
Phospholipase	Possible role in host cell lysis
ATPase	Associated with the tail sheath; presumed to be involved with the contractile process
Endonuclease V	Excision repair (see Chapter 13) of phage or host DNA
alt Function	Alteration of the host RNA polymerase

Adapted from Mathews, C.K. (1977). Reproduction of large virulent bateriophages, pp. 179–294. In: Fraenkel-Conrat, H., Wagner, R.R. (eds.) Comprehensive Virology, vol. 7. New York: Plenum Press.

or **plaque,** in an otherwise confluent bacterial lawn. For obvious reasons this technique is called a plaque assay. Each plaque is assumed to represent a phage-infected original cell, and therefore each plaque is considered analogous to a bacterial colony for counting purposes. Four specialized techniques have been developed for obtaining more specific types of information about the events that occur during phage infection: the one-step growth experiment; the single burst experiment; the premature lysis experiment; and the electron microscopic observation experiment.

The **one-step growth experiment** was developed by Ellis and Delbrück and depends on the production of a synchronous phage infection (all phage-infected cells at the same stage in the infectious process at the same time). The synchrony is accomplished by either (1) limiting the attachment of the phage to the bacterial cell to a short period of time or (2) first treating the bacterial culture with a reversible metabolic poison such as potassium cyanide and then adding the virions to the culture. In the latter case, the phages go through the early stages in their life cycle, but the cyanide prevents any macromolecular synthesis from taking place. In either case, after a suitable period of time, any phages that have not attached to the cells are eliminated by either diluting the culture to the point where collisions between phages and bacteria become improbable or by neutralizing with phage-specific antiserum. If cyanide is used, it is washed out of the culture and metabolism allowed to start. The time of washing or dilution becomes the zero time, after which samples are removed from the culture at various time intervals, mixed with indicator (phage-sensitive) bacteria, and assayed for infectious centers (plaque-forming units, PFU).

A typical curve from such an experiment is shown in Figure 4-3. The

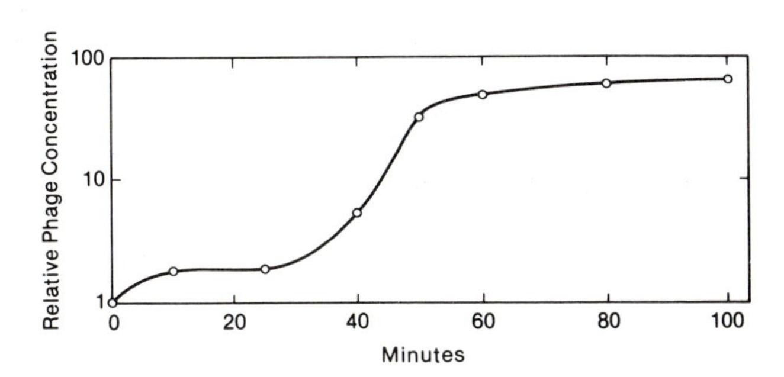

Figure 4-3. One-step growth curve. Bacteriophage T4 was added to a bacterial culture at time zero. After 10 minutes of adsorption, the entire culture was diluted 10^{-4} so that the probability of further collisions between virus particles and bacterial cells was small. When the release of new phage particles began, the culture was further diluted 10^{-1} to prevent a new step in the growth cycle. The period prior to 25 minutes is the latent phase, between 25 and 45 minutes is the rise phase, and later than 45 minutes is the plateau phase. The entire experiment was carried out at 37°C. Redrawn from Figure 3 of Ellis, E.L., Delbrück, M. (1939). The growth of bacteriophage. Journal of General Physiology 22:365–384.

curve is triphasic with an initial **latent phase,** during which the number of infectious centers is constant, followed by a **rise phase** and then a **plateau phase.** The interpretation given to these results is that during the latent phase an infectious center results from a phage-infected cell that lyses at some time after it has been mixed with indicator bacteria and immobilized in agar. The soft agar ensures that the released virions cannot diffuse more than a few micrometers, and only a single plaque is formed. During the rise phase the infected bacteria begin to lyse before the sample is removed from the culture so that an infectious center may represent an infected cell or a free virion. During the plateau phase an infectious center represents only a free virion. The ratio between the number of infectious centers in the plateau phase and the number of infectious centers during the latent phase represents the average **burst size,** the average number of phage particles released per infected cell.

The **single burst experiment,** also devised by Ellis and Delbrück, can be used to study individual infected cells rather than using the averages obtained in the one-step growth experiment. The goal of the experiment is to arrange matters so that each culture tube has only one infected cell. In such a case, after lysis is complete all phage particles found in the tube result from a single burst cell. In order to achieve this goal, a culture of bacteria is infected with a small number of phages such that the average number of phages per bacterium (the **multiplicity of infection,** MOI) is less than 1.0. Then the distribution of infected cells among small samples taken from the culture can be described by the Poisson distribution (see Appendix). If the sample size is appropriately chosen, the probability of obtaining two or more phage-infected cells per sample is low. The samples are diluted to reduce the cell density and then incubated for several hours. During the incubation any infected cells lyse, but the released phage particles are unable to infect new cells because of the low culture density (for cultures with fewer than 10^6 objects/ml, the probability of collision between any two objects is vanishingly small).

The data obtained by Ellis and Delbrück are presented in Table 4-2. Assuming one infected cell per tube, the range of individual burst sizes is enormous, varying from 1 to 190. The average burst size, as determined from the 15 infected tubes, is 883 total phage particles per 15 infected tubes, or about 59 phage particles per burst. A more accurate determination of the average burst size is possible if one takes into account the probability that there was more than one infected cell per tube. It can be done by using the zero case of the Poisson distribution to determine m, the average number of infected cells per tube (which works out to 0.47). Because 40 tubes were used in the experiment, the number of infected cells expected in the entire experiment is $40 \times 0.47 = 18.8$ infected cells, which would have been expected to be distributed among the 15 tubes that produced phage particles. The excess of the calculated number of infected cells over the observed 15 infected tubes suggests the existence of samples

Table 4-2. Single burst experiment of Ellis and Delbrück[a]

Culture	No. of plaques	Culture	No. of plaques
1	0	21	0
2	130	22	0
3	0	23	0
4	0	24	53
5	58	25	0
6	26	26	0
7	0	27	48
8	0	28	1
9	0	29	0
10	0	30	72
11	123	31	45
12	83	32	0
13	0	33	0
14	9	34	0
15	0	35	0
16	31	36	0
17	0	37	190
18	0	38	0
19	5	39	9
20	0	40	0

[a]A culture of *E. coli* was infected with a dilute phage solution, allowed to stand for 10 minutes, and then diluted more than 100-fold. Aliquots of 0.05 ml were placed in separate tubes and incubated for 200 minutes. The entire contents of each tube was then used in a plaque assay.

From Ellis, E.L., Delbrück, M. (1939). The growth of bacteriophage. Journal of General Physiology 22:365–384.

with two or more infected cells (which are expected to occur 8% of the time). The new average burst size then becomes 883 total phage particles per 18.8 total infected cells, or approximately 47 phages per infected cell. Both values for the average burst sizes are in accord with values determined in one-step growth experiments. Phage infection is apparently a complex process and subject to numerous outside influences that affect its efficiency, because even if the highest three values of Table 4-2 are discarded on the grounds that they represent multiple infections, the burst sizes still range from 1 to 83.

The **premature lysis experiment** was designed by Doermann, who took a standard one-step growth experiment and altered it so that the samples removed at various times were treated to lyse all the cells in the culture. The lysates can then be assayed for the presence of free phage particles.

The cells can be lysed by shaking with chloroform or by superinfecting the culture with a high multiplicity (~100) of T6 phages whose simultaneous attempts to inject their DNA into cells cause the cells to lyse before any T6 infection can be initiated (**lysis from without,** in contrast to lysis from within caused by a completed phage infection). The curve Doermann obtained is shown in Figure 4-4. Under the conditions of this experiment, no infectious phage particles could be detected until about 12 minutes postinfection. This particular phase of the growth cycle, a period when there are no detectable infectious phages in the culture, is referred to as the **eclipse phase.** An experiment of Hershey and Chase, using phage particles carrying different radioactive labels in the protein (^{35}S) and DNA

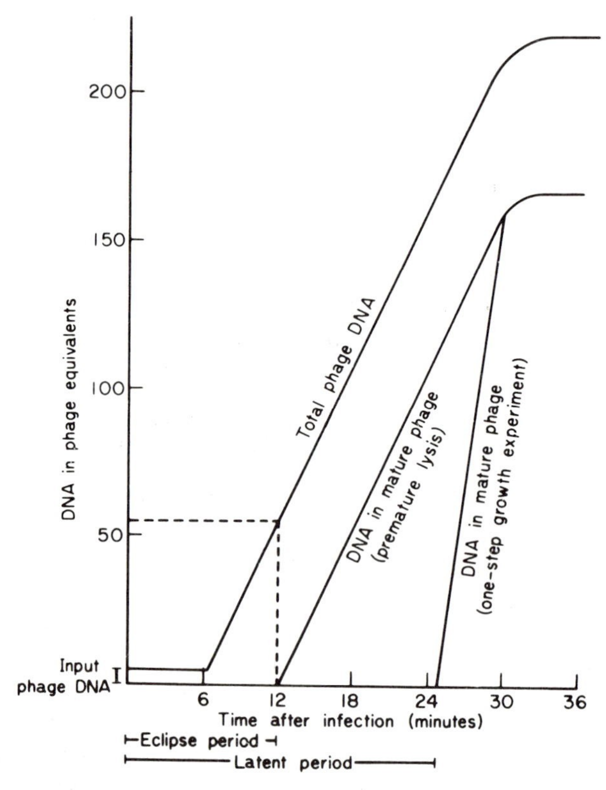

Figure 4-4. Idealized curves illustrating the kinetics of synthesis of phage DNA in bacteria infected with one of the T-even phages. The amount of DNA observed in various experiments is expressed as phage equivalents, which in the case of the DNA found in mature phage correspond to the actual number of infectious virions present. "Total phage DNA" refers to both the DNA within the virions and the DNA in the bacterial cytoplasm. "DNA in mature phage" refers to those molecules found within infectious virions. From Hayes, W. 1968. The Genetics of Bacteria and Their Viruses. Oxford: Blackwell Scientific.

(^{32}P), indicated that only the phage DNA enters the cell. The premature lysis experiment indicates that the pure phage DNA is not infectious and that the length of the eclipse phase therefore must represent the time required to form a new protein coat for the virus. When infectious phage particles do appear, the rate of their production is not exponential, as with the release of phages in the one-step growth experiment but, rather, arithmetic. The arithmetic growth rate during the remainder of the latent period indicates that infectious phage particles appear singly, as though they were produced by an assembly-line type of process rather than by some sort of fission process as would be observed in a bacterial culture in which two new cells arise from one old one.

Electron microscopy has proved to be a powerful tool for understanding the interactions between phages and bacterial cells. Two general types of procedure can be used. Free phages, or portions thereof, can be examined by negative staining techniques such as in Figure 4-1. Partially purified fractions of cells or phages, such as pieces of cell wall or phage particles devoid of DNA **(ghosts),** can be interacted with the intact attachment site to study early events in the adsorption of virions to the bacteria. Phage-infected cells can be thin-sectioned to demonstrate the internal interactions of virus components with the cell wall and membrane.

Genetic Organization of T4

Types of Mutations Observed in the T4 Genome

Most of the biochemical reactions of the eclipse phase have been genetically defined by appropriate mutations. In nearly all cases these mutations have turned out to be conditional, usually chain terminators, as the enzymatic activities are required for successful infection. These mutations, although indubitably useful to a geneticist, produce phenotypes that are frequently difficult to assay. Consequently, the earliest genetic studies relied on mutations that affected general physiologic traits and gave rise to readily observed phenotypes. These mutations are usually referred to as **plaque morphology mutations.**

The standard T4 plaque has a clear center with a halo or turbid area surrounding it. The halo is an area where lysis was not complete. Some cells lysed, but many others did not. This lack is due to a phenomenon called **lysis inhibition,** which results when a cell that has been infected by a T4 phage is reinfected (superinfected) by another T4 phage prior to cell lysis. The new phage delays (inhibits) lysis of the cell in some fashion so that the intracellular production of phage may continue for 60 minutes or more. The burst size, of course, is enormous and may be as large as 1000 phages per cell. Hershey isolated mutant T4 phages called *r* (for rapid lysis) that were not susceptible to lysis inhibition and whose plaques therefore had sharp edges (Fig. 4-5). Progeny from crosses between r^+ and *r* phages are easy to categorize phenotypically, as all that is required is an inspection of the resulting plaques. The experiments by Benzer, which are discussed below, dealt exclusively with rapid lysis mutants.

Genetic Crosses

The basic strategy for phage crosses is to simultaneously infect cells with two or more genetically distinct phages, allow the cells to be lysed, and then examine the genotypes of the progeny phages. The recombination frequency is calculated as the number of recombinants per number of minority parental phages recovered, assuming no selective pressure is applied. It is assumed that the greater the genetic distance between any two mutations, the more frequently do genetic exchanges occur between them. Therefore closely linked markers show little recombination, whereas unlinked markers approach the 50% frequency characteristic of independently inherited traits. Table 4-3 presents data from a typical set of T4 crosses. The three mutations, *r47, r51,* and *tu41* (a temperature-sensitive *r* mutation), appear to be situated near one another and therefore to comprise one **linkage group** (a group of markers that tend to be coinherited). The gene order for the linked mutations is *r47 - r51 - tu41,* with a greater genetic distance between *tu41* and *r51* than between *r47* and *r51*. Each member of this linkage group recombines at equal frequency with *tu44,* which must therefore be considered to lie in a second linkage group. If it did not, either *r47* or *tu41* would be expected to show a lower recombination frequency than the rest of the members of the linkage group. Using the data presented it is impossible to tell whether *tu44* lies to the right or to the left of the other mutations.

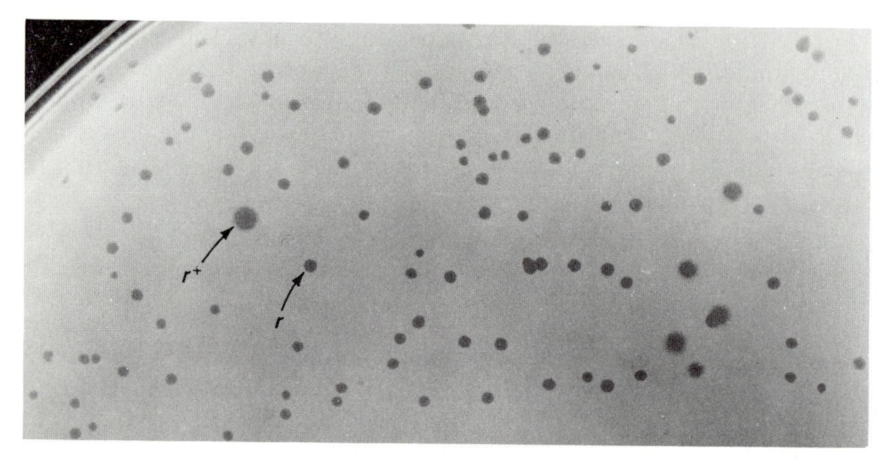

Figure 4-5. The *r* phenotype of phage T4B. Note the several large plaques with fuzzy edges that were produced by wild-type phages. The small sharp-edged plaques were produced by *rIIA* phages.

Table 4-3. Results of some T4 crosses[a]

Cross	Percentage recombination	Cross	Percentage recombination
r47 × *r51*	5.4	*tu44* × *r47*	36.1
r51 × *tu41*	23.7	*tu44* × *tu41*	37.0
r47 × *tu41*	25.0	*tu44* × *r51*	38.8

Deduced Map Order

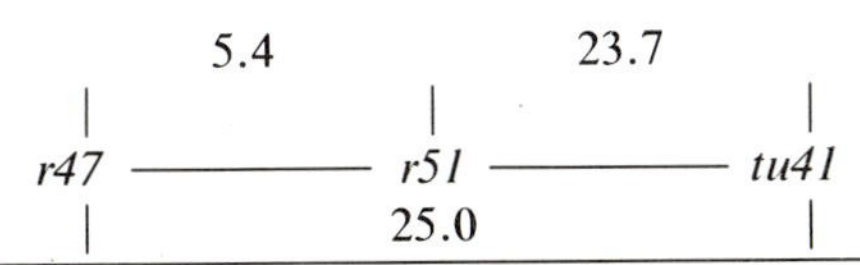

[a]Two *r* mutants and two turbid halo (*tu*) mutants were subjected to pairwise crosses and the percentage of recombinant progeny determined by examining the plaques formed by the progeny phages. It is not possible to use these data to determine the map position for *tu44* (see text).

Data are from Doermann, A. H., Hill, M. B. (1953). Genetic structure of bacteriophage T4 as described by recombination studies of factors influencing plaque morphology. Genetics 38:79–90.

Two important problems may arise when certain types of mutant are used in phage genetic crosses. One problem is that many mutant phenotypes involve a series of nonallelic genes, meaning that a mutation at any one of several independent sites produces the same phenotype. A good example is the case of the *r* mutations discussed above. Although *r47* and *tu44* are unlinked (and therefore not identical), they confer the same phenotype. The second problem is that using mutations affecting the adsorption characteristics of a phage may result in phenotypic mixing. **Phenotypic mixing** results in a virion whose DNA does not code for all of the virion proteins encasing it. Such mixing occurs because virions are assembled from pools of constituents; therefore in a cell that has been infected with two genetically distinct phages (a mixed infection), it is possible for DNA that codes for one phenotype to be accidentally packaged into a phage particle with a second phenotype. If the resultant particle is taken through more than one cycle of infection, as in a standard plaque assay, the discrepancy between genotype and phenotype makes no difference, because after the next infection cycle the DNA will be correctly packaged and have the appropriate phenotype for the virion. However, if the experimental protocol calls for a phenotypic selection during the first growth cycle, it is possible that phage particles of the correct genotype will, nonetheless, not be recovered because they have the wrong phenotype.

Using recombination frequencies calculated as in Table 4-3, it would be possible to construct a genetic map for T4, but pairwise matings between all mutations in a given region would be required in order to determine the map order. Before such extreme measures became necessary, Benzer developed a mapping system that could be used to reduce the number of pairwise matings to few or none and yet allow the unambiguous ordering of each mutation on the genetic map.

Benzer began by adapting the standard **cis-trans test** of eukaryotic genetics for use with bacteriophages. He infected bacterial cells with different mutant phages having the same phenotype (one that would normally prevent their growth) and investigated whether phage infection was successful (Fig. 4-6). This step was the *trans* portion of the test, because the mutations were located on different DNA molecules. The *cis* portion of the test was a control experiment that consisted in infecting a cell with one phage that carried both mutations and another phage that carried none. In the *cis* case the wild-type phage would be expected to provide the gene product(s) (RNA or protein) that was defective in the mutant phage. This product would diffuse throughout the cytoplasm and permit both types of phage to grow and lyse the cell. In the *trans* case, if the mutations affected different genes, each phage would contribute a different functional gene product (i.e., **complementation** would occur), and the phages should carry

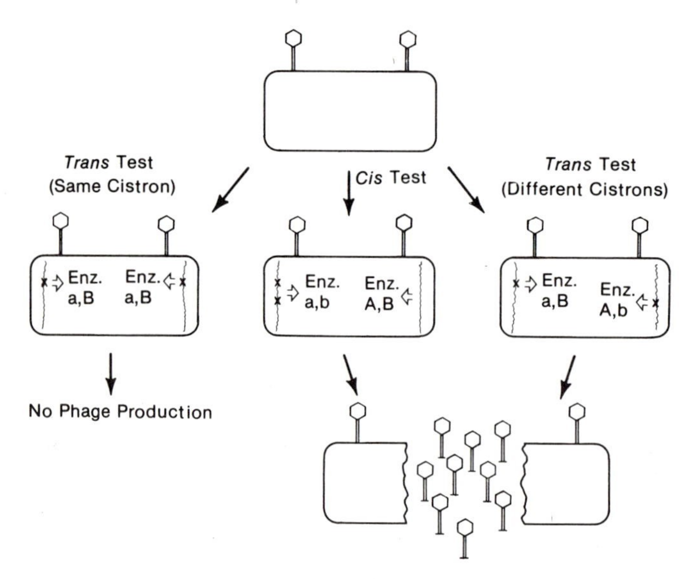

Figure 4-6. Possible results of a *cis-trans* test. A cell is initially infected with two genetically distinct phage particles. In the second row are shown three arrangements of mutations among the injected DNA molecules, and in the third row are the two possible results of the simultaneous infection. The site of a mutation on the phage DNA is indicated by an X. Mutant proteins are indicated by lower-case letters, and functional proteins are indicated by upper-case letters. It is assumed that the protein products manufactured from the genetic information are freely diffusible within the cell and that functional A protein and functional B protein are both required for production of new virions and cell lysis.

out a successful infection. However, if both mutations affected the same gene, the same gene product would be defective from both phages, and neither phage particle would be able to grow. If complementation occurs, the *cis-trans* test is considered positive; if it does not occur, the test is considered negative. To the extent that recombination occurs between the two mutations in the *trans* test, the results may be somewhat muddled but can be sorted out by an examination of the phenotypes of the progeny phages. Complementation gives rise only to parental types of phages, whereas recombination produces some wild-type phages. Therefore the fundamental *cis-trans* test assumes that no recombination occurs.

The results of a large number of *cis-trans* tests were sufficiently unambiguous for Benzer to propose a new term, **cistron,** to describe the region of the genome within which all mutations give negative *cis-trans* tests. The term cistron is much more precise than the term gene, as the latter has no operational definition. Nevertheless, the term gene has not disappeared and probably will not. In most cases when the term gene is used, it is possible to substitute the term cistron, and it is in this way that the two terms are used in the remainder of this book. Thus the "one gene–one enzyme" hypothesis of Beadle and Tatum becomes the "one cistron–one polypeptide" hypothesis, etc. The term gene is reserved for poorly defined segments of DNA, such as those used for gene splicing (see Chapter 14).

Once mutations had been assigned to specific cistrons, the next problem was to order the mutations within each cistron. When demonstrating this procedure, Benzer dealt specifically with the *r*II region of T4, which he showed to consist of two cistrons, but the technique has general applicability. All that is required is a large number of deletion mutations in the region of interest. Deletion mutations can be readily identified because they can never revert to wild type because they suffer from a lack, rather than an alteration, of genetic information. If two deletion mutations overlap, i.e., have lost some of the same genetic material, it is impossible to recombine the two deleted DNA molecules so as to produce a wild-type DNA molecule. Therefore in a cross between two *r* deletion mutants that overlap no r^+ progeny would be expected, whereas if there were no overlap r^+ progeny would be expected. By a similar argument, when a point mutation (missense, nonsense, frameshift) is crossed to a deletion mutation, if the point mutation lies within the deleted region no r^+ recombinants are possible. If it lies outside the deletion, r^+ recombinants are produced at a frequency that depends on the distance between the point mutation and the nearest end of the deletion: the greater the distance, the more recombinants expected.

Benzer's deletion-mapping protocol requires a series of overlapping deletions whose map order has been determined by crossing them to known point mutations and to each other in a manner such as that described in Table 4-3 (realizing that it is impossible to distinguish left from right because there is no landmark such as a centromere on the DNA molecule). The particular set of deletions used by Benzer is displayed in Figure 4-7. The upper group of deletions divided the two cistrons of the *r*II region into seven discrete segments. The lower group of deletions could be used to subdivide any particular segment as required. The crosses of the point mutations to the deletions could be carried out quickly by means of a spot test. All *r*II mutants grow on *Escherichia coli* strain B but not on *E. coli* strain K-12 (λ^+) (which is a phage lambda lysogen). All r^+ phages grow on either strain. Various T4 phage lysates prepared on strain B can be tested in a pairwise fashion by spotting phage carrying a point mutation and phage carrying a deletion mutation together on a plate with a lawn of *E. coli* K-12 (λ^+). If recombination yields r^+ progeny, lysis occurs and a clear area develops in the lawn, which can be visually scored (Fig. 4-8).

Using his deletion-mapping procedure, Benzer was able to assign any given *r*II mutation to one of 47 segments on the DNA in just two exper-

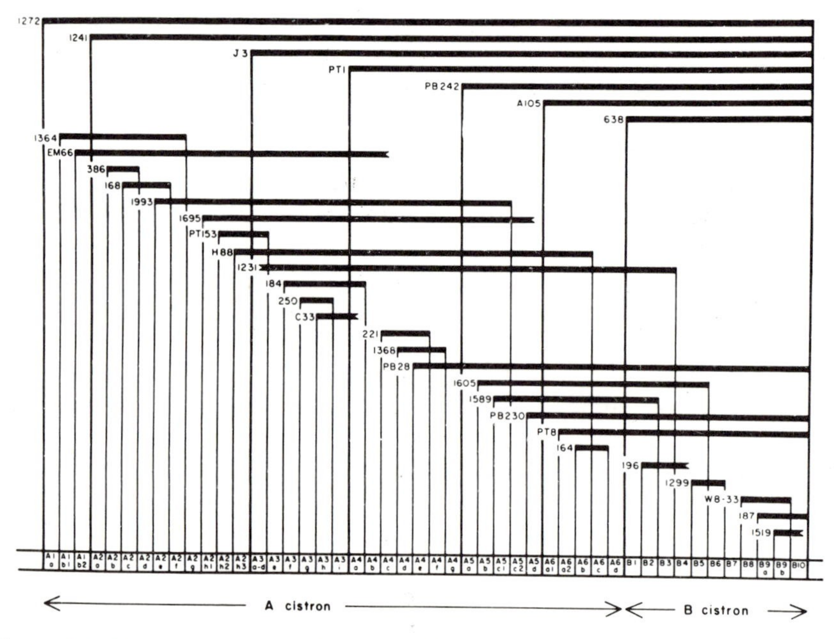

Figure 4-7. Deletions used to divide the *r*II region of the T4 genome into 47 small segments. Each horizontal line represents the material that has been deleted in a particular mutant. If the end of the deletion is shown as fluted, its position is not precisely known and it has not been used to define a segment. The *A* and *B* cistrons, which are defined by the *cis-trans* test, coincide with the indicated portions of the recombination map. From Benzer, S. (1961). On the topography of the genetic fine structure. Proceedings of the National Academy of Sciences of the United States of America 47:403–415.

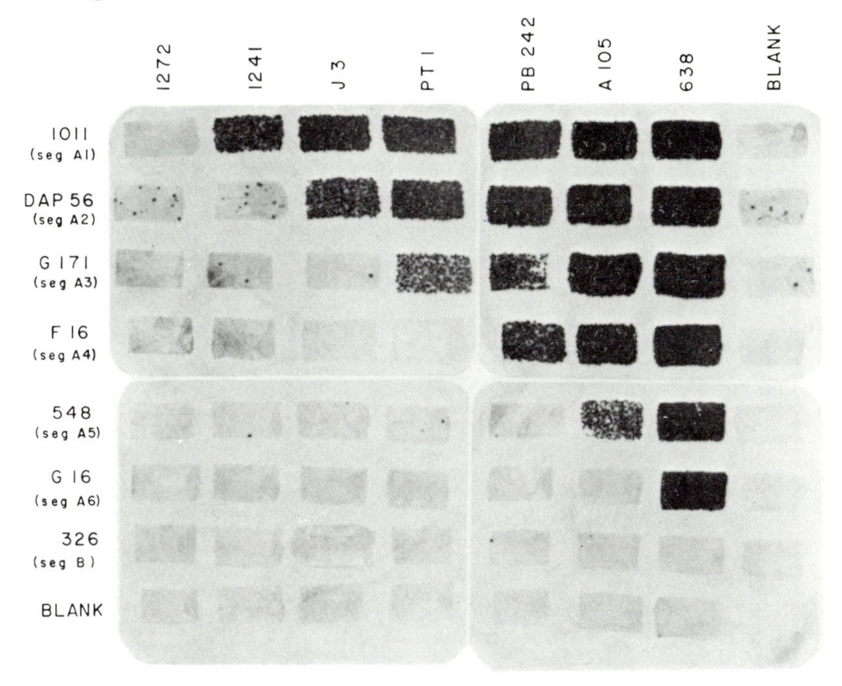

Figure 4-8. Crosses for mapping *r*II mutations. The photograph is a composite of four plates. Each row shows a given mutant tested against the reference deletions shown at the top of Figure 4-7. The test consisted in allowing the two mutants to infect *E. coli* B and then using a paper strip to spread some of the culture onto a lawn of *E. coli* K cells that had been spread on the plate. Under the conditions of the test, neither parental phage was able to grow on *E. coli* K. However, if recombination occurred during the one cycle of growth in *E. coli* B, the progeny phage would be able to grow on *E. coli* K, and would manifest themselves as a clear area on the plate. Isolated plaques appearing in the blank spaces represent spontaneous revertants in the phage stock. As an example of the interpretation of the data, mutant 1011 formed no $r\mathrm{II}^+$ recombinants with deletion 1272 but did with all other deletions tested. Therefore the mutant site in strain 1011 is included within the region uniquely deleted in strain 1272 (i.e., the region at the far left in Figure 4-7). From Benzer (1961).

imental steps by crossing it to the seven large deletions and then to some of the smaller deletions. Final map order could be obtained by two- and three-point (marker) crosses as above. By the time the work was completed, Benzer had mapped more than 2400 independently derived mutations (i.e., from more than 2400 separate cultures), representing 304 sites on the T4 genome. A map presenting the frequency distribution of 1612 spontaneous mutations is shown in Figure 4-9. These experiments were genetic analysis in the purest sense, because Benzer completed the entire study without knowing anything about the biochemistry of the gene products he was studying. In fact, more than 10 years elapsed before two

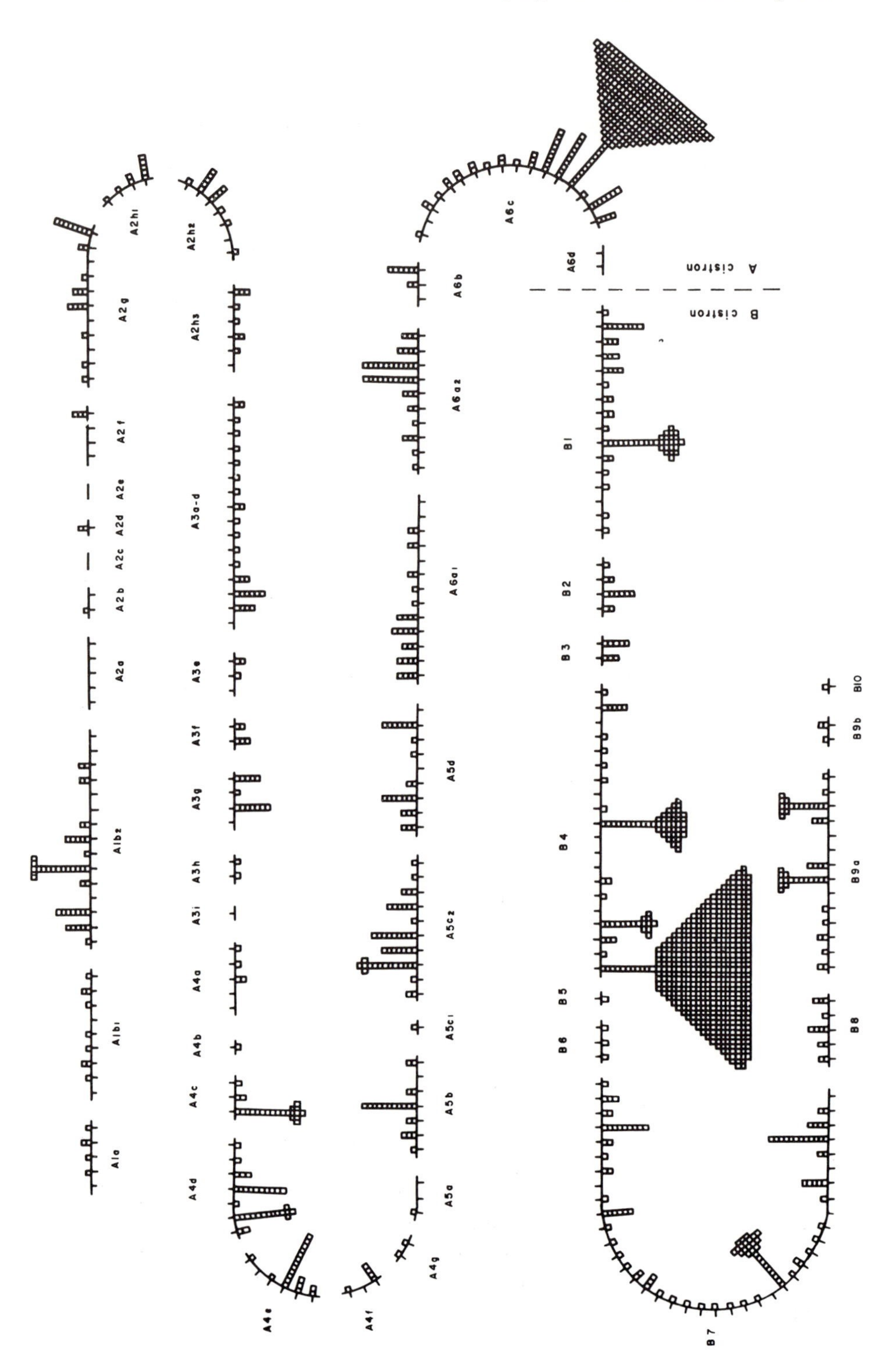
A1a
A1b1
A1b2
A2a
A2b
A2c
A2d
A2e
A2f
A2g
A2h1
A2h2
A2h3
A3a-d
A3e
A3f
A3g
A3h
A3i
A4a
A4b
A4c
A4d
A4e
A4f
A4g
A5a
A5b
A5c1
A5c2
A5d
A6a1
A6a2
A6b
A6c
A6d
A cistron
B cistron
B1
B2
B3
B4
B5
B6
B7
B8
B9a
B9b
B10

proteins appearing in the cell wall of infected bacteria were identified as the products of the *r*II cistrons.

Benzer's work still stands as the ultimate in genetic fine structure analysis. However, in addition to providing a thorough genetic map of the *r*II region, several other important concepts were developed from an analysis of his data. Although it is commonly accepted today that genetic maps are linear structures, the question of linearity was by no means resolved when Benzer began his experiments. In fact, until electron microscopy of DNA molecules came into vogue, maps such as the one shown in Figure 4-9 were the only real evidence for linearity.

By comparing the frequency distribution of mutations at different sites in the *r*II region in a fashion similar to that of Figure 4-9, Benzer showed that spontaneous mutations tend to occur at sites different from those of mutagen-induced mutations. Moreover, it is possible to test if the mutations are randomly distributed over the cistrons by comparing the Poisson distribution with the observed distribution of mutant sites. Highly mutable sites occur too frequently, indicating that mutations are not totally random, but the sites represented only once or twice in the distribution fit the Poisson distribution well. Because the sizes of the $r = 1$ and $r = 2$ classes were known, Benzer was able to use the Poisson distribution to estimate the probability of the $r = 0$ case, the number of sites within the cistrons at which no mutations had been observed. The calculated value was 28%, giving a map saturation value of 72% (i.e., 72% of all possible mutable sites had been identified). These data, taken in conjunction with some estimates of the physical size of the *r*II region, provide an experimental basis for the tacit assumption that the smallest mutable unit of a genome (a **muton**) is a single base. Moreover, they indicate that the smallest unit of recombination (a **recon**) is also a single base.

Using deletion mapping, pairwise crosses, and some physical mapping techniques (see Chapters 5 and 6), it is possible to construct a genetic map for T4 (Fig. 4-10). The zero point is taken as a junction between the *r*IIA and *r*IIB cistrons, as they have been precisely mapped. One map unit represents 1% recombination in a standard phage cross, with the provision that map distances greater than 1.0 unit (1% recombination) are generally not additive. For example, two markers (mutations) separated

Figure 4-9. Topographic map of the *r*II region for spontaneous mutations. Each mutant arose independently in a plaque of either standard-type T4B or various *r*II mutants. Each square represents one occurrence observed at the indicated site. Sites with no occurrences indicated are known to exist from induced mutations and from a few other selected spontaneous mutations. Each segment of the map is defined by combinations of the deletions shown in Figure 4-7. The arrangement of sites within each segment is arbitrary but could be absolutely established by crosses such as those shown in Table 4-3. From Benzer (1961).

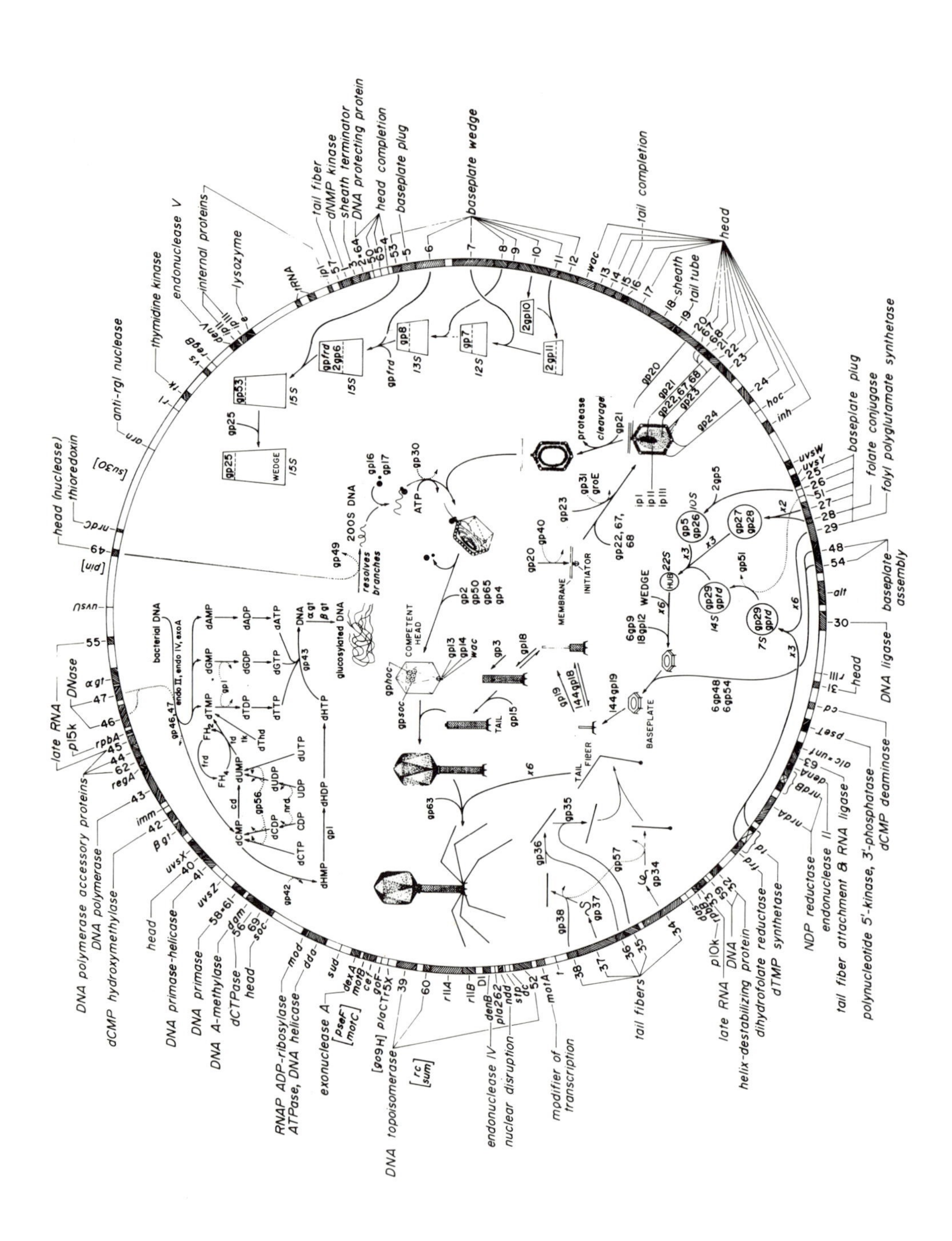
tail fiber
dNMP kinase
sheath terminator
DNA protecting protein
head completion
baseplate plug
baseplate wedge
tail completion
head
sheath
tail tube
thymidine kinase
endonuclease V
internal proteins
lysozyme
anti-rgl nuclease
head (nuclease)
thioredoxin
late RNA
DNase
DNA polymerase accessory proteins
DNA polymerase
dCMP hydroxymethylase
DNA primase-helicase
DNA primase
DNA A-methylase
dCTPase
RNAP ADP-ribosylase
ATPase, DNA helicase
exonuclease A
DNA topoisomerase
endonuclease IV
nuclear disruption
modifier of transcription
tail fibers
helix-destabilizing protein
dihydrofolate reductase
dTMP synthetase
NDP reductase
endonuclease II
tail fiber attachment & RNA ligase
polynucleotide 5'-kinase, 3'-phosphatase
dCMP deaminase
DNA ligase
baseplate assembly
baseplate plug
folate conjugase
folyl polyglutamate synthetase
bacterial DNA
glucosylated DNA
COMPETENT HEAD
MEMBRANE
INITIATOR
WEDGE
BASEPLATE
TAIL
TAIL FIBER
200S DNA
resolves branches
protease cleavage

by 9.5 map units actually have a recombination frequency of 4%. The fact that the map is circular but the DNA is linear can be explained by the physical structure of the DNA, which is discussed later. Individual proteins are often designated according to the gene (cistron) that encodes them. Thus the protein product of gene 32 is known as gp32.

The T4 data forced reassessment of the traditional genetic term allele. Two or more markers that affect the same phenotype and map at the same locus are considered allelic in both a functional and a structural sense. However, the data of Benzer clearly demonstrated that functional alleles (those occurring in the same cistron and therefore affecting the same phenotype) were not necessarily structural alleles because they could be separated by recombination.

Visconti–Delbrück Analysis

Visconti and Delbrück set out to provide a mathematic model of the recombination process in T4 as it occurs in crosses such as those described in Table 4-3. They began with four basic observations:

1. If mixedly infected cells are separated from one another in a single burst experiment, it can be shown that both recombinant and nonrecombinant (parental-type) phages can be released in the same burst. This result implies that recombination can occur after the DNA has begun to replicate, because if the two parental DNA molecules recombined in a reciprocal fashion before replication only recombinant progeny could result.
2. In a triparental cross, recombinant phages are produced that carry markers from each parent, which can result only if DNA molecules can recombine more than once.
3. If a mixed infection is set up with one parent definitely in the minority, it is possible to find more recombinant phages than there were minority parents. Once again, it suggests that genetic exchange occurs repeatedly and after DNA replication has begun.
4. If the number of recombinant phages observed after premature lysis is plotted as a function of the time at which the culture was sampled, the resulting curve shows a drift to genetic equilibrium rather than a continuous increase. Therefore recombination occurs at all times during the infectious cycle and may include exchanges between recombinant DNA molecules that regenerate parental DNA molecules.

Figure 4-10. Genomic map of bacteriophage T4. The various genes are indicated at the outer edge of the circular map. Where the function is known, it is indicated next to the locus. The interior of the circle shows the assembly pathways for the mature virion and the pattern of host DNA degradation and recycling. Courtesy of Drs. B.S. Guttman and E.M. Kutter, The Evergreen State College, Olympia, Washington.

The model developed to explain these observations assumed that the replicating (vegetative) DNA generated a large pool of T4 DNA molecules within the cell. For any given DNA molecule within the pool, three processes might occur. It might replicate again; it might be packaged up into a phage particle **(maturation)**; or it might undergo recombination before replicating or maturing. Because premature lysis experiments indicated that the pool of DNA molecules reaches a certain size and then remains constant, it was assumed that replication and maturation occurred at the same rate. It was further assumed that any recombination that occurred was random with respect to time and partner.

The previous section discussed the T4 map and pointed out that 1% recombination was equal to 1 map unit. Another way to devise a map would be to measure distance in terms of the average number of recombination events (genetic exchanges) that occur between two markers, regardless of whether the recombination resulted in a phenotypic change. Then a "true linkage" would be expressed by the parameter l, which is equal to the average number of genetic exchanges that occur between a pair of markers. If genetic exchanges are randomly distributed, their distribution can be described by the Poisson distribution. However, as Haldane pointed out, only an odd number of genetic exchanges between two markers produces a phenotypically recombinant DNA molecule, as an even number of exchanges has no effect on the linkage of flanking markers. Therefore the probability of a phenotypically expressed recombinant given two markers x and y a distance l apart is just the sum of the odd terms of the Poisson distribution. Thus when l is large, the recombination frequency K approaches 50%. However, when l is small (i.e., the expected number of exchanges is less than 1.0), most exchanges are 1.0 and K is approximately equal to l. Of course, when l is zero, K is also zero because if two mutations have occurred at an identical site they cannot recombine.

An equation can be set up to describe the number of recombinant phages inside a cell. Each time two nonidentical DNA molecules undergo a genetic exchange (a mating), either new recombinants are formed or old recombinants are lost. If there are M rounds of mating, then, in the small interval dM the increase in recombinants is $(1/2)\ (1 - R_{xy})^2 K_{xy} dM$; where R_{xy} is the proportion of recombinants for markers x and y in the population, and K_{xy} is the probability of an exchange occurring between the markers if a mating does take place. The factor 1/2 is necessary because in an equal-input cross only half of the matings are between nonidentical DNA molecules. By a similar argument, the decrease in recombinants is $(1/2)R_{xy}^2 K_{xy} dM$. Then the change in the proportion of recombinants is:

$$dR_{xy} = (1/2)(1 - R_{xy})^2 K_{xy} dM - (1/2)R_{xy}^2 K_{xy} dM. \qquad [4\text{-}1]$$

Integration of Eq. 4-1 gives:

$$R_{xy} = (1/2)(1 - e^{-K_{xy}M}). \qquad [4\text{-}2]$$

Unfortunately, the only observable quantity in the equation is R_{xy}. If the markers x and y are chosen so that they are far apart, however, K approaches 0.5 and it is possible to solve for M after measuring R_{xy}. When this step has been done, the values of M have ranged from 2 to 5. Because five to eight rounds of DNA replication are necessary to account for the observed amount of T4 DNA in a cell, the number of rounds of mating (the average number of times a particular DNA molecule can be expected to undergo recombination) approximates the number of rounds of replication.

Phage Heterozygotes

An important clue to the structural nature of T4 DNA was discovered accidentally during an examination of the progeny of r^+ and r matings. When the phenotypes of the plaques were scored, about 2% of them were mottled (parts of the margin looking sharp like that of an r mutant) and the remainder had the turbid halo characteristic of r^+. When the phages comprising a mottled plaque were retested, they were observed to be a mixture of r^+ and r, each of which subsequently bred true. The original phages that gave mottled plaques were thus behaving either as heterozygotes **(hets),** in which segregation was occurring after infection, or as cases of multiple infection of an indicator bacterium by genetically distinct phages.

The latter possibility was easily ruled out by repeating the cross and subjecting the progeny phages to various doses of UV before plating. Phages exhibit characteristic inactivation kinetics under these conditions. A particular dose of UV destroys the infectivity of a certain percentage of virions. However, if two viable phage particles are required to form a heterozygous plaque, the number of such plaques should decrease more rapidly than the total number of phages, as the inactivation of either phage particle would prevent production of a mottled plaque. In fact, the phage hets were inactivated at exactly the same rate as the rest of the population and therefore were not due to multiple infection.

A careful examination of phage hets revealed that they could occur for any pair of markers and that the length of the heterozygous region was on the order of a few cistrons. If a cross was set up in which the heterozygous locus on the genome was flanked by other markers (e.g., *a r b*), two types of hets were observed (Fig. 4-11): those that were recombinant for the flanking markers and those that were not. Some of the recombinant hets were apparently true recombination intermediates that were accidentally packaged into phage particles (strand mismatch hets). The formation of this type of intermediate is discussed in Chapter 13. However, the balance of the hets were due to an unusual feature of T4 DNA, **terminal redundancy** (e.g., if there were nine numbered cistrons comprising the T4 genome, their sequence might be 1234567891′2′).

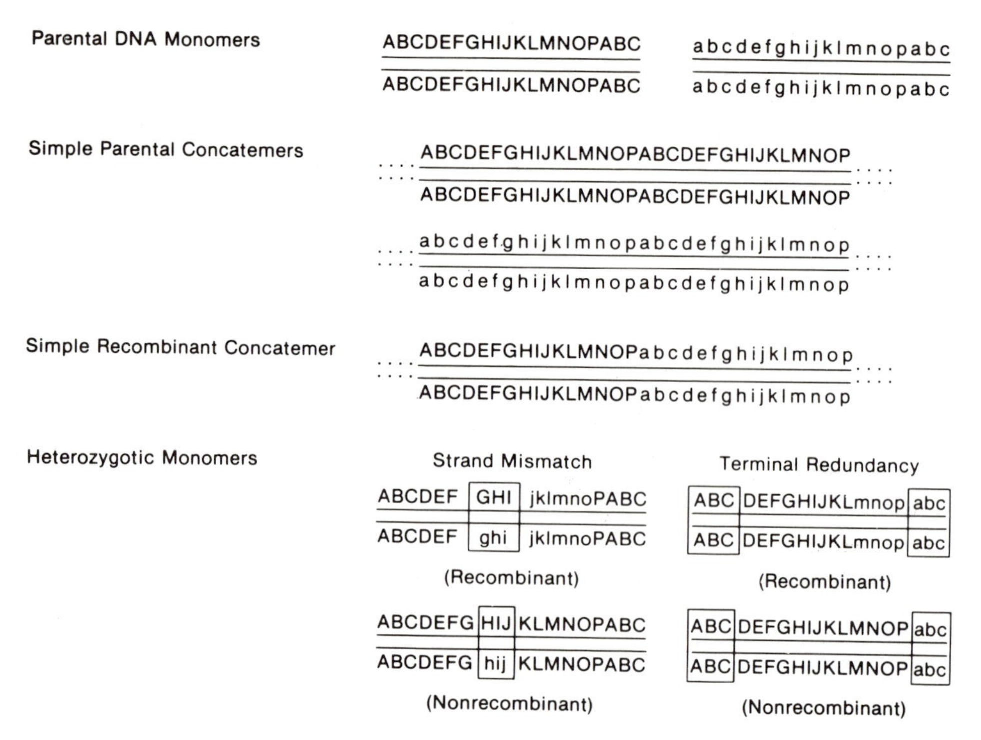

Figure 4-11. Possible T4 DNA arrangements leading to heterozygosity. The DNA monomers are shown as they would be isolated from a virion with their terminal redundancies intact. The DNA sequence of each monomer strand is indicated by letters, with lower and upper case being used to identify genetically different markers. Simple concatemers are assumed to be assembled by recombination in the terminally redundant region. Several types of recombinant concatemers are possible, although only one is shown. The heterozygotic monomers are assumed to have been cut from concatemers such as those shown above or others generated by recombination between simple concatemers. If the recombination process had not completely finished prior to the cutting of the concatemer, strand mismatch heterozygotes would be generated. If the recombination event that generated a strand mismatch consisted in the insertion of a single DNA strand (see Chapter 13), a nonrecombinant heterozygote would be generated. It is particularly important to note that in this case the term "nonrecombinant" refers not to the region of heterozygosity but, rather, to the genetic markers flanking the heterozygotic region.

It is possible to show experimentally that about 2% of the genetic information present at the left-hand end of the linear T4 DNA molecule is repeated at the right-hand end by treating T4 DNA with an exonuclease that specifically removes part of the 5′-strand from each end. If the molecules are indeed terminally redundant, the single-stranded redundant ends pair up and cause the molecule to become circular as sometimes occurs in the case of the cloning vectors discussed in Chapter 2. Such circular structures can be detected with the transmission electron microscope.

Further information about the physical structure of T4 DNA can be obtained by making random heteroduplexes. Regions of the DNA that are identical form a perfect double helix. Regions of nonidentity form various types of loop or single-stranded tails. When the techniques were applied to T4 DNA, structures such as the ones diagrammed in Figure 4-12 were obtained. The apparently random length of the single-strand tails

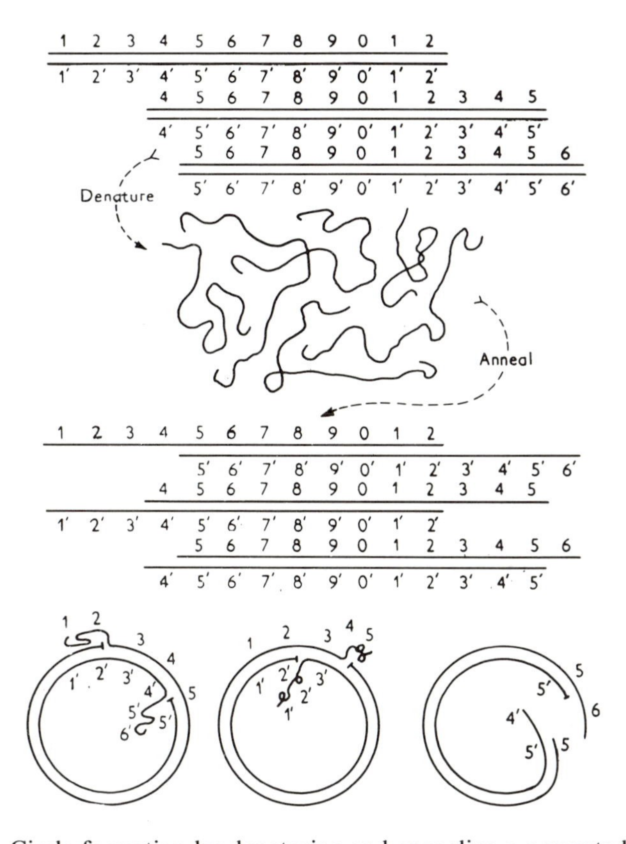

Figure 4-12. Circle formation by denaturing and annealing a permuted collection of duplexes. Each horizontal line represents a single strand of DNA. The numbers are used to indicate the arrangement of different pieces of genetic information. Note that each permutation is also terminally redundant. These redundant ends cannot find complementary partners during circle formation (*bottom*) and are left out of the resulting duplex molecule. The separation of the single strands within any circular duplex depends on the relative permutations of the partner chains. From MacHattie, L.A., Ritchie, D.A., Thomas, C.A., Jr. (1967). Terminal repetition in permuted T2 bacteriophage DNA molecules. Journal of Molecular Biology 23:355–363.

that were observed indicates that the DNA molecules were not only terminally redundant but were also circularly permuted.

Circular permutations are generated by having the sequence of cistrons or DNA bases in the form of a closed circle and then cutting the circle at different points. Each new cut generates a linear molecule representing a different circular permutation. The existence of circularly permuted, terminally redundant DNA molecules resolves the question of how a linear DNA molecule can give rise to a circular genetic map. The individual molecules may circularize, or different permutations may recombine to give longer-than-unit-length molecules. The important point is that because the DNA has no fixed endpoints, neither can the genetic map. Although it is not difficult to conceive of mechanisms by which circular permutations might be generated in vivo, it is less obvious how the terminal redundancy can be produced. The answer to this question lies in the way in which T4 DNA is replicated and packaged.

Molecular Biology of the Phage Infection

Using information derived from experiments such as those described earlier in the chapter, it is possible to present a reasonably clear picture of the course of T4 infection. These events are summarized in Figure 4-13. The activity begins when the tips of the tail fibers (gp38) attach reversibly to the core of the lipopolysaccharide located on the outer membrane of the gram-negative cell. It is theorized that the six tail fibers bind and release randomly so as to allow the virion to "walk" across the surface of the cell until an appropriate site is reached and the binding of the virion becomes irreversible. The tail fibers bend to allow the baseplate to touch the cell surface. Electron microscopy shows a **zone of adhesion** (physical connection) between inner and outer cell membranes at the site of phage attachment, although it is not known whether the zone is caused by the phage or is preexisting.

When the baseplate touches the zone of adhesion, the tail sheath contracts and the plug located in the center of the baseplate (Fig. 4-1) is removed, allowing the DNA to be ejected from the head. The ejection process is triggered when the tail core touches the cell membrane. The shape of the head is unchanged throughout, so the often-made analogy of DNA ejection to injection by a hypodermic syringe is not valid. For the viral DNA to actually enter the cell, the normal, inwardly directed proton gradient across the cell membrane must be intact. Indeed phage ghosts are capable of significantly depolarizing the cell membrane. It therefore seems that the cell cooperates in its own infection. However, phages with contracted tails can infect cells without cell walls and with no proton gradients, and it therefore appears that the proton gradient is not directly used for transport.

As the T4 phage DNA enters the cell, RNA transcription begins. The

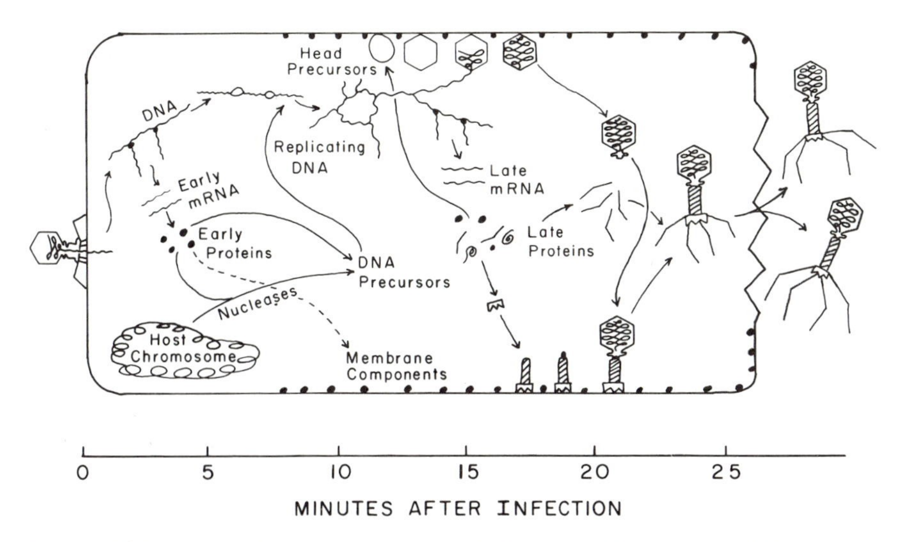

Figure 4-13. Temporal sequence of events during a T4 infection. The infecting virion is assumed to attach to the left-hand end of the cell at time zero. The position of each event or structure along the horizontal axis of the cell is reflective of the time at which it usually occurs. Vertical displacements are used to indicate the existence of separate pools of assembled head and tail substructures. From Mathews, C.K. (1977). Reproduction of large virulent bacteriophages, pp. 179–294. In: Fraenkel-Conrat, H., Wagner, R.R. (eds.) Comprehensive Virology, vol. 7. New York: Plenum Press.

phage-specific RNA molecules are assigned to four temporal classes based on the time at which abundant synthesis first occurs at 30°C: **immediate early** (30 seconds), **delayed early** (2 minutes), **middle** (6 minutes), and **late** (9 minutes). As might be expected, the early transcripts are involved with establishing the infection and the late transcripts with assembly of the progeny virions. The two groups of early RNA molecules include eight species of tRNA molecules not normally produced in large quantity by *E. coli* as well as sundry mRNA molecules. Nearly all of the early transcripts are complementary to only one strand of the phage DNA helix: the *l* strand, which is defined as that strand most readily binding to an RNA copolymer composed of guanine and uridine residues (poly UG). By contrast, late mRNA is predominantly complementary to the *r* strand. Middle mRNA is a mixture of unique transcription units and some early transcription units that can be expressed from two distinct promoters.

The difference between immediate early and delayed early RNA is one of genetic position. The genetic information for the delayed early RNA is physically located downstream (3′) to the immediate early information, and between the two groups lies one or more rho-dependent termination signals. If translation of the immediate early mRNA occurs promptly after transcription, terminator loops do not have a chance to form, and the

RNA polymerase continues into the delayed early region. On the other hand, if the antibiotic chloramphenicol is added to prevent translation of the early mRNA, only short transcripts are produced. The delayed early mRNA also has its own promoter sequences that can be separately activated after T4 protein synthesis is established.

The host RNA polymerase undergoes major changes during the growth cycle of the phage, and these changes correlate with the changes in the viral transcription pattern. Table 4-4 outlines some of these changes and indicates which viral gene products are involved in the alterations. The cumulative effect of the modifications to the host transcription system is to prevent host promoters from being recognized while making the phage middle and late promoters more easily recognized. During the transitional stages, the RNA polymerase may actually recognize more promoters than usual. The loss of the ability to recognize host promoters means that infected cells cannot express any new functions not encoded by T4 and thus cannot respond to changes in their environment in the manner discussed in Chapter 12.

Synthesis of true late mRNA requires both the occurrence of DNA replication and the presence of hydroxymethylcytosine in the DNA that

Table 4-4. T4-evoked changes in *E. coli* RNA polymerase

Change	Time[a]	Effect or function
ADP-ribosylation of one of the two α subunits	< 0.5	Lowers affinity for σ; participation in host shutoff?
Phosphorylation, adenylylation, or ADP-ribosylation of a fraction of α	< 0.5	Unknown
ADP-ribosylation of both α subunits (modification)	1.5–2.0	Lowers affinity for σ; shutoff of transcription for some early cistrons around 4 minutes; participation in host shutoff?
Binding of 10K protein	< 5	σ Antagonist
Binding of 15K protein	5	T4 cistron 60 codes for a subunit of T4 DNA polymerase and may also code for the 15K protein
Binding of gp33 (12K)	5–10	Positive control of late transcription
Binding of gp55 (22K)	5–10	Positive control of late transcription
Interaction with gp45 (27K)		GP 45 is a component of the core of the T4 replisome and is also directly involved in late transcription

[a]The time after infection at which the indicated change can be detected in the *E. coli* RNA polymerase. Where no time is given, no change occurs.

From Rabussay, D. (1983). Phage-evoked changes in RNA polymerase, pp. 167–173. In: Mathews, C.K., Kutter, E.M., Mosig, G., Berget, P.B. (eds.) Bacteriophage T4. Washington, D.C.: American Society for Microbiology.

is to be transcribed. In the absence of these two preconditions, late mRNA is produced at low levels or not at all. At least one host normal function is also required, as *E. coli* mutants that overproduce the *lit* protein do not permit late transcription of wild-type T4. A class of compensatory mutations known as *gol* can be isolated by selecting for T4 that can grow on *lit*(Con) hosts. All *gol* mutations map in the same region and affect only promoters located on the same DNA molecule, even those that are far away. In this sense they resemble eukaryotic enhancers in their mode of action.

T4-infected cells also demonstrate one other phenomenon originally thought to be unique to eukaryotic cells: RNA splicing. T4 codes for a thymidylate synthase enzyme that duplicates a function present in the host cell. It therefore is normally dispensable. Base sequence analysis of the coding region for the enzyme has shown that there is a 1017-bp intron located within the synthase coding region that contains an open reading frame of 735 bp. It is not certain that the intron protein is actually produced, as potential secondary structure might interfere (see below). The arrangement of ribosome binding sites and initiator and terminator codons is such that exon 1 is translatable even before splicing. The suggestion has been made that RNA splicing acts as a sort of "switch" mechanism to allow production of a longer protein at later stages of the infection.

The thymidylate synthase RNA transcript is an example of a self-splicing reaction in which internal guide sequences located within 220 bp of the ends of the intron cause the 3′-end of exon 1 to align with the 3′-end of the intron so as to allow excision of the intron in either a linear or a circular form. This type of reaction is characteristic of class I splicing systems such as those found in *Saccharomyces, Neurospora*, or *Tetrahymena* mitochondria, and there is significant homology between the T4 intron and the introns of the other organisms. The self-splicing reaction apparently requires some specific protein(s) for in vivo functionality because addition of chloramphenicol (which blocks protein synthesis) to the infected culture also blocks mRNA splicing. Among the enzymes that have been shown NOT to be required are the host enzymes RNase II, RNase P, and RNase E and the phage proteins RNA ligase and polynucleotide kinase. A guanosine nucleoside is required for splicing. Another T4 cistron, *nrdB*, has been reported to contain an intron. Like the thymidylate synthase mRNA, the *nrdB* mRNA can be posttranscriptionally labeled by GTP and therefore is also a class I self-splicing RNA.

After mRNA molecules have been produced, their code still must be translated into protein before they can have any effect on metabolism. It can be easily shown that various mRNA molecules may be present in the cytoplasm of a T4-infected cell and nevertheless not be translated even though they are readily translated in an uninfected cell. It is certainly true for some *E. coli* transcripts and has also been shown to be true for at least some other phages when the cell they infected becomes **superinfected** by T4 (the already-infected cell is infected by another virion). There is substantial disagreement as to the molecular mechanisms underlying the phenomenon. Some workers suggest a direct interaction between T4 proteins and the ribosomes or their protein cofactors, but other workers say they are unable to verify the association.

In the case of the cistron coding for T4 lysozyme, the basis for translational control does seem to be evident. Appropriate mRNA molecules for lysozyme are present in large quantity during both early and late transcription, but under normal circumstances only the late mRNA is translated. The complete nucleotide sequence for the region is now available, and it is possible to compare the sequences of the early and late mRNA transcripts with a view toward developing a model. As shown in Figure 4-14, the early mRNA is part of a larger transcript that is predicted to have the ability to form an internal hairpin loop than embeds both the Shine-Dalgarno sequence and the first codon for translation of the lysozyme within the duplex region. A fundamental principle of translation is that ribosomes are capable of disrupting mRNA secondary structure only after they have already bound to the appropriate initiator sites and are actively translating the mRNA. Therefore the mRNA transcribed early is predicted to be untranslatable unless some outside influence disrupts the hydrogen bonds within the stem. On the other hand, the late mRNA transcribed from eP1 or eP2 should be incapable of forming the stem-and-loop structure and consequently be readily translated.

One T4 protein that does seem to have a major effect on translation is

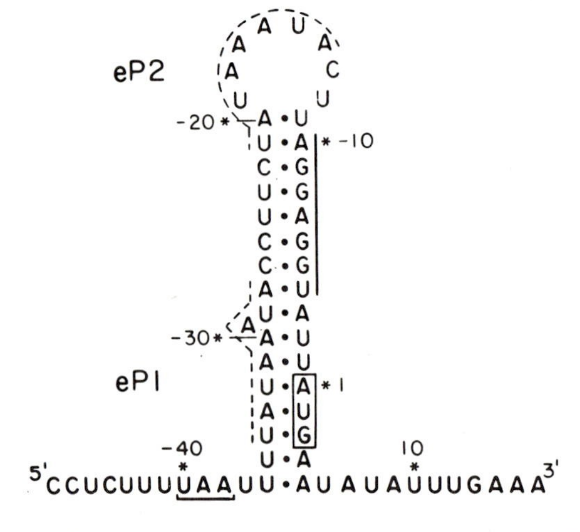

Figure 4-14. Proposed secondary structure of early lysozyme mRNA. The stem-and-loop structure that is predicted to form in early lysozyme transcript(s) is shown. Dots indicate Watson-Crick base-pairing. The initiator AUG is boxed, and the Shine-Dalgarno sequence is indicated by a line next to the bases. The two sequences homologous to the conserved T4 late promoter sequence are indicated by dashed lines and labeled eP1 and eP2. The termination codon of the upstream open reading frame is bracketed. From McPheeters et al. (1986).

the *regA* protein. The observation is that *regA* mutant strains generally overproduce a wide variety of T4 proteins, even though their transcription rates are essentially unaltered. This finding suggests that *regA* acts as a **translation repressor,** a protein that prevents mRNA molecules from attaching to ribosomes. Fitting in with this hypothesis are the observations that most if not all of the mRNA transcripts subject to *regA* control have the same or similar base sequence in the region of the ribosome binding site, whereas those mRNA transcripts that are not under *regA* control do not carry the sequence. Some host proteins are also regulated by *regA*. The common element in regulated mRNA molecules seems to be a uridine-rich region.

Many of the T4 proteins make fundamental changes in the metabolism of the host cell. Enzymes are produced to degrade cytosine and uridine triphosphates and diphosphates to their monophosphoric derivatives (nucleoside di- and triphosphatase); produce hydroxymethylcytosine (deoxycytidylate hydroxymethylase); produce thymine, guanine, and hydroxymethylcytosine triphosphates (deoxynucleoside monophosphate kinase); glucosylate the hydroxymethylcytosine residues (glucosyltransferases); and methylate certain adenine residues (DNA adenine methylase). Also produced are eight endonucleases, many of which are involved in DNA repair, and a new DNA polymerase molecule. An interesting assortment of enzymes whose roles in the infective process are uncertain is also produced: DNA ligase (necessary but normally supplied by the host cell), RNA ligase (a dual function protein that attaches tail fibers to the virion and also can join single-stranded polynucleotides), polynucleotide kinase, several DNA phosphatases, and a type II topoisomerase.

All the events described above occur prior to the end of the eclipse phase. They result in accumulation of the building blocks necessary for construction of new virions. The actual assembly of infectious phages signifies the end of the eclipse phase. The assembly proceeds in a linear manner and is discussed below.

DNA Replication and Maturation

Replication

It is obvious that normal replication, as envisioned by Watson and Crick, does not suffice to produce DNA molecules with the properties attributed to T4. From a conceptual standpoint, the easiest way to produce terminal redundancy is to have a DNA molecule that consists of two or more genomes linked end to end (a concatemer) (Fig. 4-11) and then to cut the appropriate sized piece of DNA to fit the phage head from the concatemeric structure. This mechanism is apparently the correct one, as the size of the T4 genome is 166 kb, but the size of the packaged DNA molecule is 169 kb (a 3 kb redundancy).

The problem then becomes how to produce a concatemeric structure. One method known to be used by viruses is the rolling circle DNA replication mechanism (see Chapter 5) that is used by phages f2, φX174, and lambda. However, because it can be shown that T4 DNA molecules do not need to circularize in order to replicate, this mechanism is considered unlikely.

An alternative model developed by Mosig and her co-workers to describe how linear concatemers can develop is shown in Figure 4-15. Basically the model assumes that there is a single replication origin that starts the usual bidirectional replication. Unlike the case with a circular molecule, when the replication fork reaches the end of the DNA molecule it is not

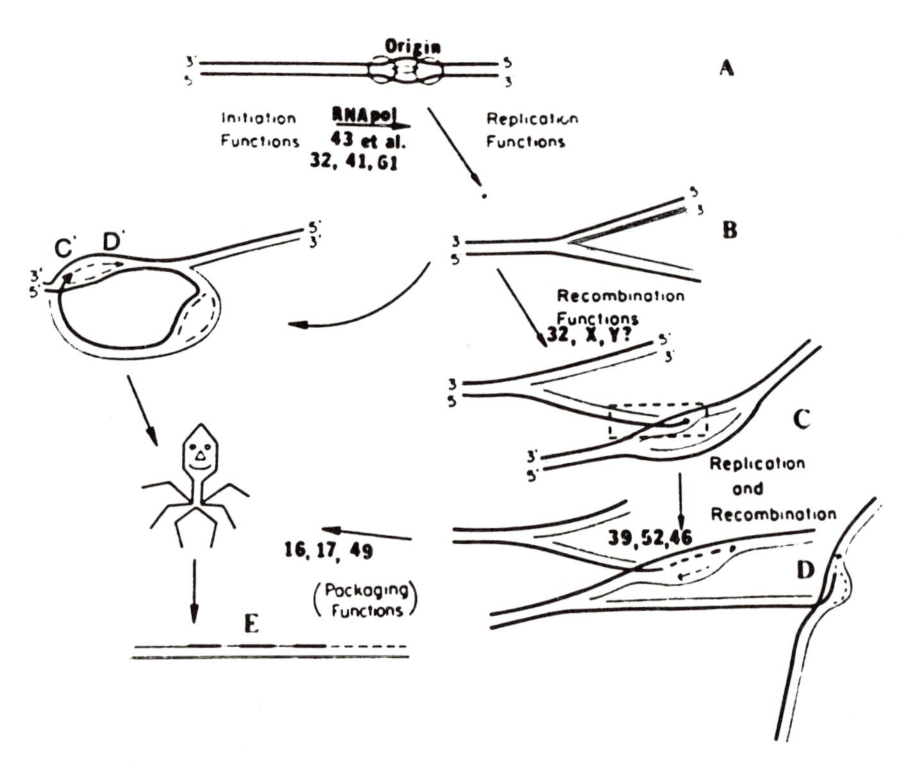

Figure 4-15. Model of T4 DNA metabolism. **(A)** Replication initiates at a unique origin and proceeds bidirectionally. **(B)** As the replication fork runs off the end of the molecule, the lagging strand is left as a single-stranded region. **(C)** The single-stranded region initiates a recombination event with a homologous region on a different DNA molecule. Self-closure of a single DNA molecule using the terminally redundant areas is possible but unlikely. **(D)** Multiple recombinations soon yield concatemeric molecules. The sites of recombination act as initiators of additional rounds of DNA replication. From Mosig, G. (1983). Relationship of T4 DNA replication and recombination, pp. 120–130. In: Mathews, C.K., Kutter, E.M., Mosig, G., Berget, P.B. (eds.) Bacteriophage T4. Washington, D.C.: American Society for Microbiology.

possible to complete replication of the lagging strand. This single-stranded region is recombinogenic (see Chapter 13) and stimulates exchange between two DNA molecules. The various breaks left by the recombination enzymes act to prime additional replication forks, and a large, multiply-branched concatemer begins to develop.

T4 DNA replication begins about 6 minutes after infection at 30°C, with the first concatemeric structures appearing several minutes later. The initial priming event is catalyzed by the host RNA polymerase, but the modifications to that polymerase noted above require that other mechanisms be invoked for late replication initiations. The RNA primers for the Okazaki fragments are synthesized by the combination of RNA primase (gp61) and DNA helicase (gp41) using the template sequences GTT or GCT. The primers are four to five bases long and always start at the middle base of the recognition sequence. The number of phage genome equivalents in the cell increases to 40 to 80 by 12 minutes, at which time maturation begins. Equilibrium is quickly achieved, so the rate of maturation is approximately equal to that of replication, causing the size of the DNA pool to remain constant even though the number of infectious phage particles is increasing.

Morphogenesis and Maturation

Three biochemical pathways generate heads, tails, and tail fibers, which are then joined to make a virion (Fig. 4-10). The process of assembly of each component is similar to that of crystallization. A cluster of proteins forms the starting structure (a nucleus), and other subunits spontaneously arrange themselves around the nucleus in an orderly fashion. At least two host-cell genetic functions, *groEL* and *groES*, are necessary for this process, but their actual biochemical role is uncertain. If one of the viral cistron products is missing or nonfunctional, the assembly process stops at the point at which that product is required, and all proteins that would have been added subsequently remain in solution. If a mutation alters but does not remove certain proteins, long structures called polyheads (Fig. 4-16) or polytails may develop. Chain terminator mutations in the same proteins involved in polyhead formation may lead to the formation of normally shaped "giant" or "petite" heads, which can produce infectious phages carrying correspondingly greater or lesser amounts of DNA.

DNA maturation requires the presence of a mature prohead, which consists of the outer proteins visible in electron micrographs as well as a complex of internal "scaffold" proteins (products of cistrons *22*, *III*, *II*, and *I*). The mature prohead is some 16% smaller than the head of a viable phage particle. The increase in size is associated with the cleavage of gp 23 by a T4-specific protease and may occur in the absence of any DNA. The process is shown in Figure 4-17.

DNA packaging occurs by the "headful" method and requires ATP and DNA ligase. Restriction digests of T4 DNA isolated from virions in-

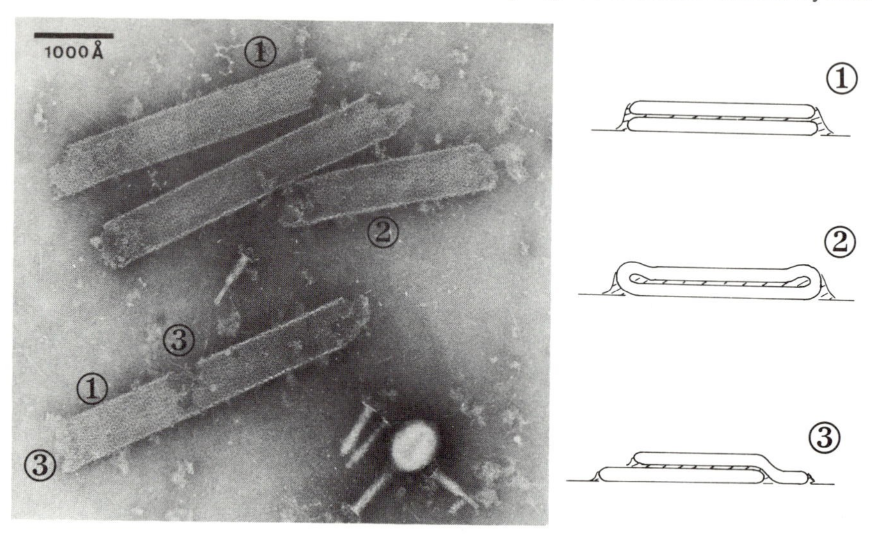

Figure 4-16. Coarse polyheads. These elongated structures result from an amber mutation in gene 24. Their sizes should be compared to the normal-sized T4 virion and the isolated tails, which are also present in the micrograph. The numbers near the polyheads refer to the sketches at the right and indicate the authors' interpretation of what was observed. From Steven, A.C., Aebi, U., Showe, M.K. (1976). Folding and capsomere morphology of the P23 surface shell of bacteriophage T4 polyheads from mutants in five different head genes. Journal of Molecular Biology 102:373–407.

dicate that there is no preference to the end sequences as is seen in the case of phage P22 (see Chapter 6), and therefore the initial cut is made randomly by an endonuclease. This random cutting gives rise to the observed circular permutations of the genome. Electron microscopic experiments in which a beam of ions is used to erode away the capsid show that the DNA is arranged in a spiral-fold model such as that shown in Figure 4-18. In this model the DNA is packed from inside to out, leaving both ends free so that either end can be the first ejected. DNA is added to the prohead until it is full. Because the head can hold more than a genome equivalent of DNA, terminal redundancy is ensured. It is presumed that after the initial packaging the endonucleolytic cut that serves to release the first virion also serves as the initiator cut for the next prohead.

Tails are attached to the filled heads, and tail fibers are then added. Interestingly, the enzyme catalyzing the addition of the tail fibers is gp63, which also functions as an RNA ligase. It is therefore a single protein that appears to have two discrete functions.

All that now remains is for the phage to lyse the host cell. This step is accomplished by a phage-specific lysozyme (product of cistron *e*) in conjunction with gp5 (located in the baseplate of the virion). When the primary particles have been released, the infectious cycle can begin again.

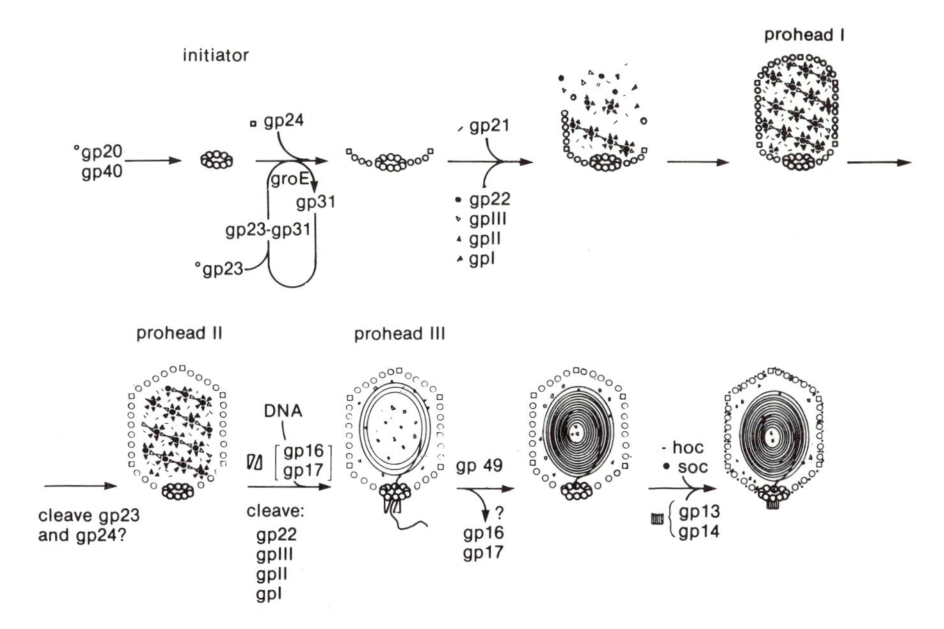

Figure 4-17. Morphogenetic pathway of T4 capsids. The various gene products are represented by geometric figures (circles, squares, triangles, etc.) shown beside the gene product designation. A dash between two components indicates association between them. The DNA is shown as a thin line that begins to fill prohead III. In phage T4, tails are synthesized in an independent pathway and then attached to complete heads. From Murialdo, H., Becker, A. (1978). Head morphogenesis of complex double-stranded deoxyribonucleic acid bacteriophages. Microbiological Reviews 42:529–576.

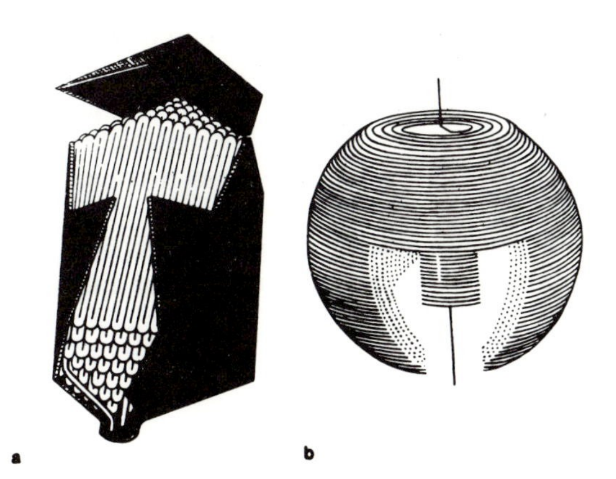

Figure 4-18. Models for the arrangement of DNA inside the head of phage T4. **(a)** Spiral fold model described in the text. **(b)** An older, concentric shell model in which the DNA is wrapped like a ball of string beginning at the center. From Black, L.W., Newcomb, W.W., Boring, J.W., Brown, J.C. (1985). Ion etching of bacteriophage T4: support for a spiral-fold model of packaged DNA. Proceedings of the National Academy of Sciences of the United States of America 82:7960–7964.

Summary

T4 phage is a large and structurally complex virus. During initiation of infection, tail fibers form the attachment to the cell wall, first reversibly then irreversibly. DNA is ejected from the virion through the tail core as the tail sheath contracts and is transported into the cytoplasm of the cell concomitantly with depletion of the proton gradient across the cell membrane. Once inside the cell, the virus is in the eclipse phase, as no infectious virus is present when cells are prematurely lysed. Phage-specific protein synthesis begins immediately, producing enzymes to initiate replication, expand transcription of the viral DNA, and supply proteins needed to modify certain cellular components. Both transcription and translation are highly regulated: the former by modifications to the RNA polymerase holoenzyme complex and the latter by mRNA structure as well as ribosomal alterations. Most proteins synthesized late in the infection are structural components of the virion. DNA synthesis begins at about 6 minutes at 30°C and generates a large concatemeric structure by multiple recombinations. At 12 minutes condensation of heads, tails, and tail fibers begins. When mature proheads are ready, an endonuclease makes a random cut in the DNA, and the phage head is stuffed with DNA until full. The "headful" mechanism leads to production of a DNA molecule that is linear, circularly permuted, and terminally redundant.

Genetic crosses with T4 phages are carried out by infecting *E. coli* cells with two or more genetically distinct phages (a mixed infection). The easiest phenotype to study is one involving plaque morphology, but more than 130 genetic loci affecting all types of functions have been mapped. The most carefully mapped portion of the genome is the *r*II region, which was intensively studied by Benzer. From his work comes the conclusion that many mutations are random in their distribution and that the unit of mutation and recombination is a single base. Roughly 2% of all progeny phage formed in a cross are found to be heterozygous for one or more traits. Some of these hets are due to terminal redundancy, while others seem to be normal recombination intermediates. The way in which recombination occurs during T4 infection has been the subject of a mathematic model prepared by Visconti and Delbrück.

References

Generalized

Cech, T.R. (1986). RNA as an enzyme. Scientific American 255(5):64–75.

Maley, F., Maley, G.F., West, D.K., Belfort, M., Chu, F.K. (1986). RNA processing in a structural gene from bacteriophage T4. Biochemical Society Transactions 14:813–815.

Mathews, C.K., Kutter, E.M., Mosig, G., Berget, P.B. (1983). Bacteriophge T4. Washington, D.C.: American Society for Microbiology.

Specialized

Chu, F.K., Maley, G.F., Maley, F. (1987). Mechanism and requirements of in vitro RNA splicing of the primary transcript from the T4 bacteriophage thymidylate synthase gene. Biochemistry 26:3050–3057.

Furukawa, H., Kuroiwa, T., Mizushima, S. (1983). DNA injection during bacteriophage T4 infection of *Escherichia coli*. Journal of Bacteriology 154:938–945.

Hall, D.H., Povinelli, C.M., Ehrenman, K., Pedersen-Lane, J., Chu, F., Belfort, M. (1987). Two domains for splicing in the intron of the phage T4 thymidylate synthase (td) gene established by nondirected mutagenesis. Cell 48:63–71.

Kalinski, A., Black, L.W. (1986). End structure and mechanism of packaging of bacteriophage T4 DNA. Journal of Virology 58:951–954.

Kao, C., Gumbs, E., Snyder, L. (1987). Cloning and characterization of the *Escherichia coli lit* gene, which blocks bacteriophage T4 late gene expression. Journal of Bacteriology 169:1232–1238.

Kuhn, A., Keller, B., Maeder, M., Traub, F. (1987). Prohead core of bacteriophage T4 can act as an intermediate in the T4 head assembly pathway. Journal of Virology 61:113–118.

McPheeters, D.S., Christensen, A., Yound, E.T., Stormo, G., Gold, L. (1986). Translational regulation of expression of the bacteriophage T4 lysozyme gene. Nucleic Acids Research 14:5813–5826.

Miller, E.S., Karam, J., Dawson, M., Trojanowska, M., Gauss, P., Gold, L. (1987). Translational repression: biological activity of plasmid-encoded bacteriophage T4 RegA protein. Journal of Molecular Biology 194:397–410.

Riede, I., Drexler, K., Schwarz, H., Henning, U. (1987). T-even-type bacteriophages use an adhesin for recognition of cellular receptors. Journal of Molecular Biology 194:23–30.

Rojiani, M., Goldman, E. (1986). Dependence of MS2 and T4 phage growth upon host amino acid biosynthesis during infections of *Escherichia coli*. Virology 150:313–317.

Sjöberg, B-M., Hahne, S., Mathews, C.Z., Mathews, C.K., Rand, K.N., Gait, M.J. (1986). The bacteriophage T4 gene for the small subunit of ribonucleotide reductase contains an intron. EMBO Journal 5:2031–2036.

Chapter 5
Genetics of Other Intemperate Bacteriophages

Bacteriophage T4 has probably been the most intensively investigated intemperate virus, but there are many other viruses that have also been the subject of considerable study. In this chapter descriptions of selected bacteriophages are presented to illustrate the high degree of genetic diversity available to the bacterial geneticist and to provide comparisons of these phages to each other and to T4. To facilitate these comparisons, the physical properties of each phage discussed in this chapter are summarized in Table 5-1.

Other Members of the T Series

Delbrück's group isolated seven bacteriophages that were placed in the T series. Closer examination of these phages revealed that all of the even-numbered phages were similar to each other but were different from the odd-numbered phages. Electron microscopic studies have shown that all even-numbered phages possess contractile tails and elongated heads, but the T-odds all have noncontractile tails with octahedral heads. Phages T3 and T7 are similar, but phages T1 and T5 are unlike T7 or each other.

Bacteriophages T2 and T6

Bacteriophages T2 and T6, the remainder of the T-even series, are similar to T4. They have similar physiology and similar genetic composition, although they do differ in the way in which the DNA is glucosylated. T2 and T6 have 25% of their hydroxymethylcytosine (HMC) residues unglucosylated, whereas T4 has none. Nearly all of the glucosylated T2 HMC

Table 5-1. Properties of some common intemperate bacteriophages

		Nucleic acid			Virion		
Phage	Usual host	Type	Molecular weight $\times 10^{-6}$	Topology	Morphology	Dimensions (nm)	Related phages
T4	*E. coli*	2-DNA[a]	130	Linear, circularly permuted, terminally redundant	Oblong head, contractile tail	80 × 120 113 × 20	T2,T6
T1	*E. coli*	2-DNA	31	Linear, unique sequence [b] terminally redundant	Octahedral head, noncontractile tail	50 150 × 10	
T5	*E. coli*	2-DNA	75	Linear, unique sequence, terminally redundant	Octahedral head, noncontractile tail	90 200	BF23
T7	*E. coli*	2-DNA	26.3	Linear, unique sequence, terminally redundant	Octahedral head, noncontractile tail	60 15 × 15	T3,φII
SP01	*Bacillus subtilis*	2-DNA	91	Linear, unique sequence, terminally redundant	Icosahedral head, contractile tail	100 200 × 20	SP8,SP82, φe,2c
φ29	*Bacillus subtilis*	2-DNA	11	Linear, unique sequence	Oblong head, noncontractile tail	32 × 42 32 × 6	φ15,Nf, GA-1,M2Y
f1	*E. coli* F^+	1-DNA	1.90	Circular	Filamentous	870 × 5	fd,M13, HR,Ec9, AE2,δA (the Ff group)
φX174	*E. coli* strain C	1-DNA	1.7	Circular	Icosahedral	25	S13,φR, G4
R17	*E. coli* F^+	1-RNA	1.1	Linear, unique sequence	Icosahedral	25	fr,f2, MS2,M12, Qβ

[a]The number in front of the nucleic acid refers to the number of strands in the molecule.
[b]As contrasted to a circularly permuted sequence.
Adapted from Strauss, E.G., Strauss, J.H. (1974). Bacterial viruses of genetic interest, pp. 259–269. In: King, R.C. (ed.) Handbook of Genetics, vol. 1. New York: Plenum Press.

residues have only one glucose attached in the α-configuration, whereas in the case of T6 phage the HMC residues are diglucosidic, having first an α-linkage and then a β-linkage.

The degree of genetic homology (the extent to which the DNA sequences are identical) between the T-evens can be readily demonstrated by the techniques of heteroduplex mapping, described in Chapter 2. Kim and Davidson used them to compare all of the T-even genomes. If DNA carrying a deletion is heteroduplexed to nondeleted DNA, the reannealed structure has a characteristic loop of single-stranded DNA originating from a point on the nondeleted DNA strand that represents the site of the deletion on the corresponding molecule (Fig. 5-1). Regions of nonhomology (different base sequences) between the two phage DNA molecules result in the production of two single-stranded loops (which may not be of identical size). By carefully measuring the distance between deletion loops whose map position was known from genetic crosses and the unknown loops along the DNA molecule, and knowing the final magnification of the molecule that was measured, it is possible to determine the distance between various points on the phage DNA, in terms not of recombination units but of physical units. Frequently these physical distances are expressed as a percentage of the total genome length, as it simplifies the calculations by eliminating the necessity for exact measurement of the magnification. Kim and Davidson estimated the overall homology between phages T2 and T4 at 85%, with the late cistrons being more homologous than the early ones. Similarity of sequence implies similarity of virion (late) proteins, which accounts for the large amount of immunologic cross-reactivity among the T-even phages.

Some of the early work with T2 involved a class of mutations that did not occur in T4. These mutations were designated *h* (**host range**) and affected the tail fibers and their ability to attach to certain bacterial cell walls. *E. coli* can mutate so normal T2 no longer infects the cell owing to a change in the cell wall surface receptor site. However, the appropriate type of tail fiber mutation will once again permit attachment of the T2 tail fiber to the cell and a successful infection of either mutant or wild-type bacteria. The *h* phenotype is another example of a phenotype that can be scored by plaque morphology. However, unlike the *r* phenotype, it is necessary to use a mixed indicator. When dealing with *h* or h^+ virions and using a mixture of normal and phage-resistant bacteria as indicators, two types of plaque are possible. One type is perfectly clear, meaning that both types of indicator bacteria had lysed and the phage was an *h* mutant. The other type of plaque is turbid, meaning that the phage-resistant cells had survived the infection and the phage was not an *h* mutant.

The existence of host range mutants has led to the suggestion that an endless cycle of bacterial resistance and phage compensatory mutation is possible. In fact, the *h* mutants are best looked on as extended host range mutants: They still infect the wild-type cells and can be found in only certain viruses. Moreover, in the case of phages T3 and T7 (see below),

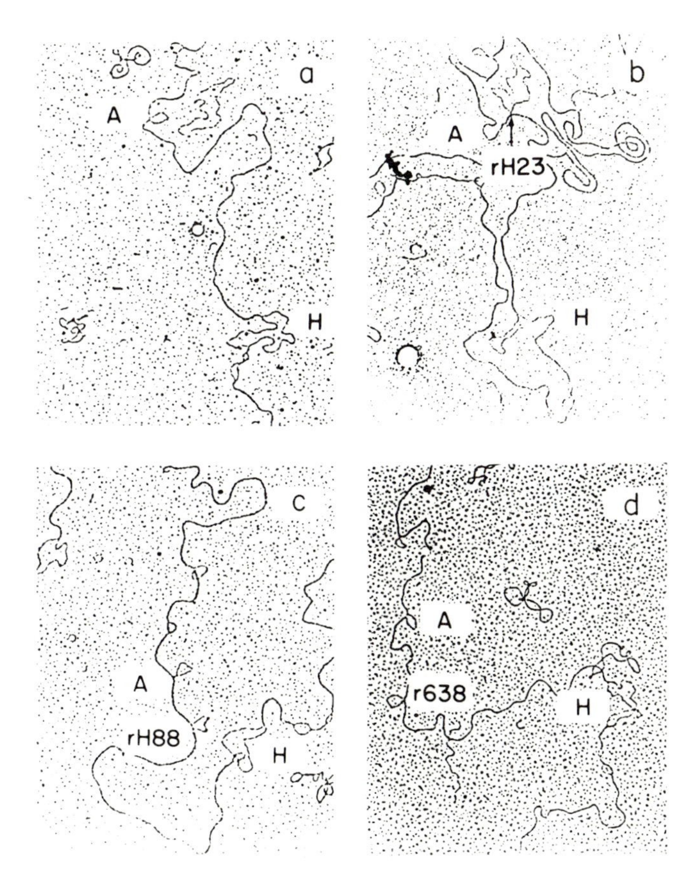

Figure 5-1. Electron micrographs of **(a)** T2/T4B, **(b)** T2/T4B*rH23*, **(c)** T2/T4B*rH88*, and **(d)** T2/T4B*r638* heteroduplex DNA molecules around the *r*II region. In this region T2 and T4 DNA molecules have a short segment in which their base sequences are drastically different (a substitution) and a short segment present in T4 DNA but deleted in T2. These differences give rise to the substitution loop H and deletion loop A that are seen in **a**. Each of the *r*II mutations is also a deletion, and in **b,c,d** the extra deletion loop observed in the heteroduplex is also labeled. The distance between loops A and H in **a** is 11,500 bp. The distances from loop A to the deletion loops *rH23*, *rH88*, and *r638* correspond, respectively, to 400, 1100, and 2800 bp. The small circular DNA molecules seen in the background are double-stranded ϕX174 DNA used as a size standard. From Kim, N. and Davidson, N. (1974). Electron microscope heteroduplex study of sequence relations of T2, T4, and T6 bacteriophage DNAs. Virology 57:93–111.

bacteria that are resistant to host range mutants can be obtained, and the phages do not produce additional compensatory mutations.

Some unusual discoveries have emerged from studies dealing with T2 morphogenesis. Drexler and co-workers have shown that the tail fiber protein gp37 is normally processed posttranslationally by the removal of about 120 amino acids from the carboxy-terminus. If a mutation is introduced into cistron 37 that deletes 87 amino acids from the region normally removed, protein processing is not possible, and the net effect of the deletion of 87 amino acids is to produce a finished protein product that is longer than the original. It is, of course, the tail fibers that determine the host range of the virus by recognizing specific receptors on the cell surface. Each of the T-even viruses has its own specific tail fibers, and each makes gp38, which is required for attachment of gp37 to the virion. The gp38 protein from one phage or mutant can function with a heterologous gp37 from another phage, but it interacts with the gp37 in some manner to imprint the host range characteristic of the phage that provided the gp38 onto the phage that provided the gp37. It thus represents an unusual sort of phenotypic mixing.

Bacteriophage T1

Although T1 was one of the first phages used in bacterial genetics, it has not been thoroughly studied. One reason is its incredible persistence. Once the phage has been brought into a laboratory, it is difficult to eliminate, as it is capable of surviving for years on laboratory surfaces and forming stable aerosols. This activity is in sharp contrast to that of a phage such as T6, which survives only a few hours in a desiccated state. Nevertheless, a few laboratories have studied phage T1, and some facts are known about its life cycle.

The phage particle itself is somewhat smaller than T4 and belongs to the phage lambda morphologic group (see Fig. 6-1). It contains a double-stranded DNA molecule that has definite endpoints and a long terminal redundancy equal to 6.5% of the DNA molecule. This figure is dramatically higher than the 2% terminal redundancy of T4. Unlike T4, the T1 DNA molecule contains no modified bases and circularizes in order to replicate, presumably by recombination between the redundant ends.

Infection begins with adsorption of the phage to specific cell receptors controlled by the *Escherichia coli* cistron *fhuA* (formerly *tonA*). The protein is a ferric ion siderophore that is associated with the peptidoglycan layer of the cell wall and spans the outer membrane. The same receptor is also used by phages T5 and ϕ80 as well as colicin M (see Chapter 11). There has been a report that T1 can act as a transducing phage (see Chapter 7). Like *r*II mutants of phage T4, phage T1 also grows poorly on lambda lysogens, although the magnitude of the effect is less.

Bacteriophage T5

The structure of bacteriophage T5 is basically similar to that of the T1 virion except that it has four tail fibers instead of one (Fig. 5-2); there are also several differences in their DNA molecules and metabolism. The T5 virion has a linear DNA molecule of unique sequence with no unusual bases but with a large terminal redundancy, amounting to 9% of the total

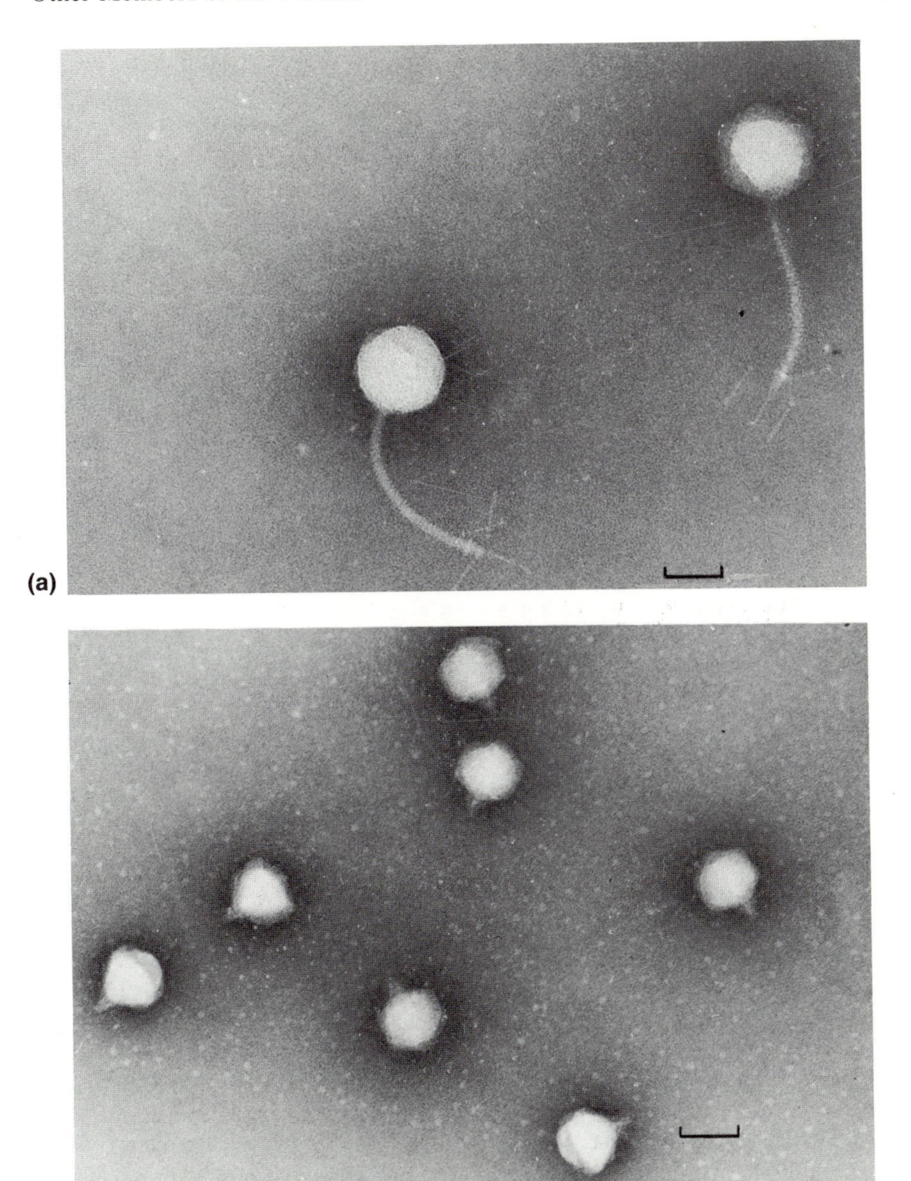

Figure 5-2. Electron micrographs of some T-odd bacteriophages. **(a)** Phage T5. **(b)** Phage T7. The length of the bar in each micrograph is 50 nm. (Courtesy of Robley C. Williams, Virus Laboratory, University of California, Berkeley.)

DNA. The DNA also contains a series of four or five **interruptions** (nicks in the phosphodiester backbone of one DNA strand) that are reparable by the enzyme DNA ligase. They occur at definite positions on the DNA molecule, are all on the same strand of the double helix (Fig. 5-3), and have no known function. The frequency of individual nicks can be increased or decreased by mutation without affecting phage activity, but no mutants are known that have lost all of their nicks.

After attachment of the phage to the *fhuA* protein, ejection of the DNA is triggered, and about 8% of the phage genome (the left end in Figure 5-3) penetrates the cell wall and membrane regardless of the presence of

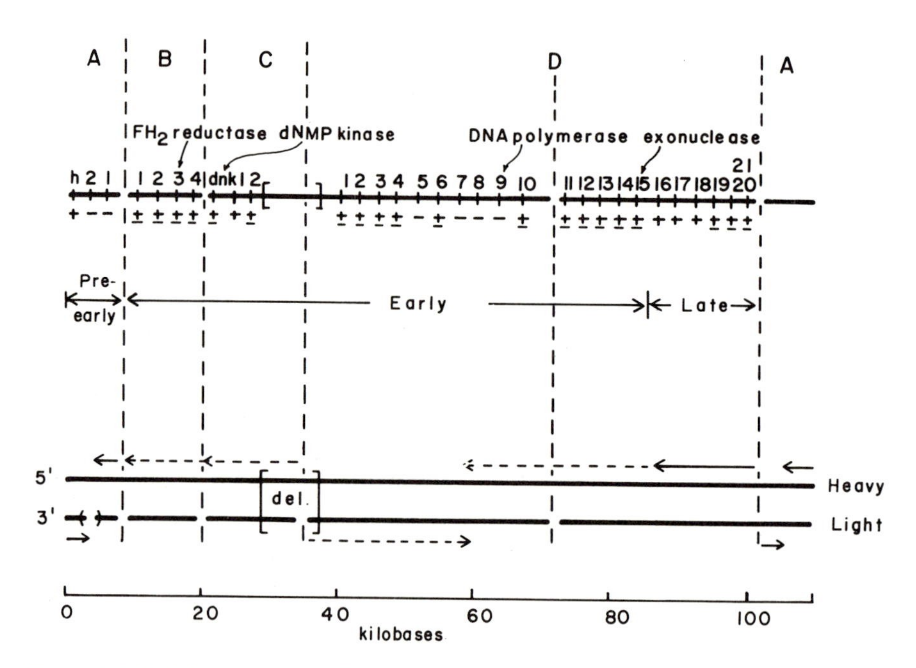

Figure 5-3. T5 phage genome. The capital letters at the top and the vertical dashed lines denote the four linkage groups. The upper heavy line denotes the genetic map, with the approximate position of each cistron given, along with the identification of four protein products. Below the map is an indication of whether mutants in each cistron synthesize DNA normally (+), synthesize no detectable DNA (−), or exhibit abnormal DNA synthesis (±). Next, the time of expression of each group of cistrons is shown. The heavy double line represents the DNA molecule, showing the positions of single-strand interruptions (the position of the leftmost nick is variable); the DNA deleted in heat-stable mutants of T5; and the direction of transcription of each group of cistrons (firm assignments are indicated with solid arrows). From Mathews, C.K. (1977). Reproduction of large virulent bacteriophages, pp. 179–294. In: Fraenkel-Conrat, H., Wagner, R.R. (eds.) Comprehensive Virology, vol. 7. New York: Plenum Press.

ionophores, the absence of a capsid, or a lack of DNA superhelicity. The only property that does seem to be required is appropriate membrane fluidity. The DNA injection process then stops at a region that has the potential to form various stem-and-loop structures and waits for the expression of certain "first step transfer" (FST) cistrons contained within the already injected DNA. The mRNA produced by these FST cistrons is referred to as pre-early mRNA by analogy to T4. After translation of the mRNA, DNA injection resumes, apparently due solely to the presence of the newly synthesized viral proteins, as removing the head and tail proteins of the phage particle after the initial attachment has occurred does not prevent DNA entry. Under normal conditions the entire injection process is complete within 1 minute. There is a transient reduction in active transport by the host cell following infection.

As the infection proceeds, the host RNA polymerase is modified by the addition of two pre-early proteins. Additional modifications seem possible, as there are three classes of mRNA produced: class I, which codes for the FST synthesis; class II, which codes for viral metabolic proteins; and class III, which codes for proteins needed for continued DNA synthesis, virion assembly, and cell lysis. Host DNA is degraded by one of the FST proteins. Assembly of the virions is probably accomplished by a mechanism more like that used by T7 than that used by T4 because, like T7 DNA, T5 DNA has a unique sequence. This property of T5 also makes it much more suitable for studying the mechanism of formation of recombinant hets because nonrecombinant hets are no longer possible.

Some of the T5 promoters are remarkably efficient, and it is possible to clone them onto plasmids carrying a selectable marker that lacks its normal promoter. The T5 promoters thereby obtained are so efficient that they can outcompete all of the normal host cell promoters to give preferential transcription of cloned DNA. Bujard and co-workers have shown that the T5 promoters are distinctive for their 75% (A+T) content and for the ability of at least some of them in vitro to accept 7-methyl guanosyl triphosphate instead of adenosine triphosphate as the initiator base for transcription. This ability allows the production of capped mRNA molecules from cloned eukaryotic DNA.

Bacteriophages T7 and T3

Bacteriophages T7 and T3 are the smallest of the T phages (Fig. 5-2) and, aside from the T-evens, the best studied. Both infect species of *Escherichia, Shigella, Salmonella, Klebsiella,* and *Pasteurella.* The primary member of the group is T7, but results obtained using T3 are discussed whenever the two phages are known to differ. The T7 DNA molecule contains no unusual bases, has a unique sequence, and has a short terminal redundancy of 160 bp (less than 1% of the total DNA). The redundancy for T3 DNA is 230 bp, and sequence analysis suggests that it is related

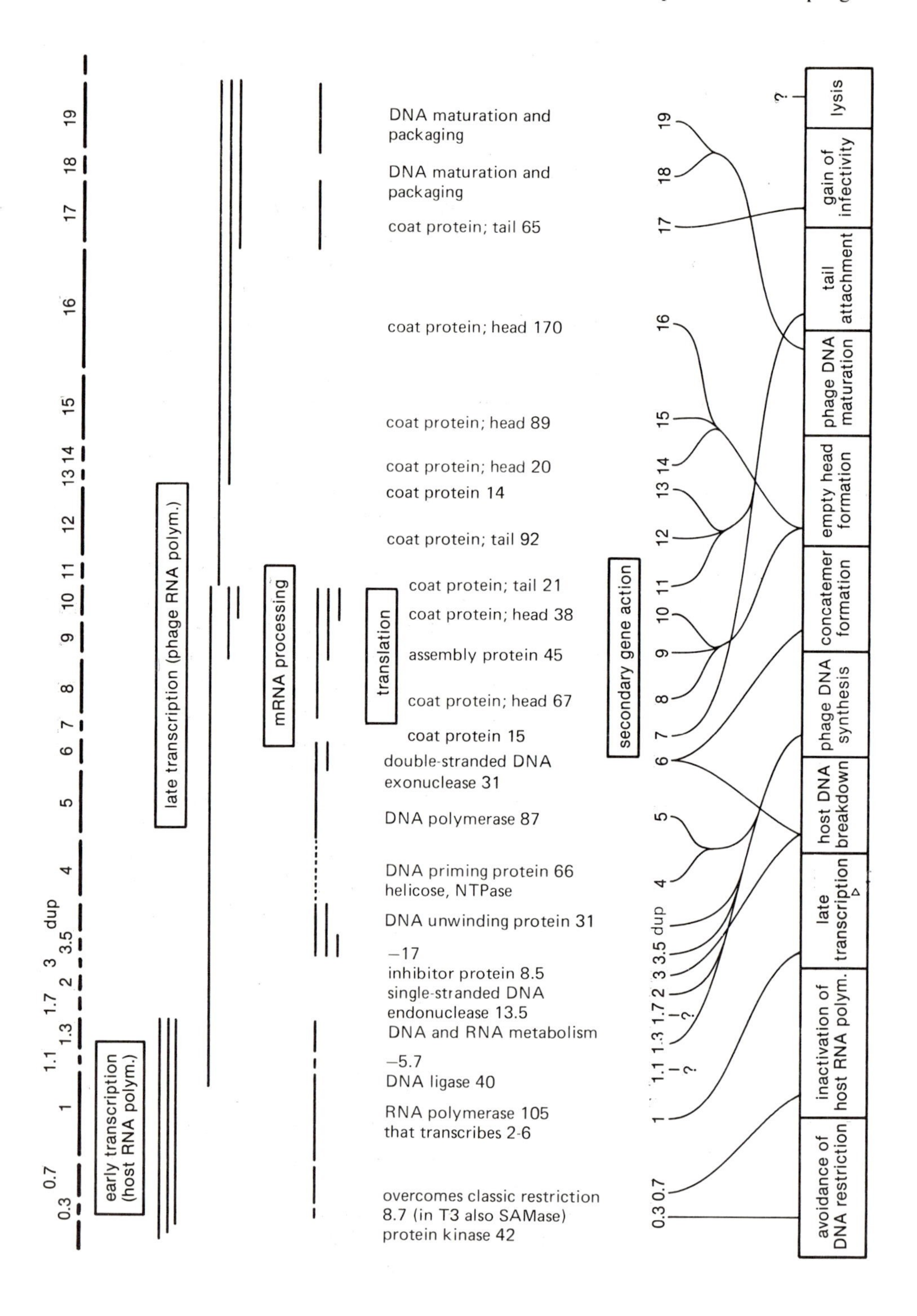

0.3 0.7 1 1.1 1.3 1.7 2 3 3.5 dup 4 5 6 7 8 9 10 11 12 13 14 15 16 17 18 19
early transcription (host RNA polym.)
late transcription (phage RNA polym.)
mRNA processing
translation
DNA maturation and packaging
DNA maturation and packaging
coat protein; tail 65
coat protein; head 170
coat protein; head 89
coat protein; head 20
coat protein 14
coat protein; tail 92
coat protein; tail 21
coat protein; head 38
assembly protein 45
coat protein; head 67
coat protein 15
double-stranded DNA exonuclease 31
DNA polymerase 87
DNA priming protein 66 helicose, NTPase
DNA unwinding protein 31
–17
inhibitor protein 8.5
single-stranded DNA endonuclease 13.5
DNA and RNA metabolism
–5.7
DNA ligase 40
RNA polymerase 105 that transcribes 2-6
overcomes classic restriction 8.7 (in T3 also SAMase) protein kinase 42
secondary gene action
lysis
gain of infectivity
tail attachment
phage DNA maturation
empty head formation
concatemer formation
phage DNA synthesis
host DNA breakdown
late transcription
inactivation of host RNA polym.
avoidance of DNA restriction

to T7 and to *Klebsiella* phage 11. All three phages seem to have terminal redundancies based on the progenitor sequence TTAACCTTGGG.

Although T3 and T7 seem to be morphologically similar, they are in fact discrete entities. Their infections are mutually exclusive, so that a cell infected with T7 cannot be infected with T3 and vice versa, whereas cells infected with other intemperate phages can generally be superinfected with the same phage. Like T5, T7 DNA enters the cell in stages, first 10%, then 50%, then the remainder. Once again there is coupled transcription and entry for the stage 2 DNA, but the final 40% of the DNA can enter the cell only if translation is permitted.

A reasonable amount of information is available about T7 DNA replication. The replication process begins at a definite point of origin that is located 17% of the distance from the left end of the map (Fig. 5-4) using phage-specific proteins that often share some homology with corresponding *E. coli* proteins. The phage replication proteins include the cistron 5 product that is a DNA polymerase when combined with host thioredoxin, the cistron 4 and 1 products that make RNA primers, and a viral protein that acts as a single-stranded DNA binding protein. The host contributes DNA ligase and DNA polymerase I. It is not necessary for the phage DNA to circularize in order for replication to proceed because at least the first two rounds of replication are bidirectional. (If replication were not bidirectional, only part of the linear molecule would be replicated.) Concatemeric structures are formed by a mechanism other than simple recombination, and DNA replication is not a prerequisite for late transcription as it is in T4.

Transcription of T7 DNA always uses as a template the right strand of the DNA helix (preferentially binds poly-UG). This polarity leads to an automatic suquencing of events during infections, as the left end of the DNA (Fig. 5-4) is always injected first. The early mRNA molecules are produced by the host RNA polymerase and are processed by the host RNase III to yield up to five smaller mRNA molecules. At 6 minutes after infection at 30°C, late mRNA synthesis begins using a phage RNA poly-

Figure 5-4. T7 genome and steps of its expression. The thick interrupted line at the top represents the genetic map of T7. Most known cistrons are marked by their numbers. The length of the cistron-representing bars is proportional to the molecular weights of the corresponding polypeptides. Each thin horizontal line represents one species of mRNA. Transcription is always from left to right. Where possible, the names and molecular weights ($\times 10^{-3}$) of the primary gene products are given. To demonstrate schematically their mode of action, these products are represented by the number of the corresponding cistrons. Adapted from Hausmann, R. (1977). Bacteriophage T7 genetics. Current Topics in Microbiology and Immunology 75:77–110. It also includes information from Krüger, D.H., Schroeder, C. (1981). Bacteriophage T3 and bacteriophage T7 virus-host cell interactions. Microbiological Reviews 45:9–51.

merase that consists of a single protein molecule. This molecule must functionally differ from the elaborate *E. coli* RNA polymerase complex, as the T3 and T7 polymerases work with either phage DNA but do not transcribe λ, T4, or *E. coli* DNA. Further transcription by the *E. coli* polymerase is prevented by phosphorylation of the β′ subunit catalyzed by the protein kinase from cistron 0.7 and by the inhibitor protein from cistron 2.0. Many late transcripts are also processed by RNase III (Fig. 5-4).

Relations between the host cell and phages T3 and T7 are also unusual. T3 but not T7 grows on *E. coli* strains harboring an F plasmid (Hfr, F′, or F^+), with the difference between the phages lying at the end of gene 1. Failure to grow is due to an inability of the infected cells to synthesize any macromolecules after the time at which class III mRNA synthesis normally begins. A similar abortive infection can be observed if an *E. coli* strain B cell that is a P1 lysogen (see Chapter 6) is infected with T7. Although phage T3 is not restricted by F^+ cells, it can, under starvation conditions, replicate so slowly that the host cell survives to divide and grow, yielding a pseudolysogen (see Chapter 6). It can be shown that the inhibitory effect on phage T7 is mediated via suppression of an ochre codon. A ribosomal mutation that increases the fidelity of translation reduces the inhibitory effect, whereas the presence of an ochre suppressor mutation increases it.

Maturation of T7 begins at about 9 minutes postinfection and proceeds in a manner similar to that of T4. The major unresolved problem in T7 virion assembly is how to develop a series of DNA molecules that have a unique sequence rather than a circular permutation as in the case of T4. The conventional headful mechanism does not suffice, as that involves random cuts in the concatemeric DNA. It is possible that the mechanism is similar to that used by phage lambda (see Chapter 6 and Figure 6-2) in which closely offset nicks are made at specific points within the redundant DNA region as would be done by a restriction enzyme. The overhangs can then be filled in by an enzyme such as DNA polymerase I to yield a fully double-stranded DNA molecule with unique sequence and a terminal redundancy. Empty proheads are necessary for cutting of the concatemer, as in the case of T4. The burst size is approximately 200.

The ability of the T7 RNA polymerase to respond solely to its own promoters has led to its frequent use in cloning experiments. The T7 promoters are highly conserved in the region from base −17 to base +6 and are therefore unlikely to occur by chance. DNA sequences cloned downstream from a T7 promoter remain untranscribed until the T7 RNA polymerase is provided. The technique can be used in vitro to generate RNA probes for Northern blotting, although a certain amount of caution is necessary. In the presence of 3′-overhangs generated by restriction enzymes, the T7 RNA polymerase can initiate away from its promoter and transcribe incorrect regions or even the incorrect strand of DNA. The problem does not arise if the 3′-overhangs are digested away. The T7 polymerase cistron has also been spliced into vaccinia virus and used to control expression of DNA cloned into other vectors in mammalian cells. Transcription of the desired DNA sequence occurs only after infection by the modified vaccinia virus.

Bacteriophages Containing Single-Stranded DNA

Bacteriophages Belonging to the Ff Group

The Ff group is a large one composed of filamentous phages (Fig. 5-5) that contain a circular, single-stranded DNA (1-DNA) molecule of 6408 nucleotides. A typical genetic map is shown in Figure 5-6. The protein coat consists of about 2700 major subunit protein molecules encoded by cistron VIII and minor subunits encoded by cistrons III, VI, VII, and IX. The rod-like shape of the virion enforces a similar configuration on the DNA molecule, which folds upon itself. The ends of the DNA molecule thus defined can be differentiated by the proteins that are bound to them, the products of cistrons VII and IX at one end and the products of cistrons III and VI at the other end. A number of morphologic variants are known. Miniphages of less than normal length have the replication origin together with a variable amount of the remainder of the genome. Diploid phages occur about 5 to 6% of the time and are twice the normal length. They contain two complete, circular DNA molecules. Longer polyphages contain increased numbers of DNA molecules in proportion to their length.

The process of infection begins with the specific attachment of virions to the cell. Electron micrographs have shown the phage specifically attached to the tips of F pili (Fig. 5-5) or I pili. Because these types of pili are produced only by male cells (see Chapters 9, 10, and 11), the Ff phages are described as being **male specific.** Despite their mode of attachment, there is some question as to whether the viruses actually release their DNA into the pili. It seems more likely that infection occurs via the base of the pilus after the pilus is retracted into the cell, drawing the virus down to the surface.

Once the virion has attached to the cell, it enters the eclipse phase. Physically, the eclipse means that part of the protein coat opens up, partially releasing the DNA and making it susceptible to nucleolytic attack. The released DNA apparently penetrates through the cell membrane into the cytoplasm of the cell provided the host *fii* membrane protein is normal. The next steps in the infectious process are analogous to those of phage T5. The cistron III protein and *E. coli* RNA polymerase initiate synthesis of a complementary DNA strand, after which the rest of the DNA molecule enters the cell as DNA polymerase III completes the synthesis.

Despite the binding of RNA polymerase to the single-strand viral DNA,

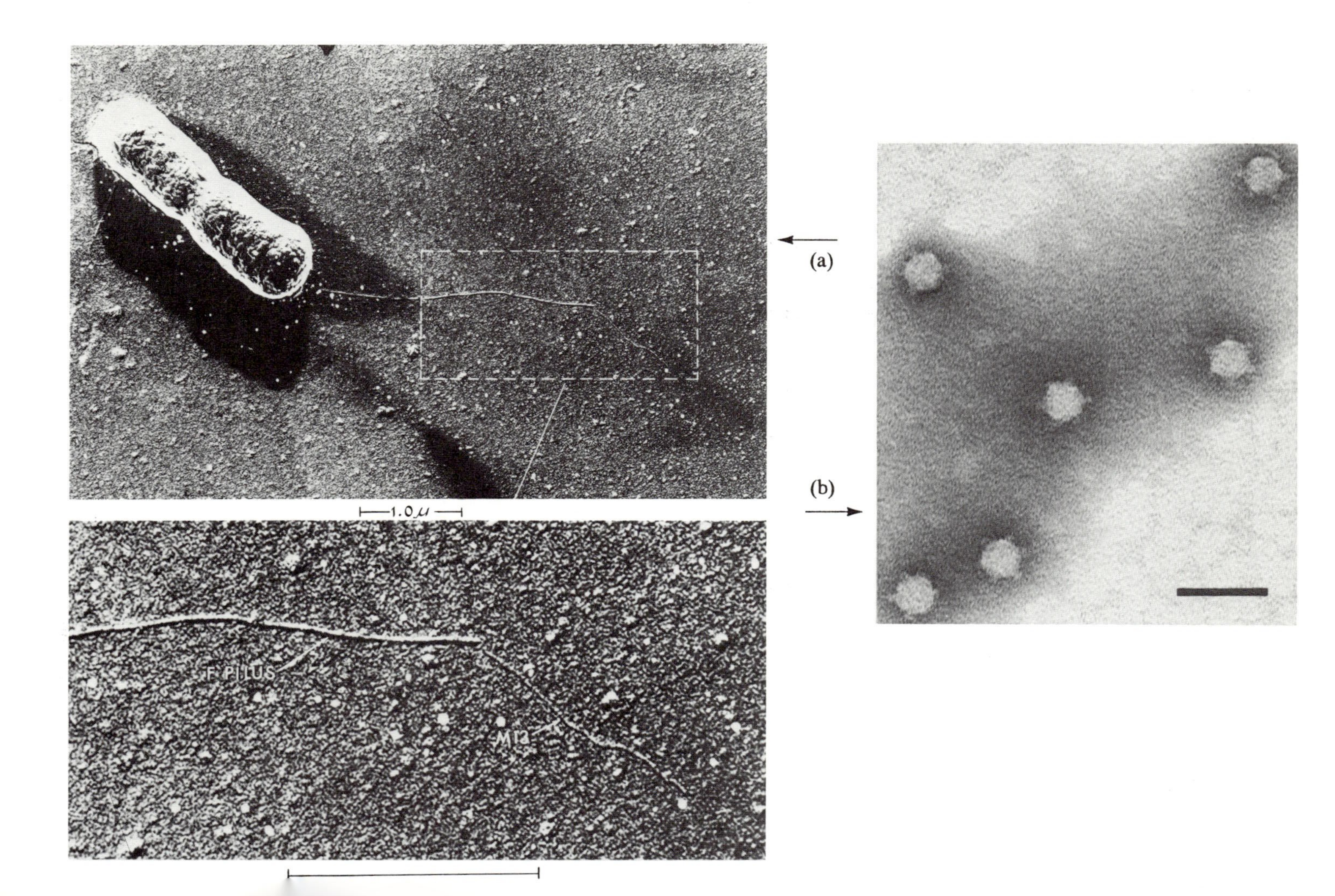
(a)
(b)
1.0μ
F PILUS
M13

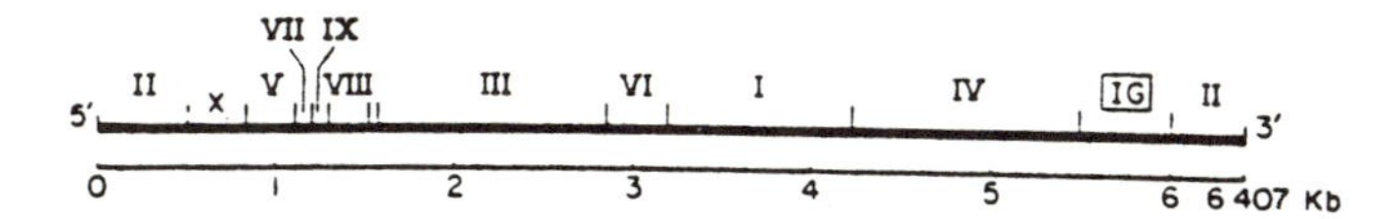

Figure 5-6. Genetic map of Ff phage. The circular genome is presented in linear form as if it had been opened at the unique *Hin*dII restriction site. Roman numerals refer to the cistrons. IG is the intergenic region. *X* refers to the part of cistron II that codes for the X protein. The direction of transcription and translation is from left to right. Wavy lines indicate mRNA transcripts. The bottom line is a scale marked in kilobases. Adapted from Zinder and Horiuchi (1985).

all the mRNA molecules are synthesized using the new complementary DNA strand as a template. The viral DNA is therefore considered plus-strand DNA, as its sequence corresponds to the mRNA, and its complement is minus-strand DNA. Promoters are positioned to allow transcription to begin at cistrons II, X, V, IX, or IV (Fig. 5-6). Some processing of the transcripts occurs using host enzymes. An unusual feature of the genetic map is that cistron X is actually the carboxy portion of cistron II. They are translated in the same reading frame, and therefore the cistron X protein is a shortened version of the cistron II protein. It is also significant that the intergenic (IG) region is not transcribed at all. Figure 5-7 shows that there is an enormous amount of potential secondary structure within the intergenic region. There are five major loops that can be reasonably predicted from the known base sequence. As outlined below, these loops seem to be associated with both transcription and replication. The lack of transcription can be attributed to loop [A], which functions as a ρ-dependent terminator of cistron IV transcription. The loops also prevent binding of single-strand binding protein (ssb) to the IG region and thereby emphasize the replication and transcription start sites.

The mechanism of replication for the Ff phages must be unusual, as conventional replication models do not generate single DNA strands. DNA replication begins with production of the complementary DNA (minus-strand) during DNA entry. Only host proteins are necessary for this reaction; they include ssb, DNA polymerase III, RNA polymerase, DNA ligase, and DNA polymerase I. A 30-base RNA primer is laid down near the base of loop [C] in Figure 5-7. The primer is extended by DNA polymerase III holoenzyme, and the primer is removed and the nick sealed

Figure 5-5. **(a)** Electron micrograph of bacteriophage M13 attached to the tip of an F pilus of *E. coli*. The bottom photograph is an enlargement of the area indicated by the dotted lines. **(b)** Electron micrograph of bacteriophage ϕX174 negatively stained with phosphotungstate. The bar represents a length of 50 nm. **a:** From Ray, D.S. Replication of filamentous bacteriophages, pp. 105–178. In: Fraenkel-Conrat, H., Wagner, R.R. (eds.) Comprehensive Virology, vol. 7. New York: Plenum Press. **b:** From Denhardt, D.T. (1977). The isometric single-stranded DNA phages, pp. 1–104. Op. cit.

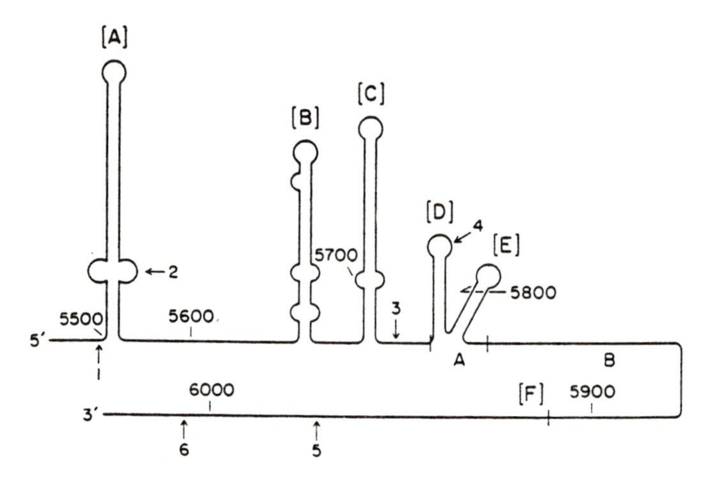

Figure 5-7. IG region of Ff phage. Secondary structure of the viral strand DNA. [A], [B], [C], [D], and [E] represent self-complementary sequence elements. [F] is an approximately 150-base-long, AT-rich sequence without self-complementarity, extending downstream from hairpin [E]. Shown by arrows are: (1) end of cistron IV; (2) rho-dependent termination of cistron IV mRNA; (3) initiation site of minus-strand primer synthesis; (4) initiation site of plus-strand synthesis; (5) initiation site of cistron II mRNA; (6) start of cistron II. Domains A and B of the plus-strand replication origin are indicated. Four-digit numbers represent nucleotide numbers. From Zinder and Horiuchi (1985).

as in conventional replication. The resulting double-strand DNA molecule is nicked in the position indicated by the number 3 in Figure 5-7, which is where the RNA primer is located. After removal of the primer in the usual way, the covalently closed, circular, relaxed DNA molecule is designated a replicative form (RF) DNA, specifically RF IV DNA (Fig. 5-8). This completes stage I replication. RF IV DNA is converted to RF I (supercoiled) by DNA gyrase.

Although the RF I DNA should in principle be able to replicate in the usual way, it does not. Instead it replicates via a **rolling circle** mechanism that was first proposed by Gilbert and Dressler for concatemer production in phage T4. This model seems to be adequate to account for all DNA replication in the single-strand coliphages and for some replication in λ and other temperate phages. Ironically, it does not seem to apply to T4.

During stage II replication, RF I replicates to make more RF I. The process begins with a nick introduced into the plus-strand by the cistron II protein at site 4 of the IG (Fig. 5-7) to yield RF II DNA. If both strands should be nicked, the resulting linear molecule is designated RF III. The 3′-OH group located at the site of the nick serves as a primer for DNA synthesis by host proteins DNA polymerase III holoenzyme, *rep*, and ssb. As new bases are laid down, the old plus-strand is displaced (Fig. 5-

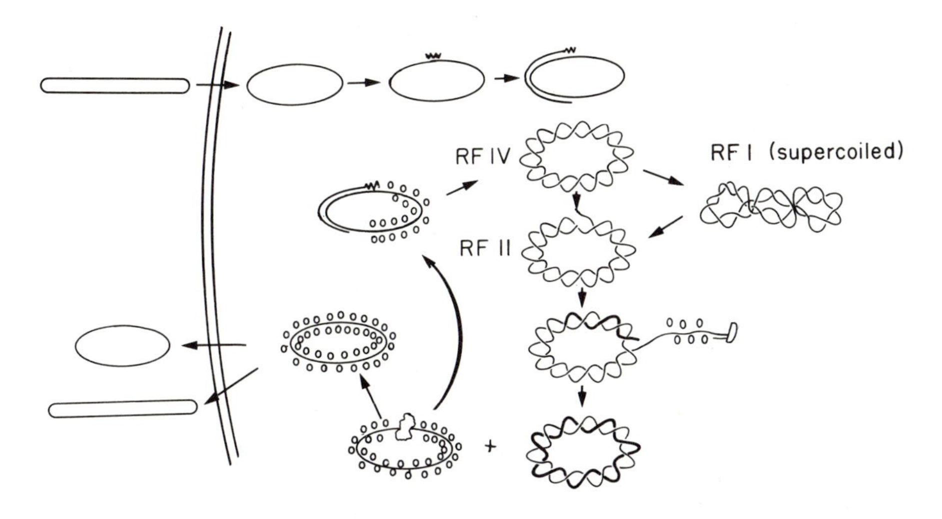

Figure 5-8. Rolling circle model for Ff phage DNA replication. The phage is represented as a long cylinder. The phage DNA enters the cell at the upper left as a single-strand, circular molecule. Priming with RNA (wavy line) initiates synthesis of a complementary DNA strand to yield an RF IV molecule that is covalently closed and circular. It can be converted to RF I by supercoiling. Both RF I and RF IV can be used to generate RF II by introducing a nick at a specific site. The nick serves to prime stage II DNA synthesis in which the complementary strand is used to synthesize another viral strand (heavy line) while the original strand is displaced. The displaced strand is coated with protein (small ovals) to stabilize it. Recircularization of the displaced strand is accomplished by intramolecular pairing facilitated by the cistron II protein. During stage II replication that protein is ssb and can be displaced by standard lagging strand synthesis to yield more RF IV. For stage III replication the displaced strand is coated with cistron V protein, which prevents synthesis of the complementary strand. As the DNA strand migrates through the cell membrane, coat protein subunits replace the cistron V protein coating the DNA, and a new virus particle is formed. Some DNA also is extruded without being coated.

8) and coated with ssb. Complementary minus-strand synthesis occurs either after the old plus-strand is released in circular form (with the ends originally paired by loop [D] and sealed by the cistron 2 protein) or in a fashion analogous to synthesis of the lagging strand in conventional DNA replication. The RF IV molecules are then converted to RF I. Most replicating molecules synthesize only a unit length of DNA and then terminate at a site in domain A of Figure 5-7. Domain A is essential for replication and includes the initiation, termination, and cistron II protein binding sites.

Domain B functions in a manner similar to a eukaryotic enhancer sequence and is dispensible in phages with a compensatory mutation in cistron II.

Stage III replication results in the production of only plus-strand DNA suitable for packaging into mature virus. The switch from stage II to stage III replication is controlled by the concentration of the cistron V and X proteins, both of which have an affinity for single-strand DNA. As the concentration builds up, the cistron V protein may coat the plus-strand DNA instead of ssb, and DNA thus coated is available to be packaged into virions. Stage III replication occurs only after stage II replication has become established but does not completely replace it, although the presence of protein V also inhibits the synthesis of protein II. Thus the virus continues to replicate within the cell at all times.

The Ff phages do not cause cell lysis but, rather, leak into the medium without significantly damaging the cell membrane. As a result, infection by Ff phages slows the growth of a culture but does not kill it. The "plaques" that are observed with these phages are actually due to the difference in growth rates between infected and uninfected cells and tend to disappear if incubated for longer periods of time. Maturation of the virus occurs as an extrusion process of the DNA through the cell membrane. The capsid proteins are synthesized as procoat molecules with 23 extra amino acid residues located at the amino-terminus. These extra amino acids constitute a **signal peptide** (hydrophobic tail) that causes the protein to be cotranslationally inserted into the cell membrane. During or after the insertion, the signal peptide is cleaved from the procoat molecule by a leader peptidase. As the circular phage DNA molecules are transported across the cell membrane, the coat proteins in combination with the host *fip* protein attempt to displace the cistron V protein from the DNA and form the mature virus. However, some viral DNA molecules are also extruded without any coat. Exceptionally high titers of intact phage (circa 10^{12}/ml) of Ff phage can be routinely obtained.

The Ff phages have assumed an important role in molecular biology thanks to the efforts of Messing. He constructed a series of DNA cloning vectors based on phage M13, the M13mp series, which now numbers some 19 members. The phages carry a portion of the β-galactosidase code (see Chapter 12) to serve as a control marker and a specially constructed DNA insert that contains multiple restriction enzyme sites within a short space (a polylinker). This insert is located so it lies close to the origin of replication for the minus-strand. Therefore by using an appropriate primer, it is possible to set up a replication system in vitro that preferentially copies the cloned DNA. This single-strand DNA replication mechanism is ideally suited for use in the Sanger dideoxy DNA sequencing technique. The size of the DNA spliced into the polylinker is not critical because of the method of virion assembly. The longer spliced DNA is simply coated with additional coat protein subunits, up to a point.

Bacteriophage φX174

Bacteriophage φX174, as its name implies, was the 174th isolate in group 10 of a large series of bacteriophages. It is the principal representative of a group of phages that are simple icosahedrons with 5-nm spikes extending from all 12 vertices (Fig. 5-5b). The capsid is composed of 60 molecules of F protein with vertices constructed from five molecules of protein G plus one molecule of protein H. Also associated with the capsid are 30 to 50 molecules of protein J, one of A* (the carboxy-terminal portion of A), and the polyamines spermidine and putrescine. The DNA molecule is once again single-stranded and circular. As in the case of the Ff phages, φX174 eclipses outside the cell, protruding some portion of the DNA. If the DNA is not converted to a double-strand form, the infection is abortive.

Priming of stage I replication occurs via a primosome such as that seen for bacterial chromosome replication. More than one initiation site is apparently possible. Extension of the primer is via the same mechanism as the Ff phages. Complementary strand synthesis is initiated at a stem-and-loop structure and generates an RF molecule.

Stage II replication is initiated by a nicking protein, protein A, which attaches covalently to the nicked DNA strand via a tyrosine residue. About 30 bp of φX174 DNA are necessary and sufficient to initiate rolling circle replication when cloned into a plasmid. The general outline of rolling circle replication is the same as for the Ff phages. Termination of the synthesis of the plus-strand DNA is via the attached protein A. As protein C concentration increases, it inhibits the initiation or reinitiation of rolling circle replication by binding to the RF-protein A-*rep* complex in the presence of ATP. The new complex is a substrate for the packaging system when proheads and protein J are supplied. Protein A* also assists in the switch to stage III replication.

Unlike the Ff group, φX174 infections cause cell lysis, which is dependent on cistron E function. The E protein activates the normal cell autolytic enzymes, and the cell wall is degraded. Studies with inhibitors of autolysis show that a lytic infection is possible only when normal autolysis can occur, so it is mainly dependent on host functions. For example, the normal proton gradient must be present across the cell membrane before lysis can occur. However, penicillin-tolerant (and therefore defective in parts of the autolyic process) cells can still be lysed by the phage, indicating that at least some functions are not in common.

There has long been a problem concerning mRNA synthesis and translation in φX174-infected cells due to a peculiarity of the genetic map. Based on the number and size of the proteins resulting from an infection, it has been estimated that 6100 nucleotides would be required to code for all the necessary amino acid sequences. The genome of the virus, however, consists of only 5386 nucleotides. This discrepancy was clarified by the discovery that three of the cistrons are embedded in other cistrons (Fig. 5-9). Cistron B lies at the end of cistron A, and cistron E lies at the end of cistron D. Cistron K translation begins at an overlap of the two terminator codons for cistron B and spans the last 86 bases of cistron A and the first 89 bases of cistron C. The proteins produced by the **embedded cistrons,** unlike the case for the A and A* proteins, do not resemble the proteins encoded by the large cistrons because the mRNA molecules are translated in different reading frames.

Genetic relatedness is often determined by codon analysis, which is based on the principle that where several codons are possible for a single amino acid an organism may characteristically use only one or two types. In the case of leucine residues in protein G, for example, six codons are possible, but GAT is never used and GAA is used 50% of the time. An analysis of the codons comprising the embedded and nonembedded cistrons suggests that cistron A was formerly shorter but lost its terminator signal and now reads through the B cistron in a different reading frame.

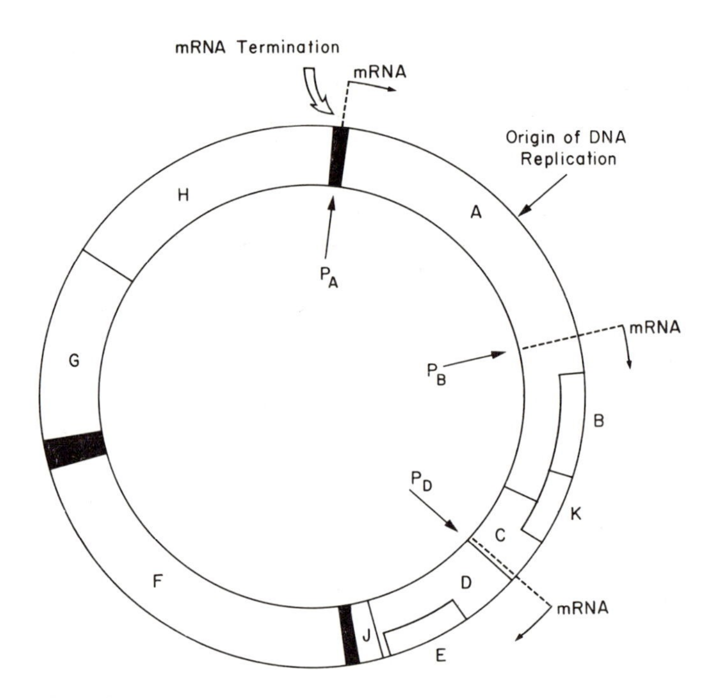

Figure 5-9. Genetic map of φX174. Ten cistrons are drawn, separated by solid lines. The 11th is A*, which is an internal restart (i.e., an alternative ribosomal binding site) within cistron A. The intercistronic spaces are marked in black. P_A, P_B, and P_D are the three known promoters (RNA polymerase binding sites) and are shown by dotted lines. The mRNA termination site is also indicated. Adapted from Godson, G.N. (1978). Bacteriophage G4. Trends in Biochemical Sciences 3:249–253.

The amino-terminal portion of A and all of the B cistron use codons that are typical for φX174, but the region of the A cistron that overlaps B employs unusual codons. By contrast, all of cistron E uses codons that are rarely used anywhere else in the φX174 genome, even though cistron D uses normal codons. This finding suggests that the E cistron evolved from a preexisting D cistron by formation of a new translation start signal, and that D was always as large as its present size. Codon analysis for cistron K suggests that its origin is similar to that of E.

Obviously, most mutations in an overlapping region affect both proteins. Nevertheless, it is theoretically possible to mutate one protein without changing the other. This feat has been accomplished by Gillam and co-workers, who used site-directed mutagenesis to make a T→A transversion that created an amber codon with respect to cistron K but retained a leucine codon in the A reading frame. The mutated phage was still viable but had a reduced burst size.

RNA-Containing Bacteriophages

The RNA phages are small icosahedral viruses that contain a molecule of linear, single-stranded RNA inside a capsid consisting of subunits of coat protein plus one molecule of maturation protein. They can be subdivided into four groups based on criteria such as the immunologic cross-reactivity of the coat protein, the buoyant density of the virion, the ratio of adenine to uracil residues in the RNA molecule, and the amino acids that are not utilized in the synthesis of coat protein. Each group also seems to produce a specific enzyme for RNA replication that does not replicate the RNA from any other group. Commonly encountered group members are (I) f2, MS2, R17; (II) GA; (III) Qβ; and (IV) SP, FI.

The RNA phages are further examples of male-specific phages. They infect Hfr, F^+, or F′ *E. coli* cells or cells of any other genus into which the F plasmid has been transferred (e.g., *Salmonella, Shigella,* and *Proteus*). The site of attachment of the phage is the side of the F pilus. Therefore male cells that have been depiliated by shearing in a blender are resistant to phage infection because they lack receptors. Cells that have been infected or superinfected with a DNA phage are also resistant to RNA phage infection.

Once the phage RNA reaches the cytoplasm of the cell, it serves as its own mRNA (it is a plus-strand virus). The viral RNA contains sufficient nucleotides to code for at least four proteins: the coat protein, the A (maturation) protein, the replicase protein, and the L (lysis) protein. In Qβ, a fifth protein, A1, has been observed that results from inefficient termination at the end of the A cistron and consequent read-through into the first portion of the coat protein cistron. This situation is another example of overlapping cistrons.

Genetic maps of the RNA are difficult to prepare, as recombination has not been detected in RNA phages. The major reason is that reversion rates for mutations in RNA phages approach 0.1%, which is a much higher frequency than that generally observed for DNA recombination. In turn, the high reversion rate may be due to the much less stringent accuracy requirements for RNA synthesis in contrast to DNA synthesis in an *E. coli* cell. The genetic sequence was finally determined by biochemical techniques in which the RNA molecule was fragmented in a known way, and then fragments were translated using an in vitro protein-synthesizing system. The genetic map for the MS2 group is shown in Figure 5-10. Note that two reading frames are used and that the L protein is the product of an embedded cistron.

The complete sequence of the RNA molecule from phage MS2 has been determined by Fiers and co-workers. Some in vitro protein-synthesizing systems cannot translate the replicase cistron without first translating the coat protein. This phenomenon is known as **translational coupling** and has been attributed to the existence of secondary structure in the RNA molecule. In order to reflect possible intramolecular structure, the base sequence of the RNA can be presented so as to maximize the number of intramolecular hydrogen bonds. Such a structure is shown in Figure 5-11. The highly convoluted loops are frequently referred to as "flower structures," and Fiers and co-workers described the entire molecule as a "bouquet." Note that the start codon for the replicase protein is predicted to be masked in a double-stranded region of the molecule that would make it unavailable for the 30S ribosomal subunit and therefore block translation. This prediction has been confirmed by Berkhout and van Duim. Because translation of the coat protein allows translation of

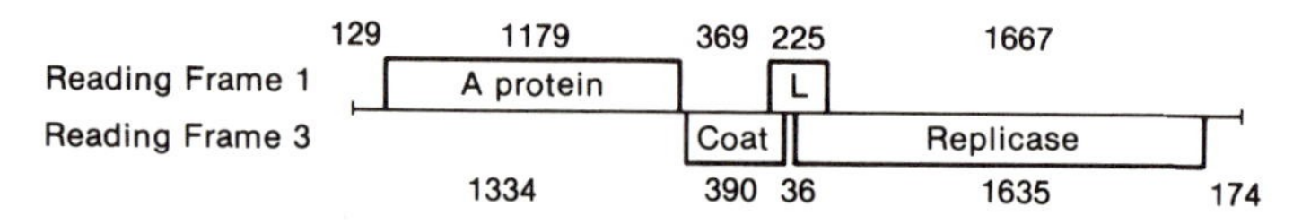

Figure 5-10. Genome of MS2. The four cistrons are shown as rectangles; untranslated regions are indicated by narrow lines. The 5′-end of the RNA molecule is shown at the left. Reading frame 1 begins at the first base, and reading frame 3 begins at the third base. The numbers above and below the diagram indicate the number of bases in a particular segment of the reading frame. In each case the initiator codon is taken to be part of the cistron, and the terminator codon is taken to be part of the untranslated region. At the present time no proteins are known to be read using reading frame 2. Redrawn from Fiers, W., Contreras, R., Duerinck, F., Haegeman, G., Iserentant, D., Merregaert, J., Min Jou, W., Molemans, F., Raeymaekers, A., Van den Berghe, A., Volckaert, G., Ysebaert, M. (1976). Complete nucleotide sequence of bacteriophage MS2 RNA: primary and secondary structure of the replicase gene. Nature 260:500–507.

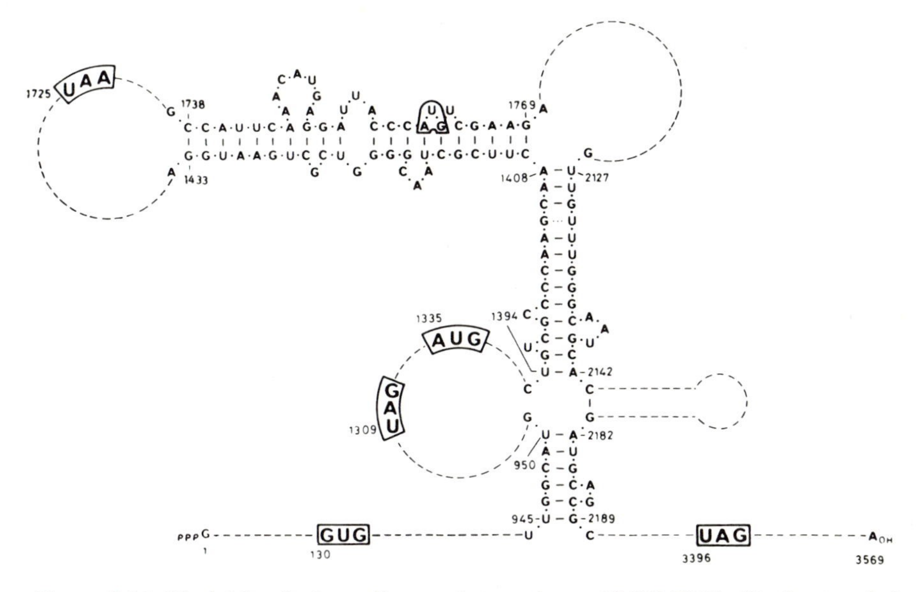

Figure 5-11. Model for the long-distance interactions of MS2 RNA. Single-stranded RNA molecules usually exhibit a high degree of folding, and the MS2 RNA is no exception. This skeletal outline shows the type of folding predicted from the base sequence of the RNA. The initiator and terminator codons for three proteins are shown by boldface type and emphasized by a line around them. The number scheme for the nucleotides is the same as that used in Figure 5-10, reading frame 1. The A protein initiator is an unusual one, GUG, and is found in a single-stranded region. The coat protein initiates with the more usual AUG codon, which is also found in a single-stranded region. The L protein initiator (not shown) occurs in the single-stranded loop at the upper left. However, the replicase protein is initiated by an AUG codon that is part of a region predicted to be double-stranded. Such a region could not be translated by a ribosome. Translation of the replicase could occur if the double-strand structure were disrupted by translation of the coat protein cistron. This prediction is in accord with observations made with in vitro protein synthesizing systems. From Fiers et al. (1976).

the replicase protein, it seems that a ribosome can disrupt hydrogen bonds in an mRNA molecule once it has attached but cannot disrupt them in order to effect its own attachment.

Another interesting point about translation of the MS2 RNA molecule is that the ribosome binding site upstream from the region coding for the L protein (and therefore part of the coat protein coding sequence) is normally blocked by a hairpin loop. Once again translational coupling is observed, this time between the L and coat proteins. However, an added complication is that the two proteins are in different reading frames (Fig. 5-10). Van Duim and co-workers demonstrated that ribosomes translating the coat undergo occasional frameshifts that result in premature termi-

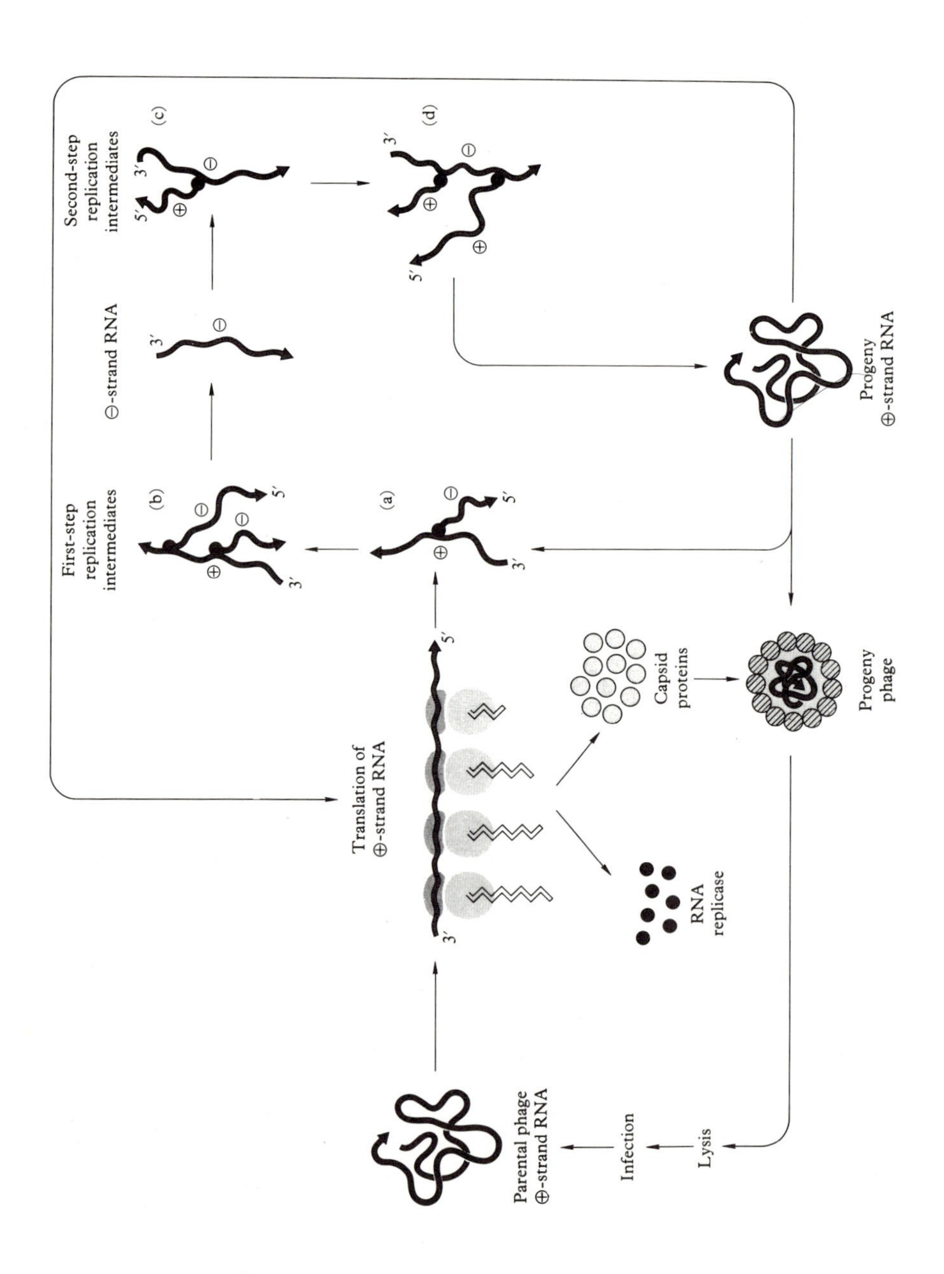
Parental phage
⊕-strand RNA
Infection
Lysis
Translation of
⊕-strand RNA
3′
5′
RNA
replicase
Capsid
proteins
Progeny
phage
First-step
replication
intermediates
(a)
(b)
⊖-strand RNA
Second-step
replication
intermediates
(c)
(d)
Progeny
⊕-strand RNA

nation of the coat protein in the region upstream from the L protein coding sequence. The ribosomes then restart in the new reading frame and synthesize L protein. The frequency of the frameshifting can be affected by changing the base sequence upstream from the ribosome binding site for the coat protein or by altering the translational fidelity of the ribosome.

Replication of the RNA is a rather involved process (Fig. 5-12). It requires a complex of four proteins designated by Greek letters, three of which are provided by the host and which, surprisingly, are part of the RNA translation system. These are ribosomal protein S1 (α) and the two elongation factors for protein synthesis, Tu (γ) and Ts (δ). The fourth protein is, of course, the viral RNA replicase (β). The complex forms an RNA-dependent RNA polymerase. The viral plus-strand uses this complex to produce minus-strands, and the minus-strands then serve as templates for the production of more plus-strands. In vivo few duplex RNA molecules can be detected, a condition that leaves the plus-strands free to be translated or packaged into phage particles.

If an in vitro reaction for Qβ replicase is set up but no template RNA is provided, the enzyme apparently generates RNA molecules spontaneously. Should one of those molecules accidentally prove to be a suitable substrate for further replication, it will reproduce rapidly and account for most of the RNA in the tube. Comparisons between reaction mixtures indicate that each time the reaction is run different RNA molecules are obtained. In a similar vein, normal phage RNA can be added to an in vitro system under conditions where only the replicase protein must be transcribed and translated in order for RNA replication to occur. If the reaction mixture is diluted at regular intervals, a premium is placed on minimizing the size of the RNA molecule to be replicated, and self-shortening of the RNA molecule is observed (but always preserving the replicase

Figure 5-12. General scheme of intracellular reproduction of an RNA phage. Upon infection, the parental RNA plus-strand is released from the capsid and translated to form capsid proteins and the RNA replicase. The replication of the phage RNA can now begin. At the initial stage of the replication process, the single-strand parental plus-strand serves as the template for the synthesis of complementary minus-strands. It results in the formation of first-step replication intermediates (*a* and *b*), open structures in which the template plus-strand and the replica minus-strand do not form an RNA double helix. In the next stage of the replication process, the single-strand minus-strand serves as the template for plus-strand synthesis. It results in the formation of second-step replication intermediates (*c* and *d*), which are similar in structure to the first-step replication intermediates except that their full-length template is a minus-strand rather than a plus-strand. The replica plus-strands are encapsulated by the capsid proteins to form structurally intact progeny phages, to be released on lysis of the infected cell. (Throughout the scheme shown here all complete plus- and minus-strands have the same length.) From Stent, G.S., Calendar, R. (1978). Molecular Genetics, 2nd ed. San Francisco: W.H. Freeman.

function). This system therefore provides an in vitro model for a type of evolution.

Maturation proceeds in much the same way as in the DNA phages. The coat protein binds to the plus-strand RNA molecule and removes it from the possibility of further replication or translation, thereby serving the incidental function of turning off the expression of the viral replicase. A terminator mutation in the coat protein results in the accumulation of greater-than-normal amounts of replicative RNA molecules. The maturation protein is necessary for phage infectivity. If a phage carries a terminator mutation in the A gene, phage particles of normal appearance are produced but they are noninfectious.

Bacteriophages Infecting *Bacillus subtilis*

Bacteriophage SP01

Bacteriophage SP01 (and the similar SP82) has the same general morphology as T4 but is somewhat larger. The linear DNA molecule within the virion has a unique sequence with 12.6-kb direct terminal repeats and a unique chemical composition. Instead of thymine (5-methyluracil, MdUMP), the DNA contains the modified base 5-hydroxymethyluracil (HMdUMP), which serves as a chemical label to permit restriction enzymes to distinguish between host and viral DNA. The phage codes for enzymes that degrade thymidine and uridine deoxyribonucleotide triphosphates (dTTP and dUTP) to their respective monophosphates and that convert deoxyuridine monophosphate to HMdUMP and then to HMdUTP so it can be used in DNA synthesis. The enzymes that break down dTTP and dUTP are not essential for phage growth, but in their absence up to 20% of the HMU residues are replaced by thymine.

Infection occurs by slow injection of DNA into the cell, a process that requires several minutes to complete. Crude genetic maps of the phage genome can be prepared by the simple expedient of violently shearing the virion and uninjected DNA away from the cell with a blender. The cells are then superinfected with various defective phages and observed to see if cell lysis occurs. The results are analyzed as for a *cis-trans* test. As might be expected, the cistrons controlling DNA metabolism are the first to be injected.

After injection of the phage DNA, the physiologic pattern is similar to that of T4. Host DNA replication shuts down within 6 to 8 minutes postinfection at 37°C, although little or no DNA degradation is observed. The shutdown of replication is independent of the enzymatic breakdown of dTTP. Host rRNA synthesis continues until cell lysis. However, host mRNA synthesis is rapidly replaced by viral RNA synthesis due at least in part to the extensive modifications to the RNA polymerase complex.

Three subunits are produced that confer different template specificities on the RNA polymerase holoenzyme by replacing the normal sigma factor. Transcription itself is subdivided into the temporal groups early, middle, and late (*e, m,* and *l*). Early mRNA is produced by host polymerase transcribing primarily from the terminal repeats. Middle mRNA requires the viral cistron 28 protein, and late mRNA requires the cistron 33 and 34 proteins. Six subclasses of mRNA have been identified on the basis of time of appearance and time of shutoff measured from the start of infection. They are *e,* 1 to 5 minutes; *em,* 1 to 12 minutes; *m,* 4 to 12 minutes; m_1l, 4 minutes until lysis; m_2l, 8 minutes until lysis; and *l,* 13 minutes until lysis.

DNA replication in SP01-infected cells produces concatemeric structures presumably analogous to those of the T phages. The maturation and packaging systems are assumed to be similar to those of phage T5. The onset of sporulation in the host cell effectively blocks further development of SP01 and prevents cell lysis. Consequently it is possible to have endospores that carry viral genomes and release phage particles during outgrowth. The mechanism of inhibition is thought to involve further modification of the RNA polymerase holoenzyme so it no longer transcribes SP01 DNA efficiently but transcribes the bacterial *spo* cistrons.

Bacteriophage φ29

Bacteriophage φ29 is the smallest double-strand DNA phage of any considered in this chapter, in terms of both head size and DNA size (Table 5-1, Fig. 5-13). The DNA molecule is unusual in that it is circularized by a protein (product of cistron 3) that is covalently bonded via a serine residue to the 5′-ends of the linear DNA, which contain short inverted repeats. Removal of this protein dramatically reduces the efficiency of transfection (see Chapter 8). Infection by φ29 does not significantly affect the level of macromolecular synthesis by the cell prior to the time of lysis.

DNA transcription occurs in the familiar fashion. The early mRNA molecules are transcribed from the L strand, whereas late mRNA transcription uses the H strand (Fig. 5-14). The time lag between infection and start of major mRNA synthesis is 6 to 8 minutes, which is longer than for most phages. The shift from host cell to viral transcription occurs as the result of the synthesis of a new polypeptide, which tends to replace the host sigma factor in the RNA polymerase holoenzyme. It is interesting to note that cistrons 10 through 14 seem to be transcribed in vivo both early and late and, therefore, from opposite strands (Fig. 5-14). This phenomenon has been observed in several other viruses and bacteria (see Chapter 12), but its significance in this instance is uncertain.

The DNA replication system for φ29 has been studied extensively in the phage and the cloned phage DNA. No RNA primer is needed for φ29-specific replication to be initiated, but protein 3 must be covalently joined

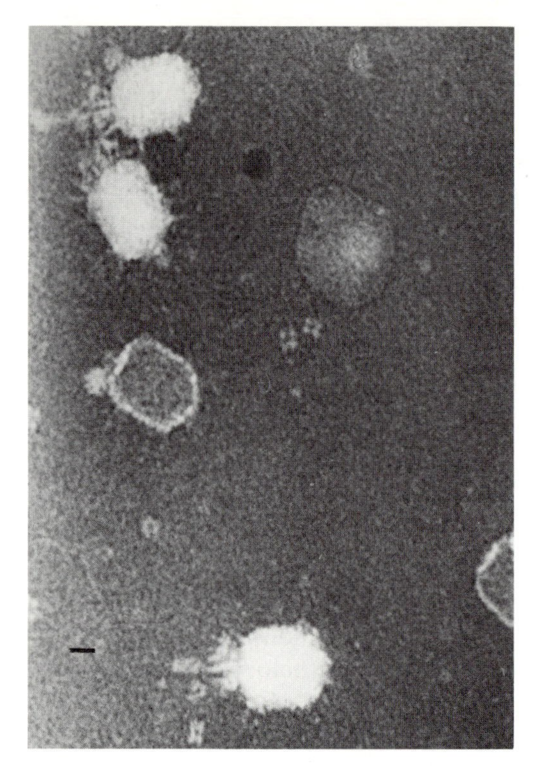

Figure 5-13. Electron micrograph of phage ϕ29, negatively stained with potassium phosphotungstate. The bar represents 10 nm.

to the physical ends of the DNA molecule. Cloned ϕ29 DNA does not self-initiate unless the plasmid is linearized adjacent to the protein 3 binding sites. Protein 2 catalyzes a reaction between free protein 3 and 5′-dAMP so as to give the initiation complex. The reaction is stimulated by the presence of protein 6. The protein—AMP complex acts as a primer and is extended by the usual proteins.

Three major transcripts have been identified: two early transcripts that correspond roughly to the ends of the DNA molecule and one late transcript that corresponds to the middle of the DNA molecule. Phage proheads copurify with a small (120 bases) RNA molecule whose presence is absolutely required for packaging to occur. There is significant homology between the left end of the DNA molecule and the prohead RNA, which may assist in the actual packaging operation.

Summary

Bacteriophages are a heterogeneous group of "organisms." They may be rod-like, spherical, or complex (having a head and tail). Their nucleic acids may be single-strand RNA, single-strand DNA, or double-strand DNA.

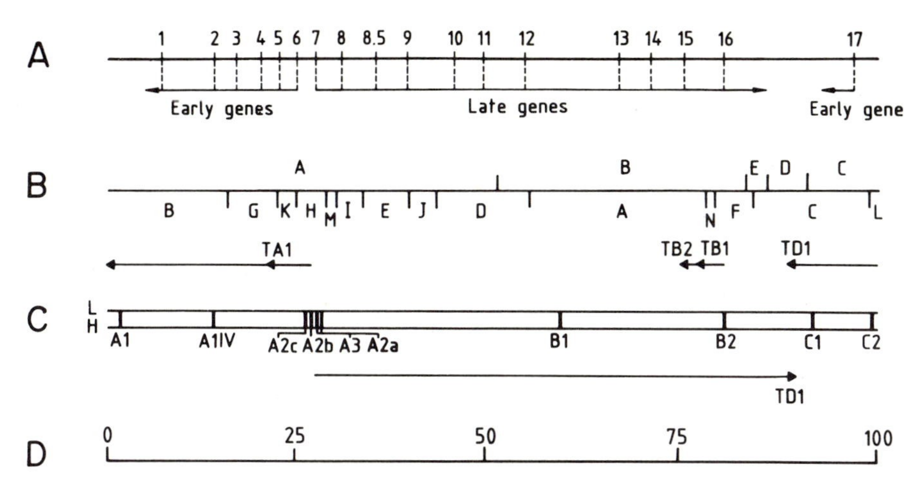

Figure 5-14. Genetic, physical, and transcriptional maps of ϕ29. **(A)** Genetic map. Arrows indicate the direction of transcription. **(B)** *Eco*RI (top) and *Hin*dIII (bottom) restriction maps. **(C)** In vivo transcriptional map. The vertical lines indicate the positions of the ϕ29 promoters. Arrows indicate the direction and extent of transcription. Arrowheads mark the termination points. Early transcription takes place from the light (L) strand of the DNA, and late transcription occurs from the heavy (H) strand. **(D)** DNA length shown in percentages. The total length is 19,285 bp. From Barthelemy, I., Salas, M., Mellado, R.P. (1987). In vivo transcription of bacteriophage Phi29 DNA: transcription termination. Journal of Virology 61:1751–1755.

As a general rule, the single-strand DNA molecules are circular, whereas all other viral nucleic acids are linear. The circularity can be attributed to the difficulty of protecting linear single-strand DNA from exonucleolytic attack.

Infection begins as the virus attaches to the cell surface. In all cases at least some of the impetus for the transfer of nucleic acid from the virion into the cell is provided by the host cell itself. Often it takes the form of a required replication or transcription step. Transcription of viral DNA is a highly regulated process, both temporally and spatially. In many instances the cistrons are arranged in the order of their use so that transcription provides an automatic sequencing of events. As in the case of phage T4, the host RNA polymerase is often modified so promoters for cistrons needed late are recognized and those for regions that have already been adequately transcribed are not.

Nucleic acid synthesis is as varied as the phages themselves. The large viruses show little dependence on host metabolism and frequently degrade the host DNA. Such phages protect their own DNA from degradation by the use of unusual bases such as hydroxymethyluracil or hydroxymethylcytosine. The smaller phages have shorter nucleic acids that lack the coding capacity to produce large numbers of polypeptides. As a result they are more dependent on host cell function. They may also exhibit extensive genetic overlap (embedded cistrons). The single-strand DNA phages convert their DNA to the double-strand form and then employ the rolling circle mode of replication, which can generate either more double-strand DNA or else single-strand DNA for the production of more virus particles.

Assembly of new virions has been studied extensively in only a few cases. Those complex phages with circularly permuted, terminally redundant DNA molecules are presumed to function similarly to T4. In cases where the DNA is not circularly permuted, the model predicts an offset cut like that of a restriction endonuclease followed by a filling in of the overhangs. The single-strand phages seem to assemble by the simple expedient of having single-strand DNA binding protein subunits attach to the nucleic acid as it is synthesized. These proteins are then later replaced by the coat proteins.

The bacteriophages have made a number of important contributions to DNA cloning procedures. M13 derivatives can be advantageously used to cloned DNA for dideoxy sequencing. Phage T5 promoters are exceptionally strong and can be used to generate large amounts of product from the cloned DNA. T7 RNA polymerase is specific for its late promoters and has found a use in regulating transcription of cloned DNA in both pro- and eukaryotic cells. The DNA remains quiescent until the polymerase is provided, at which time it is efficiently transcribed.

References

Generalized

Baas, P.D. (1985). DNA replication of single-stranded *Escherichia coli* DNA phages. Biochimica et Biophysica Acta 825:111–139.

Calendar, R. (ed.) (1988). The Bacteriophages. New York: Plenum Press.

Rasched, I., Oberer, E. (1986). Ff coliphages: structural and functional relationships. Microbiological Reviews 50:401–427.

Zinder, N.D., Horiuchi, K. (1985). Multiregulatory element of filamentous bacteriophages. Microbiological Reviews 49:101–106.

Specialized

Auyama, A., Hayashi, M. (1986). Synthesis of bacteriophage φX174 in vitro: mechanism of switch from DNA replication to DNA packaging. Cell 47:99–106.

Barthelemy, I., Salas, M., Mellado, R.P. (1987). In vivo transcription of bacteriophage Phi29 DNA: transcription termination. Journal of Virology 61:1751–1755.

Berkhout, B., van Duim, J. (1985). Mechanism of translational coupling between coat protein and replicase genes of RNA bacteriophage MS2. Nucleic Acids Research 13:6955–6967.

Biebricher, C.K., Eigen, M., Luce, R. (1986). Template-free RNA synthesis by Qβ replicase. Nature 321:89–91.

Drexler, K., Riede, I., Henning, U. (1986). Morphogenesis of the long tail fibers of bacteriophage T2 involves proteolytic processing of the polypeptide (gene product 37) constituting the distal part of the fiber. Journal of Molecular Biology 191:267–272.

Fluit, A.C., Baas, P.D., Jansz, H.S. (1986). Termination and reinitiation signals of bacteriophage ϕX174 rolling circle DNA replication. Virology 154:357–368.

Fuerst, T.R., Niles, E.G., Studier, F.W., Moss, B. (1986). Eukaryotic transient-expression system based on recombinant vaccinia virus that synthesizes bacteriophage T7 RNA polymerase. Proceedings of the National Academy of Sciences of the United States of America 83:8122–8126.

Gentz, R., Bujard, H. (1985). Promoters recognized by *Escherichia coli* RNA polymerase selected by function: highly efficient promoters from bacteriophage T5. Journal of Bacteriology 164:70–77.

Guo, P., Erickson, S., Anderson, D. (1987). A small viral RNA is required for in vitro packaging of bacteriophage Phi29 DNA. Science 236:690–694.

Gutiérrez, J., Barcía, J.A., Blanco, L., Salas, M. (1986). Cloning and template activity of the origins of replication of phage ϕ29 DNA. Gene 43:1–11.

Heusterspreute, M., Ha-thi, V., Tournis-Gamble, S., Davison, J. (1987). The first-step transfer-DNA injection-stop signal of bacteriophage T5. Gene 52:155–164.

Hunter, G.J., Rowitch, D.H., Perham, R.N. (1987). Interactions between DNA and coat protein in the structure and assembly of filamentous bacteriophage fd. Nature 327:252–254.

Krüger, D.H., Bickle, T.A. (1987). Abortive infection of *Escherichia coli* F^+ cells by bacteriophage T7 requires ribosomal misreading. Journal of Molecular Biology 194:349–352.

Moffatt, B.A., Studier, F.W. (1987). T7 lysozyme inhibits transcription by T7 RNA polymerase. Cell 49:221–227.

Perkus, M.E., Shub, D.A. (1985). Mapping the genes in the terminal redundancy of bacteriophage SP01 with restriction endonucleases. Journal of Virology 56:40–48.

Serwer, P. (1986). Arrangement of double-stranded DNA packaged in bacteriophage capsids: an alternative model. Journal of Molecular Biology 190:509–572.

Tabor, S., Richardson, C.C. (1985). A bacteriophage T7 RNA polymerase/promoter system for controlled exclusive expression of specific genes. Proceedings of the National Academy of Sciences of the United States of America 82:1074–1078.

Takagi, J., Chiura, H., Kagiyama, N., Sakaguchi, R. (1986). The behavior of the bacteriophage SP01 in adsorbing to *Bacillus subtilis*. The Journal of General and Applied Microbiology 31:569–572.

Witte, A., Lubitz, W., Bakker, E.P. (1987). Proton-motive-force-dependent step in the pathway to lysis of *Escherichia coli* induced by bacteriophage ϕX174 gene *E* product. Journal of Bacteriology 169:1750–1752.

Zavriev, S.K., Kochkina, Z.M. (1986). Bacteriophage T3 and bacteriophage T7: transcription-dependent mechanism of the transport of phage DNA into the cell during infection. Molecular Biology 20:328–334.

Chapter 6
Genetics of Temperate Bacteriophages

For all of the bacteriophages discussed in the preceding chapters, a successful phage infection always results in the immediate production of progeny virions. However, many bacteriophages are known for which there is an alternative outcome to phage infection. Instead of the customary unrestrained DNA replication and phage assembly, there is a temperate response in which the bacteriophage sets up housekeeping within the bacterial cell and maintains a stable relationship with that cell and all its progeny for many generations. The varied ways in which the temperate response can be accomplished are the subject of this chapter. The population dynamics of temperate and lytic viruses and their hosts have been analyzed by Stewart and Levin and are not covered here. The physical properties of the temperate bacteriophages discussed in this chapter are summarized in Table 6-1.

General Nature of the Temperate Response

The key characteristic of the temperate response is the modulation of phage growth. The viral DNA replicates at the same rate (on a molecule-for-molecule basis) as the host cell DNA and is distributed to both daughter cells at each cell division. However, despite the occurrence of DNA replication, most of the phage-specific proteins, especially those involved in late functions, are not produced. Because the virion structural proteins are among those not produced, there is no possibility of new phage particles being assembled, and the host cell survives the infection.

The survival of the host cell has important implications for the interaction between phage and host, as both temperate and lytic infections

Table 6-1. Physical properties of various temperate bacteriophages[a]

Phage	Usual host	DNA molecule molecular weight $\times 10^{-6}$	Topology	Virion: Morphology	Virion: Dimensions (nm)	Related phages	Prophage DNA
λ	*E. coli*	30.8	Unique sequence, cohesive ends	Icosahedral head, noncontractile tail	62 152 × 17	21, φ80, 82, 424, 434	Circularly permuted
P22	*Salmonella*	26	Circularly permuted, terminally redundant	Icosahedral head, tail of 6 short spikes about a central core	60 18		Circularly permuted
P2	*E. coli*, *Shigella*, *Serratia*	22	Unique sequence, cohesive ends	Icosahedral head, contractile tail	61 133 × 17	PK, 186	Circularly permuted
P4	*E. coli*, *Shigella* (Must be P2 lysogens)	6.7	Unique sequence, cohesive ends	Icosahedral head, contractile tail	46 133 × 17		Circularly permuted
P1	*E. coli*, *Shigella*	60	Circularly permuted, terminally redundant	Icosahedral head, contractile tail	93 220 × 18	P7	Circular, nonintegrated
Mu	*E. coli*	25	Unique sequence, ends are host DNA	Icosahedral head, contractile tail	54 × 61 100 × 18	D108	Colinear
PBS1	*B. subtilis*	190		Icosahedral head, contractile tail	120 240	PBS2, 3NT, I10	Pseudolysogen, nonintegrated

[a]Terminology is that of Table 5-1 except for "prophage DNA," which refers to whether the vegetative and prophage genetic maps have the same gene order.

Adapted from Strauss, E.G., Strauss, J.H. (1974). Bacterial viruses of genetic interest, pp. 259–269. In: King, R.C. (ed.) Handbook of Genetics, vol. 1. New York: Plenum Press.

begin in the same way. This fact means that any effect the virus has on the host cell during the early stages of infection must be either nondetrimental to cell survival or reversible. Therefore activities such as degradation of the nucleoid, which occurs shortly after T4 infection, should not be expected to occur among the temperate viruses. Neither should one expect a temperate virus to carry out wholesale modifications of the bacterial RNA polymerase as part of its regulatory system. Instead, the expected pattern should be one of careful utilization of existing host biochemistry, at least during the potentially reversible portion of the viral life cycle.

A cell that carries a temperate bacteriophage is referred to as a **lysogen,** and the quiescent phage DNA is referred to as a **prophage.** Generally a cell that is a lysogen is immune to superinfection by the same phage (homoimmune) but not by heterologous phages. As this statement implies, it is possible for a cell to carry more than one prophage, a state that constitutes multiple lysogeny. Generally, multiple lysogens involve heterologous phages, as superinfection immunity is apparently due to the presence of substances **(repressors)** that bind to the DNA and turn off viral functions in the prophage. The repressors can also act on newly injected DNA and prevent its expression, thereby conferring immunity against any phage to which the repressor may bind. Conversely, if a prophage is transferred to a cytoplasm that does not contain repressor, it can reactivate and produce a lytic infection.

The lysogenic state is not always maintained in all cells of a culture. It is possible for the prophage to revert to the vegetative state and go on to produce a lytic infection. The reasons for spontaneous reversion of the prophage to the lytic state are unknown, but it seems to occur at a rate that yields roughly 10^6 phage particles per milliliter of mid-log phase culture (about 2×10^8 cells/ml). If any lysogenic culture of moderate cell density always contains some cells that have spontaneously induced their prophages, it obviously is impossible to obtain a phage-free culture of a lysogenic bacterium. This fact can be used to identify lysogenic cultures. For many phages, treatments that damage DNA, e.g., ultraviolet radiation (UV), mitomycin C, or other mutagenic agents, result in an increased rate of conversion of the prophage to the vegetative state. This increase is termed **induction,** and viruses that can be so stimulated are considered inducible.

Occasionally a cell derived from a lysogenic culture gives rise to a cell line that does not produce infectious virus particles. This event arises from the prophage being susceptible to all of the same genetic processes as the bacterial DNA. In particular, it may undergo mutation. Simple mutations may have the effect of inactivating the prophage so it can no longer be induced, either endogenously or exogenously. More complex mutations such as deletions may result in the production of only defective phages (e.g., only tails). A completely inactivated prophage that does not produce any sort of particle is referred to as a **cryptic prophage.**

Temperate phages can be detected during routine screening because they produce turbid plaques in lawns of nonlysogenic bacteria, plaques that still contain a thin layer of growing bacterial cells. They are true turbid plaques in that some of the infected cells have lysed, and others have formed lysogens. The newly formed lysogens are, of course, immune to superinfection and continue to grow in the region of the plaque. In contrast, the turbid "plaques" produced by RNA phages are not the consequence of cell lysis but result from retardation of growth.

Bacteriophage Lambda as the Archetypal Temperate Phage

The Paris group headed by the senior Wollmans was the first to recognize that some bacterial cultures were persistently contaminated by bacteriophages and therefore must be lysogens. Later Lwoff demonstrated that lysogens were stable in the absence of inducing agents and yet were all capable of being lysed by the virus that was inside them. The major experimental effort, however, came from Jacob and the younger Wollman working with *Escherichia coli*. Although they identified a large number of distinct phages, most of their efforts were concentrated on a single phage known as lambda (λ), which had been originally identified by Esther Lederberg. Figure 6-1 is an electron micrograph of this phage, which is of average size (Table 6-1) and contains a 47-kb linear DNA molecule. At each 5′-phosphate end of the DNA molecule there is a short single-strand region of 12 bases that are complementary. These special ends of the DNA are described as cohesive because they hydrogen-bond readily in the same manner as restriction fragments and allow the DNA to circularize rapidly after injection to form nicked circles. Phage DNA that lacks one cohesive end cannot circularize and therefore cannot replicate. Phages of this type are called λ*doc* (defective, one cohesive end) and cannot reproduce themselves.

The nicked DNA circles resulting from the presence of the cohesive ends are sometimes called **Hershey circles** after their discoverer and can be permanently sealed by host DNA ligase in the same fashion as RF II is converted to RF I in the Ff phages. A careful analysis of λ virions indicates that the right-hand cohesive end of the DNA molecule (as defined in Figure 6-6, below) is always located at the head–tail junction. During abortive infection, it is the first portion of the λ DNA to become nuclease-sensitive and therefore is presumed to always enter the cell first.

Lytic Life Cycle

λ Phages have been shown to attach to the host cell in two stages: first reversibly with the tip of the tail fiber, then irreversibly with the end of the tail to a membrane structure on the surface of the *E. coli* cell that is

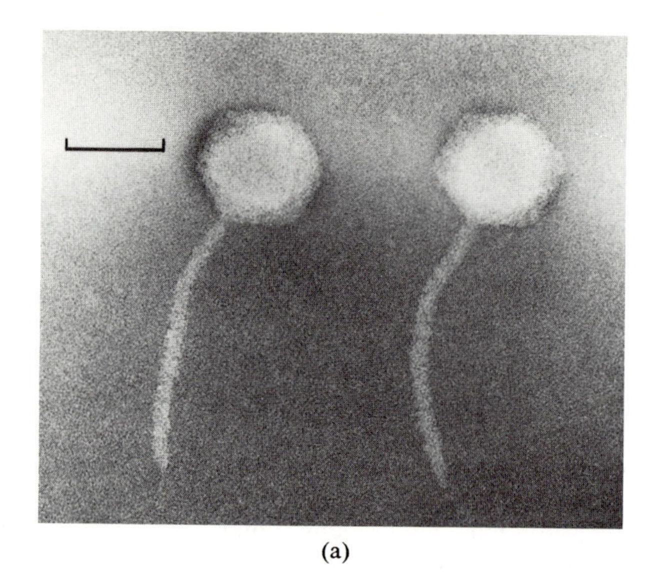

(a)

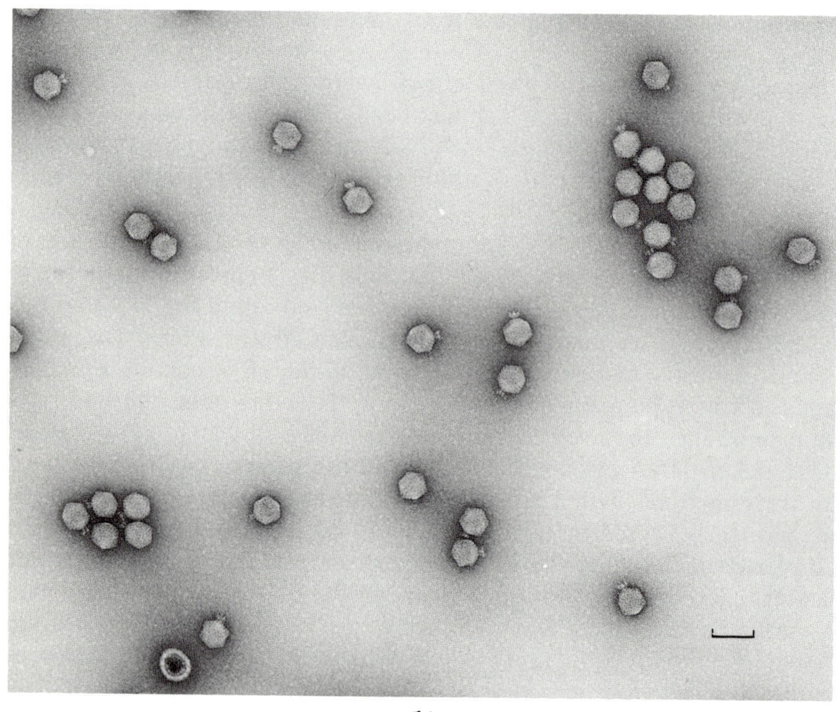

(b)

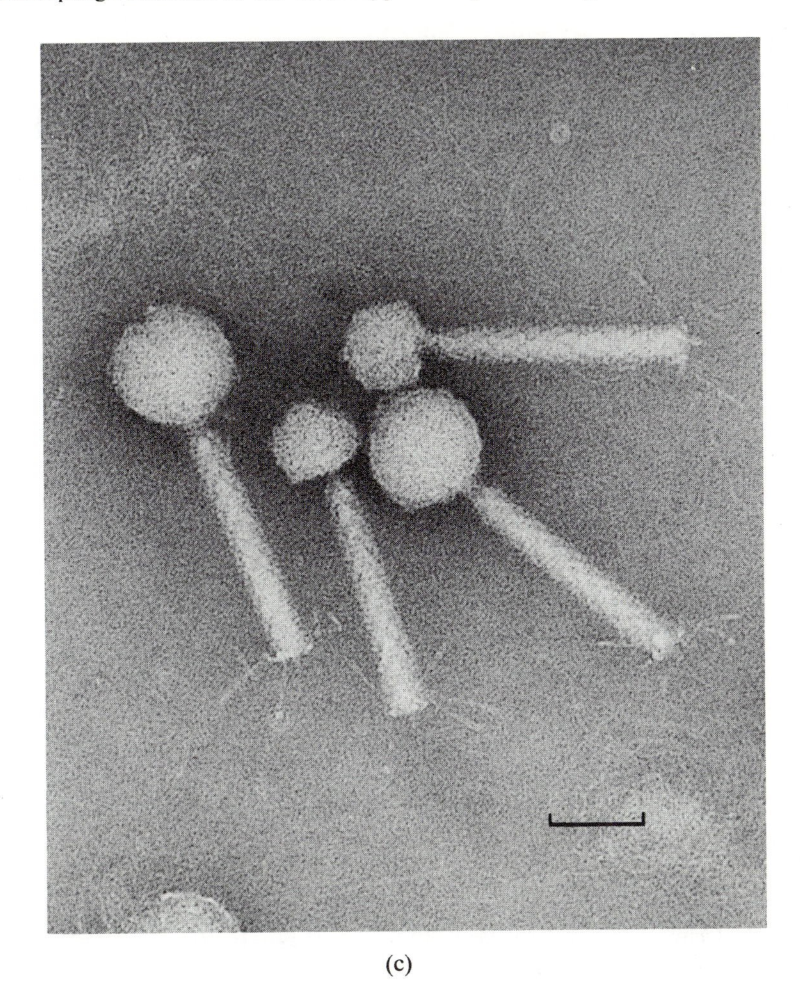

(c)

Figure 6-1. Electron micrographs of some temperate bacteriophages. **(a)** Lambda phage negatively stained with potassium phosphotungstate. The length of the bar is 50 nm. **(b)** Phage P22, also negatively stained. The length of the bar is 100 nm. **(c)** Phages P2 (large) and P4 (small) photographed by minimal beam exposure. The length of the bar is 50 nm. **b:** From King, J., Casjens, S. (1974). Catalytic assembling protein in virus morphogenesis. Nature 251:112–119. **c:** Courtesy of Robley C. Williams, Virus Laboratory, University of California, Berkeley.

a component of the maltose permeation system. Some strains of *E. coli* that are *mal* are deficient in both maltose transport and the λ receptor (see Chapter 12) and consequently are resistant to λ infection. Chemicals that uncouple energy-producing reactions from the membrane proton gradient have no effect on DNA entry, and it has been suggested that the DNA enters by simple diffusion. Consistent with that model is the observation that if artifical lipid bodies (liposomes) carrying the Mal protein are prepared, λ DNA enters the liposome until the liposome is full or the phage head is empty. Successful entry depends on the normal functioning of the host Pel protein, encoded by *manY*, which is a component of the membrane-associated phosphotransferase system. After injection of the DNA, the λ genome circularizes as described above and is ligated into a supercoiled molecule.

RNA transcription begins promptly and can be divided into three temporal classes designated immediate early, delayed early, and late. Immediate early transcription produces two short mRNA molecules that are translated to give only three proteins: N and Ral from one transcript and Cro from the other. Ral (restriction alleviation) interacts with host cell methylases to give enhanced modification of all DNA in the cell, thereby improving the chances that incoming λ DNA will avoid restriction. The functions of N and Cro are discussed below.

Delayed early transcription gives rise to substantially longer transcripts that still hybridize to the same DNA sequences as the immediate early mRNA and therefore must include within them the immediately early sequences. In fact, they are merely extensions of the immediate early mRNA. The situation is similar to that observed in the case of phage T4 where the shift between the two classes of mRNA molecules is regulated by **antitermination,** a failure to terminate the short transcript. Careful study of the antitermination function in λ has provided substantial information about RNA polymerase function.

The critical phage protein in antitermination is N. In order to function it must have host RNA polymerase with three additional host proteins, products of the *nusA*, *nusB*, and *rpsJ* cistrons. The NusA protein seems to be necessary for normal termination of transcription and serves to allow binding of the N protein to the holoenzyme. NusB protein is specifically necessary to allow antitermination, and appropriate *nusB* mutant proteins can substitute for N and NusA function. The role of ribosomal protein S10, also known as *nusE* or *rpsJ*, is still unknown. The holoenzyme is set up for antitermination at two binding sites located just downstream from the promoter. The first of these sites is called box A (sequence 5′-CGCTCTTA), and the second is called *nut* (N utilization), or box B. They are separated by one helical turn of the DNA. The *nut* sequence (5′-**AGCCCT**GAAAA**AGGGC**A) contains a short region of dyad symmetry (boldface) that should be able to form a stem and loop. The generally accepted model for N control is that RNA polymerase without N pauses at specific terminators downstream from the *nut* site and terminates transcription. In the presence of N the RNA polymerase complex pauses too briefly to allow termination, even at terminators found in other transcription units that have been joined to λ. They may be ρ-dependent or ρ-independent. Antiterminated transcription eventually stops, but the mechanism is unknown. It may involve multiple, closely linked terminators.

A similar phenomenon is observed with respect to regulation of late transcription. A protein Q that functions at a *qut* site acts to antiterminate a transcript and allow late gene expression. By 10 to 12 minutes after infection, late mRNA synthesis has entirely replaced the early mRNA synthesis, which has ceased because of the inhibitory effect of the accumulated Cro protein, a repressor of early mRNA synthesis. For additional details on the regulation of λ transcription, see Chapter 12.

This pattern of antitermination is not unique to λ but, rather, can be observed in other temperate phages such as P22 (see below). The phages remain unique entities in the sense that the N proteins are not interchangeable between phages. However, a phage mutated in *N* function can be rescued by superinfection with a normal phage that supplies protein N. The protein diffuses through the cytoplasm and causes **transactivation** of the defective DNA (it activates a DNA molecule other than the one that produced the functional mRNA). Protein Q, on the other hand, basically functions efficiently only in *cis*, primarily turning on only the DNA molecule that coded for the Q-specific mRNA. The difference in activity is thought to be due to the relative binding constants of the proteins to their specific DNA sequences. Protein N binds more strongly and hence is less influenced by the presence of extraneous DNA sequences. The weaker binding Q protein can easily attach to the site located near its cistron but loses its specificity in the face of large quantities of DNA as would be seen when acting in *trans*.

DNA replication during vegetative growth of λ phage is a complex process that begins during the delayed early mRNA synthesis. A large number of host cell proteins are required for viral DNA replication, some of which are listed in Table 1-1. There are two λ proteins, products of cistrons *O* and *P*, that are essential for replication, and *ori*, the actual initiation site, is located within the *O* cistron. As shown in Figure 6-2, *O* protein binds specifically to four 19-bp inverted repeats in the *ori* region. It is thought that the base sequence of this region imparts a curve to the DNA that is accentuated into a loop by the binding of four O protein dimers. P protein binds followed by host proteins to generate an initiator complex that gives localized unwinding of the DNA helix. Accidental initiation of replication at other sites within the DNA molecule that resemble the normal *ori* is prevented by the requirement for multiple O protein binding sites. The circular mode of replication persists for some 16 minutes and then is supplanted by a rolling circle mode.

Evidence for rolling circle replication is provided by experiments showing that concatemeric DNA molecules (two to eight times the normal genome length) are formed during replication even in the absence of recombination functions. Both types of replication start at about the same time, but significant rolling circle activity cannot begin until late mRNA synthesis commences. The problem lies with exonuclease V, which is involved in the host cell recombination pathways (see Chapter 13) and

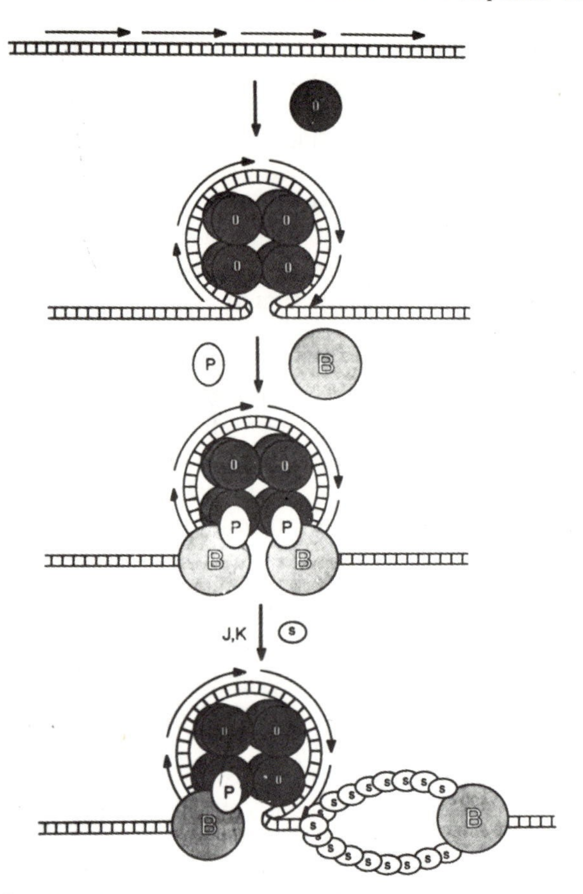

Figure 6-2. Inferred pathway to initiation of DNA replication at *ori*λ. The O protein binds to the four direct repeats in *ori*λ and self-associates to form the "O-some" (possibly along with additional O molecules). The P protein binds to O and to DnaB to generate a more complex nucleoprotein structure. The addition of DnaJ, DnaK, Ssb, and ATP allows DnaB to act as a helicase, unwinding the origin region. The locally unwound DNA, coated with Ssb, is presumed to serve as a substrate for DnaG primase to synthesize an RNA primer, as for single-strand phage replication systems. DNA polymerase III holoenzyme can elongate this primer to initiate leading strand DNA replication and thereby start normal double-strand replication. The eventual route to bidirectional replication is not known; as studied so far, the localized unwinding reaction goes only in the direction shown. From Echols, H. (1986). Multiple DNA–protein interactions governing high-precision DNA transactions. Science 233:1050–1056.

preferentially degrades single-strand DNA. The presumption is that the exonuclease would normally attack the single-strand intermediates emanating from the rolling circle. What is observed is that the phage *gam* cistron (transcribed late) codes for protein gamma that specifically inhibits exonuclease V activity. Insofar as is known, the actual molecular mechanism for the rolling circle replication is the same as that described in Chapter 5.

Maturation of λ follows a pattern similar to that of phage T4. Tails, proheads, and scaffold proteins are produced in the usual manner. The concatemeric DNA molecule is broken into unit lengths of DNA by offset nicks at specific sites called *cos* (cohesive ends). The terminase enzyme (a product of the *Nu1* and *A* cistrons) binds at one site *(cosB)* and in the presence of a host cell factor makes offset cuts at a second site *(cosN)* located about 26 bp to the left on the conventional genetic map (Fig. 6-3). Its behavior thus resembles an asymmetric type II restriction enzyme. The actual cut requires the presence of an intact prohead to neutralize the effect of a host cell inhibitor, and the left end of the DNA is inserted into the prohead first. Ion etching experiments similar to those for T4 indicated that the spiral fold model may apply to λ as well, and that the right-hand end of the DNA is located on the outside of the phage DNA mass. As the head fills, it expands slightly. Following assembly of the progeny virions, the host cell is lysed by the products of the *R, rexA,* and *rexB* cistrons, which have a lysozyme-like activity. If any lysogenic culture of moderate cell density always contains some cells that have spontaneously induced their prophages, it obviously is impossible to obtain a phage-free bacterial culture. This fact can be used to identify lysogenic cultures.

Temperate Life Cycle

The first stages of a λ phage infection that ultimately results in a temperate response are the same as those that initiate the lytic cycle. However, when a temperate response occurs, the bulk of the phage-specific RNA synthesis gradually slows to a halt at a somewhat indeterminate time after delayed early mRNA synthesis has begun. The detailed mechanism for this regulation is presented in Chapter 12.

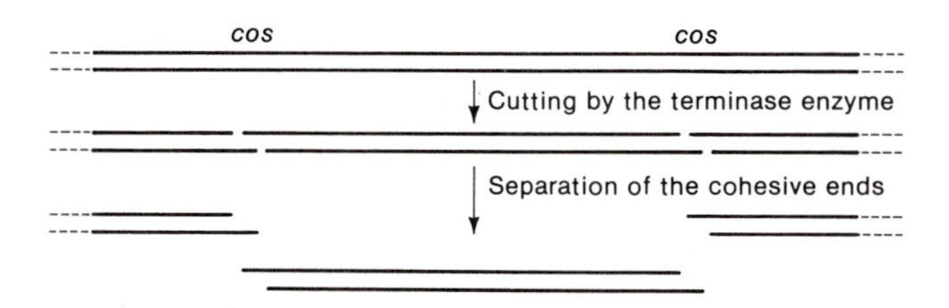

Figure 6-3. Production of the cohesive ends of λ DNA. Unit length λ DNA molecules are cut from concatemers by the terminase enzyme. The cutting takes place at *cos* sites located at the ends of the vegetative DNA molecule. In the diagram, each line represents a single strand of DNA. Only a short portion of the concatemer is shown. The cuts made by the enzyme are offset, generating single-strand DNA tails. Because all cuts occur at identical base sequences on the concatemer, the tails carry complementary sequences.

Concomitantly with the reduced mRNA synthesis, the phage DNA physically inserts itself into the bacterial DNA and becomes a prophage. This association can be shown experimentally by genetic mapping studies. The prophage behaves like any *E. coli* genetic element. It possesses a definite map position, and individual viral cistrons can be mapped like host cistrons. In many cases, however, the prophage must be induced before the genotype can be ascertained, as most of the prophage cistrons are not being transcribed. Insertion of a prophage increases the genetic distance between the host markers at either end of the prophage (the flanking markers), resulting in an increased recombination frequency. The amount of the increase is what would be expected if a piece of DNA the size of a λ genome had inserted linearly between the flanking markers. In essence, then, the bacterial and phage DNA have recombined to form a single integrated molecule.

All the experimental evidence is in accord with a mechanism for this integration proposed by Campbell (Fig. 6-4). In this model it is assumed that the λ DNA has circularized via its cohesive ends. The *E. coli* DNA is already known to be a circular molecule, so a single recombination event involving the two DNA molecules generates a single, larger circle. It can be shown that the λ prophage has a definite genetic location within the *E. coli* genome (between the *gal* and *bio* loci) and has a specific orientation (i.e., the ends of the prophage are always the same). In order to account for these observations, it is necessary that the recombination event always occurs at the same site (called *att*) on both the phage and host DNA. The two sites are designated *attP* and *attB*, respectively, and are represented in Figure 6-4 as consisting of subsites P and P′ as well as B

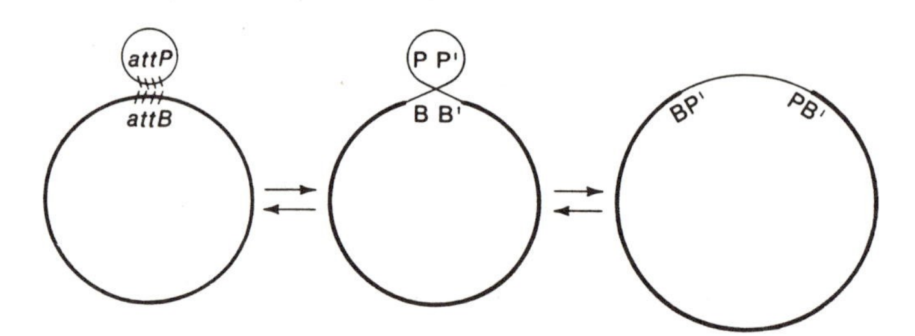

Figure 6-4. Campbell model for λ integration. The double-strand λ and bacterial DNA molecules are represented by the small and large circles, respectively. Initially these two circles associate in a region of homology designated *att*. This region can be considered to consist of two halves. A genetic exchange is assumed to occur so that the left half of the bacterial *att* region (B) is linked to the right half of the phage region (P′) and vice-versa, generating a figure-eight structure. When this structure unfolds, a larger circular DNA molecule carrying an integrated λ DNA molecule is observed. During excision of the phage DNA the entire process is reversed.

and B′. Each pair of subsites is connected by a 15-bp segment of DNA, which is the actual region of homology for pairing of the *att* sites. Other regions of partial homology are present in the *E. coli* genome, however, because if the *attB* site is deleted λ DNA can still integrate but at a much reduced frequency and at relatively random positions along the genome.

It might be presumed that the site-specific recombination event between *attP* and *attB* could be catalyzed by any recombination system including the bacterial *rec* or phage *red* systems. However, phages mutant in a cistron called *int* fail to form stable lysogens after infection, which indicates that the amount of recombination in the *att* region catalyzed by the *red* or *rec* systems is negligible. Instead, the *int* cistron codes for a 140,000-dalton **integrase** protein that is specific for the two *att* sites (see Chapter 13).

Explicit in Campbell's model is the concept that prophage excision (induction) is a simple reversal of the insertion process, which implies that *int* function is necessary and sufficient for excision as well as integration. However, biochemically it is not strictly true. Prophage excision requires two phage functions, *int* and *xis* (pronounced excise). The *xis* cistron is small, and its product is absolutely necessary for successful excision of the prophage. Apparently the integrase can recognize the BP′ end of the prophage (Fig. 6-4) but not the PB′ site. This deficiency is remedied by the xis protein. Two host cistrons, *him* and *hip,* have also been shown to be necessary for integration and excision. In combination with the *int* protein, the *him* and *hip* proteins bind to the DNA molecule to yield a nucleosome-like structure.

In order to maintain the prophage DNA in the integrated state, a protein repressor, the product of a phage cistron, is required. It can be demonstrated by conjugating a donor strain of *E. coli,* which carries a λ prophage to a recipient that is not a lysogen. When the prophage transfers from one cell to the other, induction is almost inevitable and immediate. The phenomenon is called **zygotic induction** because it occurs after a mating and is presumed to be due to the lack of repressor protein in the recipient cell.

Mutations have been isolated in λ that affect the ability of the phage to produce a functional repressor. These mutations result in the production of clear rather than turbid plaques and hence are called *c* mutations. Three classes of *c* mutations have been identified. Mutants for *c*I function never produce any lysogens, whereas mutants for *c*II or *c*III do produce occasional lysogens that are normally stable. It is reasonable to assume, therefore, that the *c*I cistron must code for the repressor protein, and *c*II and *c*III serve to enhance its expression. Certain mutations that map within *c*I result in the production of a temperature-sensitive repressor. Lysogens carrying such a mutation grow normally at 30°C but promptly undergo induction if the culture temperature is raised to 40°C. Final verification was provided by Ptashne, who isolated a 30,000-dalton protein and showed that it bound preferentially to λ DNA, which includes the regions immediately adjacent to *c*I but not to similar regions from heterologous phages such as 434.

The region of the DNA delimited by the *c*I repressor binding sites is referred to as the **immunity region** because it determines the type of superinfection immunity conferred by the prophage. It is possible to produce various types of recombinant phages that carry the structural cistrons of λ but varying immunity regions from other lambdoid phages (Table 6-1). The product of a cross between phages λ and 434 would be described as λ *imm*434, etc. A λ *imm*434 superinfects a normal λ lysogen but cannot grow on a 434 lysogen. The immunity of the lysogen is thus due to the presence of the repressor in the cell that prevents expression of all DNA molecules of the proper immunity type regardless of whether they are integrated. For further details on this interaction, see Chapter 12.

The action of the repressor is not perfectly effective, however. It is possible to obtain homoimmune (identical *c*I cistrons) double λ lysogens in which a second λ phage has infected a lysogen and integrated itself by generalized recombination into the middle of the existing prophage (Fig. 6-5). The result is two prophages in a row, but both prophages are recombinants. A second general type of double lysogen occurs when a ly-

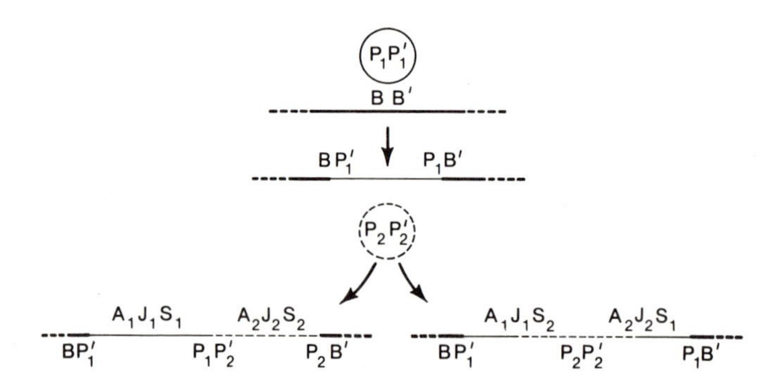

Figure 6-5. Possible mechanisms for insertion of a second phage DNA molecule into a lysogen to form a double lysogen. The small circles represent the phage DNA, and the thick lines represent a portion of the bacterial chromosome. The attachment sites are indicated as either PP′ (phage) or BB′ (bacterium). In the first step the phage DNA inserts into the bacterial DNA by an exchange event catalyzed at the attachment sites by phage integrase. The insertion of the second prophage may occur in two ways: The recombination may be catalyzed by integrase again and occur between P_2P_2' and one of the recombinant attachment sites (BP_1' or P_1B') to yield the structure shown at the left; alternatively, the recombination may occur within one of the phage cistrons such as *J*, catalyzed by a nonspecific recombination system, as shown at the right. Note that in the first case the tandem prophages are each intact, but in the second case the first prophage is separated into two pieces by integration of the second prophage. The result is still two prophages arranged in tandem, but each prophage is recombinant.

sogenic cell is infected with a heteroimmune λ such as λ *imm*434. In this case the second phage undergoes integrase-catalyzed insertion at one of the ends (equivalent to a normal *att* site) (Fig. 6-5) of the original prophage to give a tandem duplication of λ in which each prophage has the same genetic composition with which it began. This type of integration is not possible for a homoimmune phage due to the action of the λ repressor on *int*.

Genetic Map of Lambda

There are two genetic maps for lambda: one for the vegetative phage and one for the prophage. This situation arises because the *attP* site used for integration of the phage DNA is not located at one of the cohesive ends of the DNA molecule. The result is that integration of the phage DNA via the Campbell mechanism (Fig. 6-4) results in a circular permutation of the genetic map for the vegetative phage.

The vegetative map has been developed from standard phage crosses such as those used for T4. Interpretation of the data is much simpler because λ DNA is neither circularly permuted nor terminally redundant. The prophage map, on the other hand, has been derived from standard bacterial crosses using transduction (see Chapter 7) or conjugation (see Chapter 9). A simplified version of the map is presented in Figure 6-6. A more detailed map for λ is presented in Figure 12-8 as a part of the discussion on regulation of the phage.

Once again there is a distinct clustering of genetic functions. The *A-J* region and *R* and *S* represent the regions of late transcription. Although they appear separated on the vegetative map, they are actually continuous in the circular vegetative DNA molecule. The rest of the functions are arrayed about the *c*I cistron, with recombination functions to the left and DNA replication functions to the right. Located next to *c*I are the *rex* cistrons, which determine the ability of the lysogen to exclude *r*II mutants of phage T4.

It is also possible to develop so-called **physical maps** for λ in which distances are based not on recombination frequencies but on measurements of linear distance using DNA heteroduplexes and electron microscopy. The technique is similar to that discussed in Chapter 5 but is more powerful in that it yields preparations that contain nothing but heteroduplexes. In order to attain such purity, the heteroduplexes are prepared by rehybridizing single strands purified from different phage mutants using the poly(UG) binding technique discussed earlier (see Chapter 4). When the heteroduplexes are examined, regions of nonhomology appear as "loops" or "bubbles" (see Fig. 5-1). By measuring distances from the ends of the λ DNA to the various nonhomologous regions, physical distances can be estimated. Figure 6-7 shows some physical maps comparing various λ *imm* recombinants.

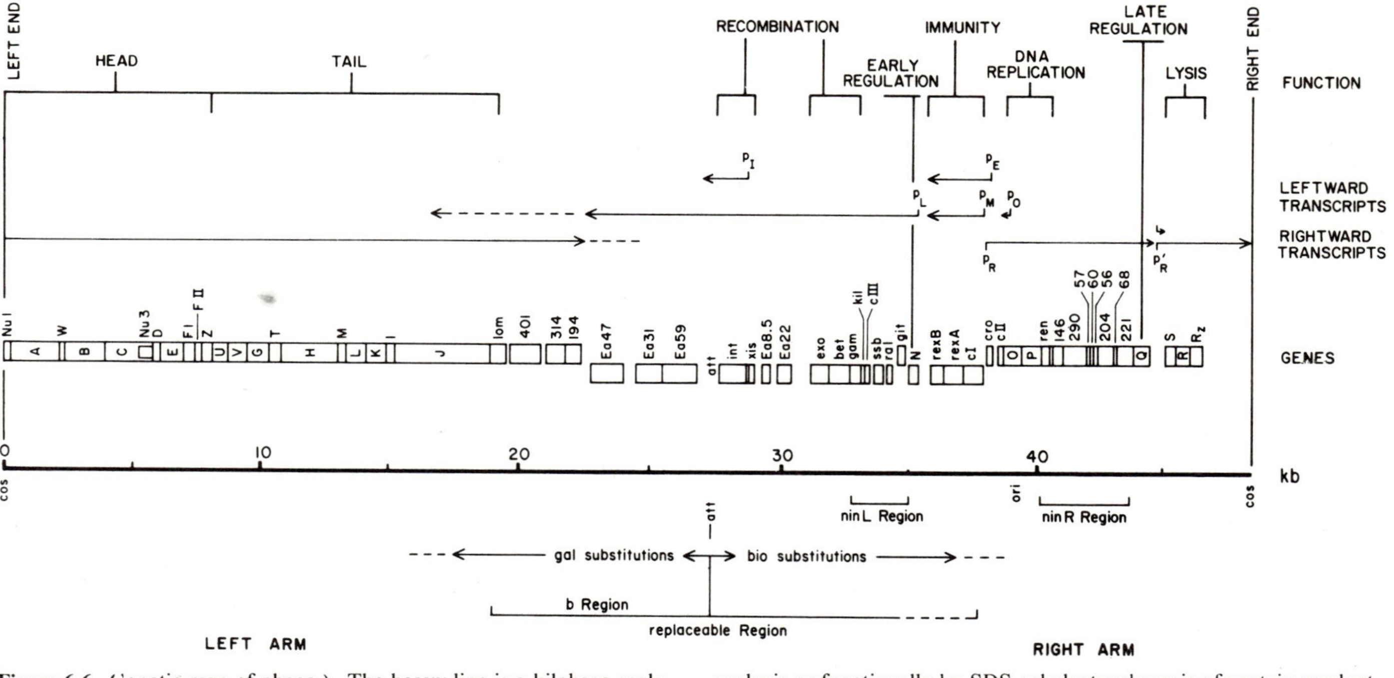

Figure 6-6. Genetic map of phage λ. The heavy line is a kilobase scale beginning and ending at a cohesive end site (*cos*), and above it is a scale drawing of the known λ genes. The brackets at the top indicate functional clusters of loci. The vertical lines show the positions of the major regulators *N* and *Q*. Known promoters are denoted by *p* with a subscript to indicate their unique points of origin. p_I = *int* protein promoter; p_E = establishment promoter for *c*I; p_M = maintenance promoter for *c*I; p_L = major leftward promoter; p_R = major rightward promoter; p_O = *oop* promoter; p_R' = late promoter. → = extent and direction of transcription; --- = readthrough transcription that can occur when termination fails. Known genes (identified either genetically by mutation analysis or functionally by SDS-gel electrophoresis of protein product, or both) are indicated with letter names. Genes known only as open reading frames (ORFs) in the DNA sequence analysis are given numbers corresponding to the predicted size of the protein. In the area below the map are indicated the major areas of substitution and deletion mutations: *att* = attachment site; *ori* = origin of replication. From Daniels, D.L., Schroeder, J.L., Szybalski, W., Sanger, F., Blattner, F.R. (1984). A molecular map of coliphage lambda, pp. 1–21. In: O'Brien, S.J. (ed.) Genetic Maps 1984. Cold Spring Harbor, N.Y.: Cold Spring Harbor Laboratory.

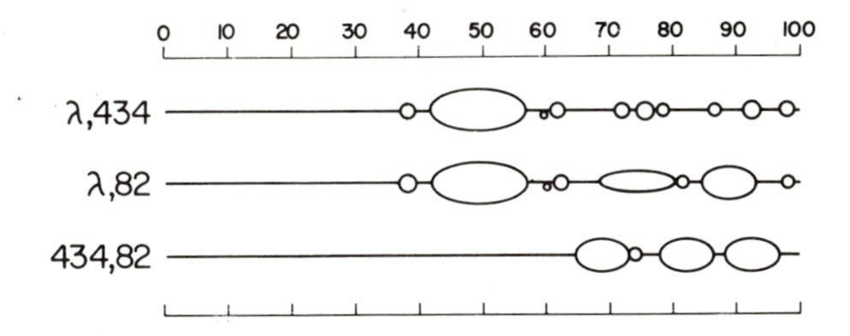

Figure 6-7. Heteroduplex analysis of hybrid phages. The lines represent stylized drawings of heteroduplex DNA molecules as observed in the electron microscope. Each substitution loop (see Fig. 5-1) represents a region of nonhomology. All possible combinations of the three lambdoid phages are shown. The immunity region is roughly between 70 and 80 (Fig. 6-6). From Campbell, A. (1977). Defective bacteriophages and incomplete prophages, pp. 259–328. In: Fraenkel-Conrat, H., Wagner, R.R. (eds.) Comprehensive Virology, vol. 8. New York: Plenum Press.

A wide variety of DNA cloning vectors based on phage λ has been prepared. Although not especially useful for DNA sequencing experiments, they can carry relatively large DNA inserts and of course offer the advantage of automatic packaging of the cloned DNA. Some examples of λ cloning vectors are discussed in Chapter 14.

Bacteriophage P22

The general morphology of phage P22 (Fig. 6-1) is roughly similar to that of phage T3, consisting of a polyhedral head attached to a six-spiked baseplate. The DNA is a linear molecule that is circularly permuted and terminally redundant like that of the T-even phages. The extent of the redundancy is about 2.5% of the genome.

When P22 infects *Salmonella typhimurium*, the DNA circularizes by recombination within the terminal redundancies, provide the *erf* (*e*ssential *r*ecombination *f*unction) gene product is present. Because it also catalyzes generalized recombination, the *erf* system is analogous to the λ *red* system. A companion cistron, *abc*, is analogous to the λ *gam* function, and a *cro* cistron functions to turn off lysogeny. During the lytic cycle, concatemeric DNA molecules arise that are probably produced by the rolling circle mechanism. The concatemers are cut to unit length and the DNA packaged by a headful mechanism similar to that used by phage T4. For additional information, see Chapter 7.

During the temperate response, P22 forms a prophage inserted between the *proA* and *proC* cistrons on the *Salmonella* genetic map (see Fig. 10-1). The Campbell model suffices to explain the mechanism, and a phage integrase is present to provide the necessary catalyst. The prophage state is maintained by two repressors, the products of the *c*2 and *mnt* cistrons. Mutations in either cistron can prevent the establishment of lysogeny. The function of the protein product of the *c*2 cistron is essentially similar to that of the repressor encoded by the *c*I cistron of λ, and the two proteins are about 45% homologous with respect to the amino acid sequence of their carboxy halves. The function of the *mnt* cistron is, however, unusual. It acts by binding to the overlapping promoters for *ant* and itself. When bound, the protein stimulates its own transcription and represses that of *ant*. The *ant* cistron codes for an **antirepressor.** In the presence of the antirepressor, the normal repressor protein molecules (from *c*2) are inactivated, and induction of the prophage occurs. Interestingly, the antirepressor is nonspecific and inactivates the repressors of other temperate phages, including some that are not normally inducible.

Induction of the P22 prophage and its genetic mapping can be carried out by the same techniques used for λ. The genetic map that has been obtained for P22 is shown in Figure 6-8. Superimposed on the P22 map is a λ map in order to emphasize the similarity of cistron arrangements in P22 and in λ. Botstein proposed that phage traits may be carried in modules that tend to recombine as entire units. Such unitary exchange could account for the remarkable similarity of regulatory functions in P22 and λ while allowing for structural diversity.

Bacteriophages P2 and P4

P2 and P4 constitute an unusual pair of bacteriophages. Morphologically they are similar (Fig. 6-1), with the exception of their head sizes, the P4 head being only about two-thirds the diameter of the P2 head. Because head size is held to reflect genome size, the size differential suggests that the P4 DNA molecule is smaller, which is in fact the case (Table 6-1). Upon analysis, P2 and P4 DNA molecules have unique sequences but with one odd similarity. Heteroduplex analysis shows that the base sequence of the DNA molecules is not similar, except for the cohesive ends, which are identical. A further analysis of the virions themselves shows that they are composed of absolutely identical subunits. The key to this similarity of proteins between the two phages can be found in the nature of their life cycles.

Phage P2 follows much the same life cycle as does λ and can insert its DNA directly into the bacterial DNA in order to form a lysogen. Like λ, P2 lysogens prevent the growth of certain other phages, in this case the T-evens and λ. In contrast to λ, however, P2 prophages have been located in at least ten sites on the *E. coli* genome. The preferred site varies with the strain used (e.g., the phage uses a different site in *E. coli* C than in *E. coli* K-12), but multiple lysogens are possible, with integration occurring at separated locations. A further contrast to λ is that the lysogens cannot

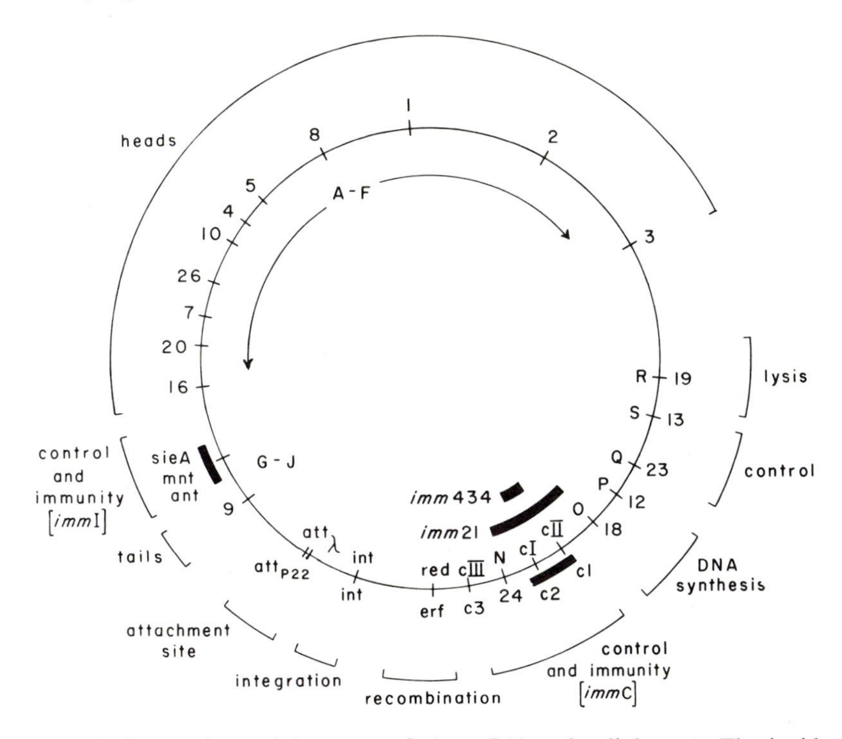

Figure 6-8. Comparison of the maps of phage P22 and coliphage λ. The inside of the circle shows the λ map and the outside shows the P22 map. A solid line connecting markers of the two phages indicates substantial similarity in genetic function. The heavy bars inside the λ map indicate the extent of material substituted in certain λ variants. The heavy bars outside the P22 map indicate the extents of the *immI* and *immC* regions. From Susskind, M.M., and Botstein, D. (1978). Molecular genetics of bacteriophage P22. Microbiological Reviews 42:385–413.

be artificially induced to give phage particles, even by zygotic induction, although spontaneous induction does occur at low frequency. The molecular basis for noninducibility remains uncertain. When the phage enters the vegetative state, it replicates its DNA by the rolling circle mode, using the host cell *dnaB, dnaE,* and *dnaG* functions. P2 undergoes maturation and assembly by mechanisms similar to those used by λ.

Phage P4 can complete its life cycle only if it infects a cell that already has a helper phage such as P2 within it or if a helper phage is supplied later. In the event that no helper phage is available, there are two possible outcomes for P4 infection. In both cases there are a few rounds of uncommitted DNA replication, but then the P4 stabilizes either as a high copy number plasmid (30 to 50 copies per cell) sometimes called a "phasmid," or as a prophage integrated at a specific site. If the repressor function

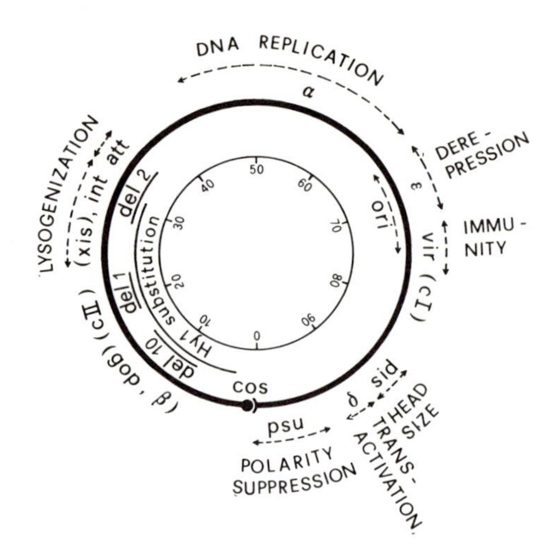

Figure 6-9. Genetic map of phage P4. The functional areas are indicated by lines with double arrows. The region at the lower left has not been well characterized and is subject to some changes. From Dehò, G. (1983). Circular genetic map of satellite bacteriophage P4. Virology 126:267–278.

is absent, those phages that attempt to follow the lysogenic pathway kill the host cell. Krevolin and co-workers have shown that virulent phage mutants replicate bidirectionally from a single point of origin using the classic replication mechanism of leading and lagging strands.

When helper phage is supplied, either in the vegetative or lysogenic state, P4 progeny are produced. A successful P4 infection requires that the helper phage be completely normal for all functions essential to the structural integrity of the virions and for lysis of the host cell. Taken together with the fact that P4 virions are composed of P2 subunits, it means that the P4 is using some of the P2 cistrons for its own purposes and is acting as a **satellite phage,** one that parasitizes another virus. However, the differences in replication mechanisms noted above indicate that P4 produces its own replication proteins and is therefore not totally dependent on P2.

The nature of the interaction between P2 and P4 phages is interesting. The helper phage is not induced during P4 infection, and if it is already in the vegetative state its DNA is not packaged. If such were not the case, it would be impossible to obtain pure stock of phage P4. The current genetic map of P4 (Fig. 6-9) shows nine identified cistrons, but the interaction of P4 and P2 revolves around only two of those cistrons, *sid* and *psu*. The remainder are involved with DNA replication and transcription of P4. Transcriptionally active P4 turns on the late functions of P2 by transactivation using Psu in much the same way that λ N protein can turn on delayed early mRNA synthesis of another DNA molecule.

The mechanism by which P2 capsid assembly is modified to produce a smaller size head for P4 has not been entirely worked out, but it is dependent on *sid* function. In the absence of *sid* activity, normal-sized P2 heads are prepared and functional P2 is produced resulting from the packaging mechanism using the λ cohesive end method so that the size of DNA is not a factor. All that is required is appropriately spaced *cos* sites, accounting for the necessity of P2 and P4 to have identical *cos* sites. The P2 enzyme system can then package P4 DNA into the smaller heads but not a complete P2 genome.

Lagos and Goldstein suggested that P4 would make a good cloning vector. Its behavior as a phage allows it to deliver cloned DNA to a cell easily. At the same time its mode of replication as a multicopy plasmid allows the experimenter to recover large quantities of cloned DNA with relatively little effort. It would be more advantageous than λ because large quantities of λ are obtained only following induction, which of course kills the cells. P4 has no such effect in its plasmid mode.

Bacteriophage P1

P1 is the largest of the *E. coli* phages and, indeed, one of the largest phages known (Fig. 6-10). It is similar to phage P7 (which was formerly known as φ-amp because its lysogens are resistant to ampicillin). Heteroduplex analysis indicates that the two phages are 92% homologous, although they are heteroimmune (i.e., they produce different repressor molecules). The terminal redundancies also differ, that for P1 being 9 to 12% whereas that for P7 is only 1%. Both phages seem to have variable assembly mechanisms, as up to 20% of the virions examined have smaller heads than usual and hence are defective. Most of the defective virions carry only 40% of the phage genome. However, because of the circular permutation of the DNA (Table 6-1), multiple infection of a single cell by defective viruses can result in progeny virus production as a consequence of complementation between defective phages.

The lytic cycle of P1 is basically similar to that of T4. Instead of tryptophan, calcium ions are absolutely required for attachment of virions to the cell. One interesting variation from the T4 pattern is that as the tail sheath contracts the tail fibers are released, a process found only in P1 and P2. Concatemeric DNA molecules are produced by recombination and are then packaged by the headful mechanism. One difference between P1 and T4 is that P1 infection does not produce extensive host DNA degradation.

The lysogenic state for P1 is unusual because the prophage is only rarely integrated into the host chromosome, generally remaining as an autonomous plasmid. Unlike P4, however, it is a low copy number plasmid of approximately one per genome equivalent. Given that P1 plasmids are circular DNA molecules, it is surprising that the genetic map for P1 (Fig.

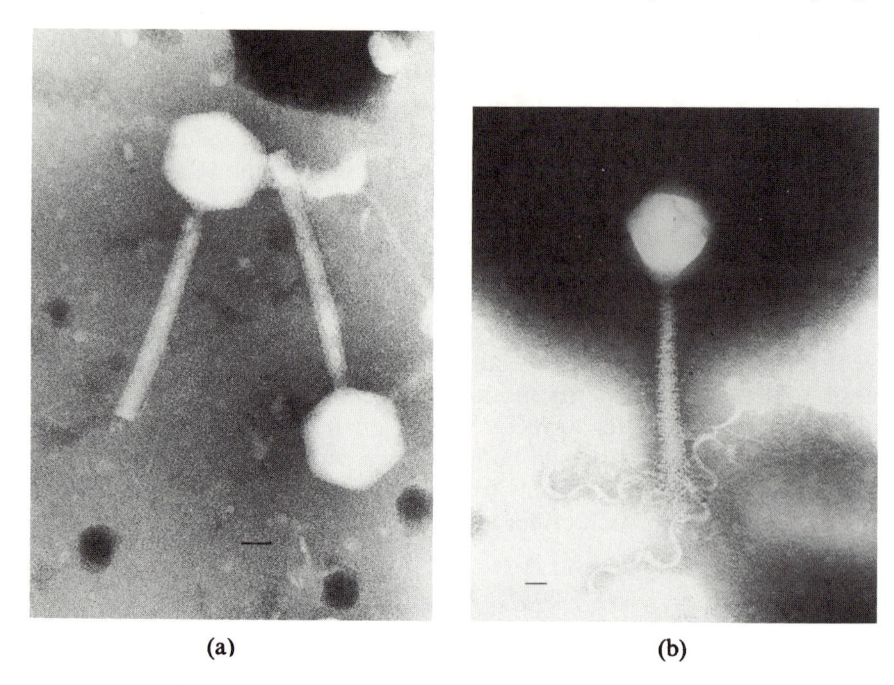

Figure 6-10. Electron micrographs of phages P1 **(a)** and PBS1 **(b)**. The phages were negatively stained with sodium phosphotungstate. The bar indicates a length of 25 nm.

6-11) is linear. The linearity is due to extensive recombination catalyzed by the *cre* protein at a site called *loxP* that lies between the "ends" of the genetic map and prevents any genetic linkage between markers situated on either side of it (recall that a recombination frequency of 50% means that it is impossible to tell how far apart two markers are situated). When integration of the prophage is found, it always occurs at the *loxP* site. A repressor protein is coded by the *c1* cistron, which in combination with the *c4* protein confers the usual superinfection immunity. Like λ, a P1 prophage is inducible.

P1 DNA must be capable of two styles of DNA replication: viral when a lytic infection is in progress and plasmid at other times. In accord with this idea, two separate but adjacent replication origins can be identified. In the case of lytic infection, the initial replication is via a circular mechanism that later converts to the rolling circle, as in the case of λ. Like most plasmids (see Chapter 11), P1 expresses an incompatability function through a series of repeated sequences designated *incA*, which prevents other P1 DNA molecules or their close relatives from replicating in a cell that already has a prophage. One model suggests that the IncA repeats bind *repA* protein, which is required for P1 DNA replication.

P1

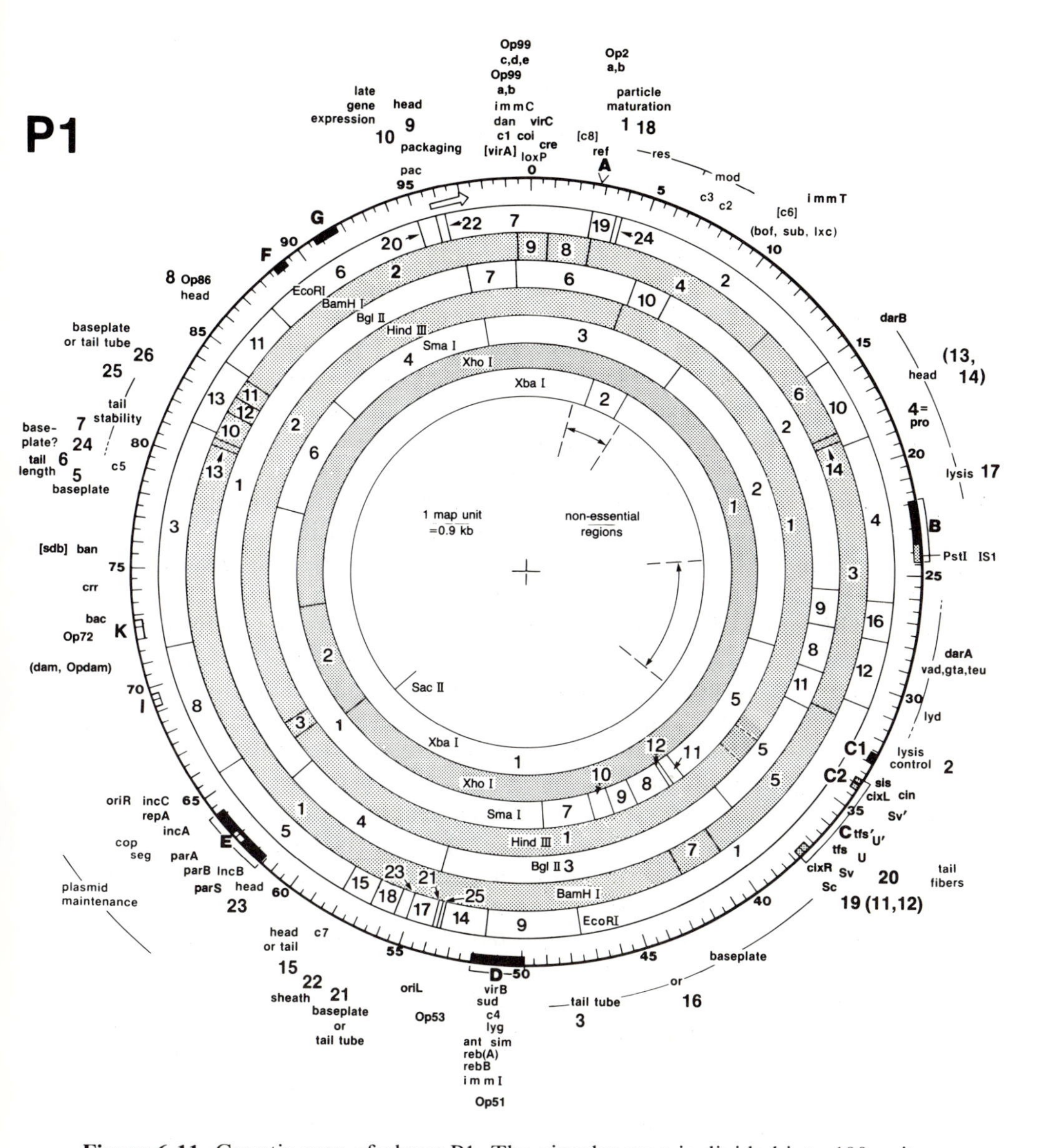

Figure 6-11. Genetic map of phage P1. The circular map is divided into 100 units with the origin defined by the *loxP* site where sufficient recombination occurs to generate a linear map. Bold-faced numbers, acronyms, or acronyms and numbers outside the circle refer to genes. Bracketed symbols indicate genes identified only in phage P7 but assumed to be present in P1. Bold-faced letters with black boxes indicate regions of substitutions in certain phages. Site A is the location of insertion of Tn*902* which codes for ampicillin resistance. Region C is the invertible segment flanked by inverted repeats indicated by stippled boxes. The inner circles are restriction maps for some commonly used enzymes. The sectors are numbered in order of decreasing DNA fragment size as seen on agarose gels. From Yarmolinsky, M. (1987). Bacteriophage P1, pp. 38–47. In: O'Brien, S.J. (ed.) Genetic Maps 1987. Cold Spring Harbor, N.Y.: Cold Spring Harbor Laboratory.

The host range for P1 is controlled by a system of invertible DNA that shows strong homology to the system used by phage Mu. It is discussed in the next section.

Bacteriophage Mu

Phage Mu is notable because it is not only a noninducible temperate virus but also the most efficient transposon known. Upon infection of a host cell, the linear Mu DNA molecule is converted to a circular form by binding a protein. Unlike the case with ϕ29, this protein is not covalently linked to the DNA. Regardless of whether the infection is destined to be lytic or temperate, the Mu DNA always integrates into the host chromosome.

The integration mechanism used is not the Campbell model as with λ but a simple transposition. The basis for the transposition can be seen in the structure of the phage DNA molecule (Fig. 6-12). The DNA is 37.5 kb and always has associated with it some host DNA sequences at both ends. The host DNA at the left, or *c*, end is short ($\sim$ 50 to 150 bp) and is always some multiple of 11 bp. At the other end the host DNA is about 1 to 2 kb. The nature of the transposition mechanism is such that there is no reason to expect any two Mu DNA molecules to have the same host sequences, and hence their presence can be readily demonstrated by heteroduplexing. To integrate the phage DNA then requires transposition from the previous host sequences (still present inside the virion) to some sequence in the new host. The choice of sequence is not entirely random, but it is close to being so. Insertion of Mu into a coding region results in the inactivation of that cistron and produces a mutation. The specific mechanism of transposition is considered in Chapter 13 as part of the discussion on recombination.

As a noninducible phage, Mu is stable following integration and gives rise to stable mutations. This situation is obviously inconvenient for the geneticist who wishes to study the phage, and so most work is done on Mu *c*ts mutants whose repressor fails to function at high temperature. After induction, Mu replicates not as an autonomous molecule but by repeated transposition. The transposase protein is the product of cistron *A* (Fig. 6-12), but maximum efficiency is obtained only when the *B* function is also supplied. As is the case with all transposons, the nature of the DNA in the middle of the molecule is immaterial, but the ends of the phage DNA must contain specific 22-bp sequences, two on the left and one on the right.

The phage life cycle is regulated in the usual way. Late transcription is activated by the *C* protein which might be an accessory transcription factor or a σ factor. Among the late functions is one called *mom, mo*dification *of M*u. In order to get transcription of the *mom* region, three GATC sites upstream of *mom* must be methylated by the host *dam* func-

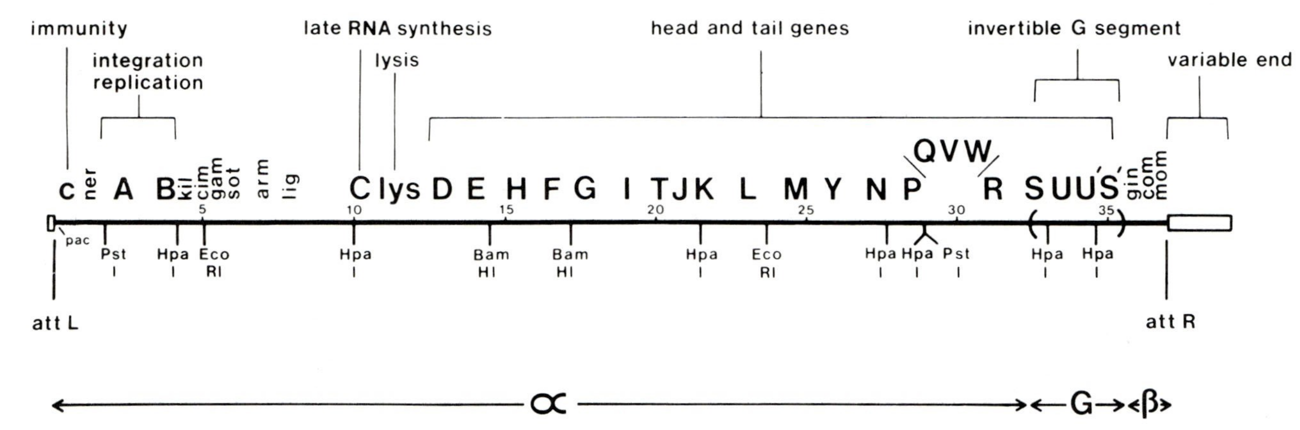

Figure 6-12. Genetic and physical map of phage Mu. The solid line represents Mu-specific DNA; the vertical line at the left end and open box at the right end represent attached host DNA sequences. Parentheses flank the invertible G segment. The functions of some of the genes are indicated above the map. Vertical tick marks above the line represent 5-kb intervals. The locations of genes between *B* and *C* and possible identity of *gam* and *sot* require additional confirmation. The terminator for the C transcript (*tc*) and the *pac* site are also shown. Below the line are indicated some of the known restriction enzyme recognition sites. The regions designated the α, G, and β segments are delineated at the bottom. From Howe, M.M. (1987). Phage Mu: an overview, p. 29. In: Symonds, N., Toussaint, A., van de Putte, P., Howe, M.M. (eds.) Phage Mu. Cold Spring Harbor, N.Y.: Cold Spring Harbor Laboratory.

tion, which is one of the few examples of regulation of transcription in prokaryotes by methylation. Mu DNA is packaged by a headful mechanism so that if the length of the DNA is changed by insertion or deletion, the left end (Fig. 6-12) is held constant and the amount of host DNA varies from its customary 1 to 2 kb. This observation suggests that the packaging mechanism begins at the left.

When Mu DNA is prepared from an induced lysogen and heteroduplexed, a bubble indicative of nonhomology is routinely observed in the region labeled G (Fig. 6-12). Careful analysis shows that the sequence has been inverted with respect to its previous orientation and is flanked by 34-bp **inverted repeat** sequences (identical sequences rotated 180° with respect to one another). The orientation of the G region correlates with the host range of the phage particle. In one orientation, G(+), the phage is able to infect *E. coli* K-12. In the other orientation, G(−), the phage is able to infect *Citrobacter, Serratia,* and *E. coli* strain C. The inversion is under the control of the *gin* cistron located in the β region. The rate of inversion is *gin*-dose-dependent and therefore usually low. Inversion rarely occurs during lytic infections, but the large number of cells and longer times involved during lysogenic growth produce substantial numbers of prophages with inverted G segments. Within the G segment are two sets of genes required for tail fiber biosynthesis oriented in opposite directions but no promoter. The promoter is located near *S* but outside the invertible region. Inversion then acts to bring various DNA segments into position for transcription.

Essentially the same phenomenon is seen with phage P1 and its invertible C segment. The central 3 kb of P1 C segment are homologous to the 3-kb G segment of Mu. Once again two sets of genes and only one promoter are used to regulate host range. An additional example of this type of regulation appears in the discussion of phage variation in Chapter 12.

Bacteriophage PBS1: Example of a Pseudotemperate Phage

PBS1 is the largest of all the bacteriophages discussed herein (Fig. 6-10). Morphologically it is complex, having both helical tail fibers and special contractile fibers which appear at the base of the contracted sheath. It infects only motile cells, as it binds first to the flagellar filament, then migrates to the base of the flagellum where DNA injection occurs. Migration of the phage requires flagellar motion, and infection with PBS1 results in a loss of motility.

In addition to its large size, the DNA of phage PBS1 is unique in containing uracil instead of thymine. As might be expected for so large a phage, a number of virus-specific functions have been identified. In addition to producing enzymes necessary to substitute uracil for thymine, the phage also seems to produce its own RNA polymerase, as it can grow in a cell that has been pretreated with rifampin to block the *Bacillus subtilis* RNA polymerase.

PBS1 produces turbid plaques, but it is not a true temperate phage. The number of free virions found in exponentially growing cultures is several orders of magnitude higher than for a normal temperate phage. Physical studies indicate that most or possibly all of the phage DNA exists in a plasmid state within the cells. It has also been observed that treatment of cultures carrying PBS1 with phage-specific antiserum results in loss of the phage, suggesting that continuous reinfection is necessary to maintain the intracellular phage. Nevertheless, the phage–bacterial association occurs over relatively long periods of time; the infected cells may survive without lysis; and the stability of the relationship depends on certain phage functions. For these reasons, the association of PBS1 and *B. subtilis* is considered an example of **pseudolysogeny.**

One interesting phenomenon that arises when studying *B. subtilis* is the effect of sporulation on a virus. For pseudotemperate viruses such as PBS1, it can be shown that the phage DNA can become entrapped within the developing spore so the pseudolysogenic state may persist during and after sporulation. When the spore germinates, the viral DNA is reactivated along with the bacterial DNA.

Summary

Temperate bacteriophages are those phages that can establish a long-term, stable relationship with a bacterium that prevents lysis of the host cell and ensures that all progeny cells also carry the virus. The combination of a cell and a virus is called a lysogen, and the viral DNA within the cell is designated a prophage. Prophages maintain themselves in a quiescent state by producing a protein repressor that prevents transcription of the DNA coding for vegetative DNA replication and for phage structural proteins. The repressor also confers superinfection immunity on the host cell. Prophages revert to the vegetative state following inactivation of the repressor, which may occur spontaneously or in response to external stimuli (in which case the process is called induction). Two types of prophage have been observed. Those such as λ, Mu, or P2 are inserted into the bacterial chromosome by either (1) a single recombination event between the circular phage DNA and the circular bacterial DNA (Campbell's model, which applies to λ or P2), or (2) transposition (Mu). Prophages such as P1 or P4 (at times) do not insert into the genome but, rather, exist as plasmids, independently replicating DNA molecules within the bacterial cell.

During lytic infections, temperate viruses follow many of the same biochemical pathways as do the virulent phages. Concatemeric DNA molecules are often produced by rolling circle replication or recombination. Packaging of DNA can be via the headful mechanism as with P1, Mu, or P22 or else via the use of specific cohesive ends that demarcate the region to be encapsidated as with λ or P2 and P4.

Many of the temperate phages show remarkable similarity in their genetic maps which may be representative of evolutionary relations within the group. Heteroduplexing is often used as a tool for identifying such relations. Phages P1 and Mu when examined in that way exhibit invertible regions that serve as host range determinants.

References

General

Botstein, D. (1980). A theory of modular evolution for bacteriophages. Annals of the New York Academy of Sciences 354:484–491.

Calendar, R. (1986). Viral transactivation. Biotechnology 4:1074–1077.

Calendar, R. (ed.) (1988). The Bacteriophages. New York: Plenum Press.

Feiss, M. (1986). Terminase and the recognition, cutting and packaging of λ chromosomes. Trends in Genetics 2:100–104.

Symonds, N., Toussaint, A., van de Putte, P., Howe, M.M. (eds.) (1987). Phage Mu. Cold Spring Harbor, N.Y.: Cold Spring Harbor Laboratory.

Specialized

Alano, P., Dehò, G., Sironi, G., Zangrossi, S. (1986). Regulation of the plasmid state of the genetic element P4. Molecular and General Genetics 203:445–450.

Brown, A.L., Szybalski, W. (1986). Transcriptional antitermination activity of the synthetic *nut* elements of coliphage lambda. I. Assembly of the *nutR* recognition site from *boxA* and *nut* core elements. Gene 39:121–127.

Casjens, S., Huang, W.M., Hayden, M., Parr, R. (1987). Initiation of bacteriophage P22 DNA packaging series: analysis of a mutant that alters the DNA target specificity of the packaging apparatus. Journal of Molecular Biology 194:411–422.

Dale, E.C., Christie, G.E., Calendar, R. (1986). Organization and expression of the satellite bacteriophage P4 late gene cluster. Journal of Molecular Biology 192:793–803.

Kamp, D., Stern, B. (1986). Receptor-independent infection of Mu G(−) phage in *Escherichia coli* K-12. FEMS Microbiology Letters 37:387–390.

Lagos, R., Goldstein, R. (1984). Phasmid P4: manipulation of plasmid copy number and induction from the integrated state. Journal of Bacteriology 158:208–215.

Loenen, W.A.M., Murray, N.E. (1986). Modification enhancement by the restriction alleviation protein (Ral) of bacteriophage lambda. Journal of Molecular Biology 190:11–22.

Schauer, A.T., Carver, D.L., Bigelow, B., Baron, L.S., Friedman, D.I. (1987). λ *N* antitermination system: functional analysis of phage interactions with the host NusA protein. Journal of Molecular Biology 194:679–690.

Seiler, A., Blöcker, H., Frank, R., Kahmann, R. (1986). The *mom* gene of bacteriophage Mu: the mechanism of methylation-dependent expression. EMBO Journal 5:2719–2728.

Stewart, F.M. Levin, B.R. (1984). The population biology of bacterial viruses: why be temperate? Theoretical Population Biology 26:93–117.

Vershon, A.K., Liao, S-M., McClure, W.R., Sauer, R.T. (1987). Bacteriophage P22 Mnt repressor: DNA binding and effects on transcription *in vitro*. Journal of Molecular Biology 195:311–322.

Wiggins, B.A., Hilliker, S. (1985). Genetic and DNA mapping of the late regulation and lysis genes of *Salmonella* bacteriophage P22 and coliphage λ. Journal of Virology 56:1030–1033.

Chapter 7
Transduction

Transduction is the term used to designate the bacteriophage-mediated transfer of DNA from one cell (a donor) to another cell (a recipient). It was first described by Zinder and Lederberg for *Salmonella* and phage P22 but has since been shown to occur in many other bacteria and to involve a variety of bacteriophages. Depending on which virus is involved, the donor cell DNA may or may not be associated with viral DNA inside the capsid of the bacteriophage. However, in all cases of transduction it is necessary for the donor cell to lyse and for the virions carrying host DNA (the transducing particles) to be capable of injecting their DNA into a new cell. A cell that has acquired a recombinant phenotype by this process is called a transductant.

Transducing particles are by-products of normal phage metabolism and are considered to be of two basic types. **Generalized transducing particles** are associated with the progeny from lytic infections involving numerous virulent or temperate phages. Almost any suitably sized portion of the bacterial genome may be found inside a transducing particle, although not all cistrons appear with equal frequency. The bacterial DNA is usually not associated with any viral DNA, so these generalized transducing particles are sometimes called pseudovirions. **Specialized transducing particles,** on the other hand, are associated with the progeny from an infection by an integrative temperate phage such as λ. The only bacterial DNA found within these particles is that immediately adjacent to one end of the prophage, and it is always covalently linked to viral DNA.

In this chapter various mechanisms for the production of transducing particles are discussed, and examples of the kinds of data obtained are presented.

Bacteriophage Lambda: A Specialized Transducing Phage

Production of Transducing Particles

A number of observations, made over a period of years, have contributed to the development of an appropriate model for the genesis of λ transducing particles. The first fact to emerge was that only two bacterial phenotypes were involved, *gal* and *bio*, which flank the *att*λ site (Fig. 7-1). As might be expected from the definition of specialized transduction, it later became apparent that any other cistrons mapping between the attachment site and *gal* or *bio* could also undergo specialized transduction with λ. However, the classic nomenclature has predominated, and all subsequent discussion deals only with *gal* or *bio*. The cistron(s) carried by a transducing particle is indicated by appending the genotype symbol to the phage, e.g., λ*gal* or λ*bio*.

The overall frequency for specialized transducing particles among normal virions produced after induction of a lysogen has been shown to be about 10^{-6}. Such a low frequency suggests that a rare event is involved in the production of transducing particles. For obvious reasons, this type of lysate is referred to as a **low frequency transducing (LFT) lysate.**

The size of the DNA molecule contained in a λ transducing particle is essentially normal. This observation is not surprising, as studies on λ deletion and insertion mutations have shown that the λ maturation process requires a DNA molecule between 75 and 109% of normal size. Molecules bigger or smaller do not permit proper encapsidation. What the size constancy does mean, however, is that there must be an approximately equal amount of viral DNA lost vis-à-vis the amount of bacterial DNA carried by the transducing particle. A loss of viral DNA implies a loss of viral function, which is indeed the case. The *gal* cistrons are somewhat farther from the phage attachment site than are the *bio* cistrons, and consequently λ*gal* particles tend to be more defective than λ*bio* particles. A defective transducing particle is indicated by adding the letter *d* to the name, e.g., λ*dgal*.

The nature of the defects in a λ transducing particle is correlated with the prophage map. A λ*dgal* phage is defective in DNA replication functions, whereas λ*dbio* is defective in various structural proteins. Examination of the prophage map (Fig. 6-6) shows that the missing viral DNA is always lost from the end of the prophage opposite the newly acquired bacterial DNA. In fact, it is as though the phage excision function occasionally cuts the prophage off center and the result is the transducing particle. It should be noted that deletions produced in this way have been extremely useful for elaboration of the λ genetic map.

A molecular mechanism that accounts for all of the above observations

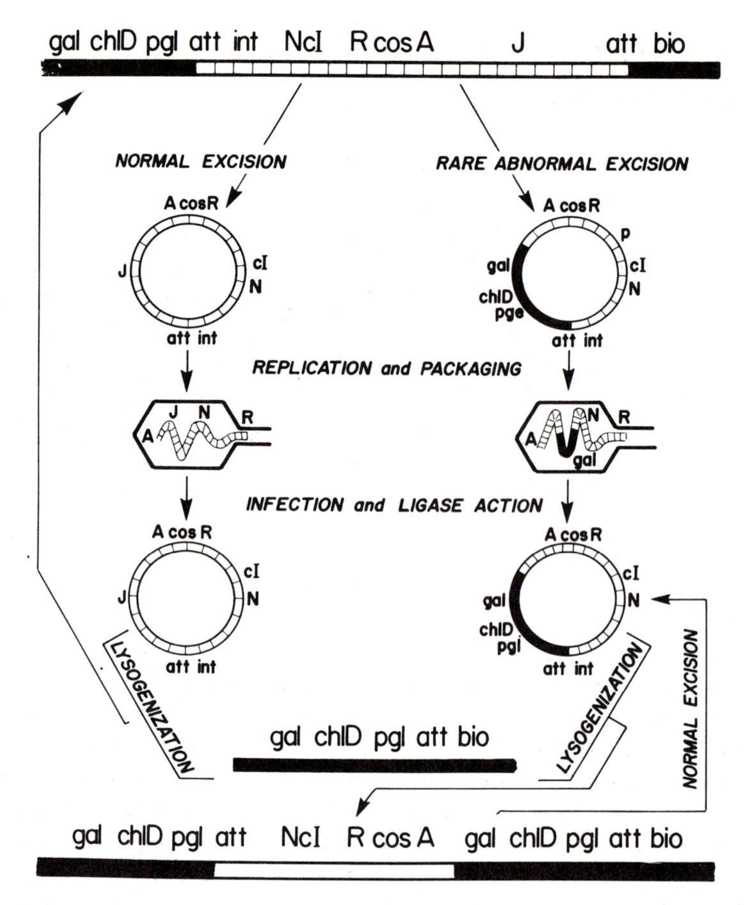

Figure 7-1. Formation of λ*gal* from a λ lysogen. The upper linear structure represents a portion of the genome of a cell that is lysogenic for λ. Each horizontal line represents a single DNA strand, and the short vertical lines represent hydrogen bonds. Both normal and abnormal excision mechanisms are shown. Note that abnormal excision may take the *bio* region instead of the *gal* region. Lysogenization following normal excision regenerates the structure at the top of the figure. Lysogenization following abnormal excision results in the bottom structure, which carries a duplication for the portion of the bacterial genome immediately adjacent to the left hand end of the prophage. From Campbell, A. (1977). Defective bacteriophages and incomplete prophages, pp. 259–328. In: Fraenkel-Conrat, H., Wagner, R.R. (eds.) Comprehensive Virology, vol. 8. New York: Plenum Press.

can be designed by a slight modification of Campbell's model for integration and excision of λ (Fig. 7-1; see also Fig. 6-4). In this revised model, excision of the prophage after induction requires that the DNA form a loop structure. If the base of the loop is not centered on the *att* sites at the ends of the prophage when the recombination event occurs, a specialized transducing DNA molecule results. An excision of this type is an example of nonhomologous recombination; other examples are discussed in Chapter 13. The size of the newly excised viral DNA molecule depends on the size of the initial loop.

There are, however, certain limitations to the displacement of the center of the loop from the center of the prophage if a viable phage particle is to be produced. If the *cos* site is lost, a λ*doc* phage results that cannot package its DNA. Furthermore, if the *ori* site should happen to be deleted, the DNA would be unable to initiate replication. It is obvious, then, that transducing phages that have lost relatively little viral DNA are less defective and are more likely to be found during a random search. In fact, genetic analysis shows that it is possible to recover λ*gal* phages in which some of the sequences lying between *gal* and the left end of the prophage have been deleted, resulting in less defective phages.

Physiology and Genetic Consequences of Transduction

The physiology of transducing phage infection depends on the genetic alteration that produced the DNA. If the phage is fully functional, events can proceed as described in Chapter 6. This situation is not necessarily as rare as might be expected from the preceding discussion, as the *b2* region, which lies near the right-hand end of the prophage (Fig. 6-6), can be deleted with no apparent effect on the phage. However, if the phage is defective, a normal helper phage is required for successful lysogeny but not for initial production of transducing particles or actual DNA injection. The independence of the initial production is due to the multiple genome equivalents present in logarithmically growing *E. coli* cells that allow intracellular production of both normal and defective phages. In the case of λ*dgal,* the helper also supplies the necessary region of homology for integration of the transducing phage by integrating itself into the bacterial genome first. This process is similar to that diagramed in Figure 6-5.

The results of infection of a cell by a transducing particle are variable. In about one-third of cases, new *gal* or *bio* cistrons are substituted for old by simple recombination, and the transductant does not become a lysogen. In the case of λ*dgal,* it is the type of transductant recovered when the cell is infected in the absence of a helper. Alternatively, about two-thirds of the transductants are observed to be lysogenic and to carry a duplication for the transduced marker (e.g., two sets of *gal* cistrons) (Fig. 7-1). In this case the transductant cells frequently segregate gal^+ and gal^- progeny because of recombination of the *gal* markers within the duplicated region that results in loss of the prophage.

Cells that are lysogenic for a specialized transducing phage can be induced by the usual techniques. If the transducing prophage is defective, it is necessary to supply the missing functions by means of a helper phage unless one has integrated with the original transducing phage to give a double lysogen. The progeny virions produced include wild-type (from the helper phage) and specialized transducing particles (from the appropriate prophage). However, the transducing particles may represent as much as 50% of the burst, and this type of lysate is called a **high frequency transducing (HFT) lysate.** The transducing particles are identical to the original phage that gave rise to the lysogenic transductant, but HFT lysates do apparently differ from LFT lysates in some fashion. When an HFT lysate is used as the source of phage, the proportion of lysogenic transductants is more than 90%, in contrast to LFT lysates for which the figure is about 70%.

Lambda Phages That Transduce Additional Genetic Markers

In its original form λ is of limited genetic usefulness because of the restricted regions of host DNA that can be transduced. There has, however, been a concerted effort to extend the range of markers that are transducible by λ because there are many biochemical advantages to obtaining a specialized transducing phage for the particular cistron under study. Chief among these advantages is that, upon induction, all of the prophage cistrons are turned on (derepressed), including the extra bacterial DNA. Therefore the amount of a bacterial cistron product in the culture can be greatly increased by attaching the appropriate cistron to λ phage DNA and inducing the resulting prophage. The result is that a given amount of product can be obtained from a substantially smaller culture than would otherwise be required (i.e., it is at a higher concentration relative to other host cell products).

One good method for constructing new transducing phages is to make them artificially by DNA splicing techniques. However, prior to the discovery of the splicing techniques, a method was developed by Signer to extend the range of transduction by randomizing the prophage location on the bacterial genome. This method requires that the normal attachment site for λ be deleted from the bacterial genome. This deletion lowers the frequency of lysogens by several hundredfold but does not completely eliminate them. Cells that do manage to form lysogens carry the prophage at secondary attachment sites distributed about the genome. Each new site offers the opportunity for obtaining a new type of specialized transducing phage. An additional method for obtaining new transducing phages is discussed below in connection with ϕ80.

Specialized Transducing Phages Other Than Lambda

Bacteriophage ϕ80

Phage λ is the best-studied member of a large group of morphologically (and sometimes serologically) similar viruses that infect *Escherichia coli.*

Several members of this group were mentioned in Chapter 6, including phages 21 and 434. Another well-studied member is bacteriophage ϕ80, isolated in Japan in 1963 by Matsushiro. This phage is heteroimmune to λ and has its own attachment site near *trp* on the *E. coli* genetic map.

The life cycle of ϕ80 is essentially identical to that of λ, including its ability to produce specialized transducing particles. The usual marker transduced by ϕ80 is *trp* in the form of a ϕ80*dtrp*. One of the genetic markers lying between the attachment site for ϕ80 and the *trp* cistrons is *tonB,* which codes for the receptor for phages T1 and ϕ80 and which can also be carried on a transducing particle. The insertion of any new DNA into the *tonB* cistron has two effects: The cell becomes resistant to phage T1, and if it was already a ϕ80 lysogen the inserted material could become part of a transducing phage particle.

Gottesman and Beckwith developed a technique to exploit the convenient location of the *tonB* cistron; they called it **directed transposition.** The first requirement is that the marker to be inserted at *tonB* is a dominant one. The marker must also be carried on a particular type of F plasmid that is temperature-sensitive for replication, which means that the plasmid DNA is capable of self-replication only at low temperatures. At high temperatures (42°C) the plasmid ceases to replicate itself and is gradually lost from the culture unless it recombines and integrates itself into the bacterial chromosome.

The F plasmid carrying the desired marker is transferred by conjugation into a ϕ80 lysogen that carries a deletion at the site on its genome corresponding to the extra marker on the F plasmid. After conjugation, the phenotype of the cell is nonmutant due to the presence of the F plasmid marker. While maintaining the selection for this phenotype, the temperature of the culture is raised. The selection forces the cell to keep the plasmid if it wishes to grow, but the high temperature prevents plasmid replication and should lead to the segregation of cells lacking the plasmid and which are therefore unable to grow. However, if the F plasmid integrates into the bacterial genome, it can be replicated by the bacterium in the same way that a prophage is replicated, and loss of the plasmid is avoided. Homologous recombination is not possible because of the deletion, so integration is at more or less random sites, including *tonB*. Cells carrying the F plasmid integrated at *tonB* can be readily selected from among the survivors of the high temperature selection by treating the culture with T1 phage. The T1 resistant cells made lysogenic with phage ϕ80 can be induced with ultraviolet radiation (UV) to give specialized transducing particles of the appropriate type.

Bacteriophage P1

Bacteriophage P1 is normally considered to be a generalized transducing phage because it produces a plasmid-type lysogen. However, it is possible to show that P1 integrates into the bacterial chromosome at a frequency of about 10^{-5}. By utilizing various combinations of bacterial and phage mutants, it can be shown that the integration of P1 may be catalyzed either by the host *rec* system or by phage Cre protein.

Upon induction, a cell carrying an integrated P1 gives rise to the usual specialized transducing particles. Although integrated P1 lysogens are difficult to detect, the LFT lysate they produce can be demonstrated in a straightforward fashion using a strain of *Shigella* as a recipient. Members of the genus *Shigella*, although similar to *E. coli*, do not utilize the sugar lactose as a sole carbon source and have no DNA corresponding to the *E. coli lac* region. When such a strain is treated with P1 from *E. coli*, Lac^+ cells can be obtained. Because there is no homology, the *lac* DNA is not integrated into the *Shigella* genome but is acting as part of a P1 plasmid, which in this instance is self-replicating, not integrated. As predicted by this model, Lac^+ transductants are generally immune to P1 superinfection and segregate Lac^- cells.

The Lac^+ transductants can also be induced with UV to produce HFT lysates, and in many cases the presence of a helper phage is not required during the induction. This independence is due to the long terminal redundancy in the P1 DNA, which allows insertion of relatively large pieces of extraneous DNA into a DNA molecule of constant size with no corresponding loss of function. The practical limitation is that enough of the terminal redundancy must remain to allow the DNA to circularize after infection.

Bacteriophage P22

As a temperate phage with a life cycle similar to that of λ, P22 can form the same type of specialized transducing phages as does λ. It is, however, more versatile than λ in that it occasionally integrates at secondary sites on the *Salmonella* genome, providing a broader range of transducible markers than does λ.

A second type of specialized transducing particle has been observed for P22. It consists of a P22 genome into which DNA coding for resistance to one or more antibiotics has been inserted by a transposon. The origin of the resistance DNA presumably is one of the R plasmids (see Chapter 11), which can coexist with P22. The DNA of P22 is terminally redundant, although not to the same extent as that of P1, and therefore the same strictures concerning the size of the inserted DNA that apply to P1 also apply to P22. If the size of the P22 DNA plus the insertion becomes too large to fit into a head, each virion is able to carry only a partial genome. However, an infection is still possible if the multiplicity of infection is more than 1.0. It has been shown that there is sufficient randomness in the way in which circularly permuted P22 DNA is produced so that two defective phage particles generally have a complete set of P22 cistrons if their DNA molecules are joined by recombination (Fig. 7-2).

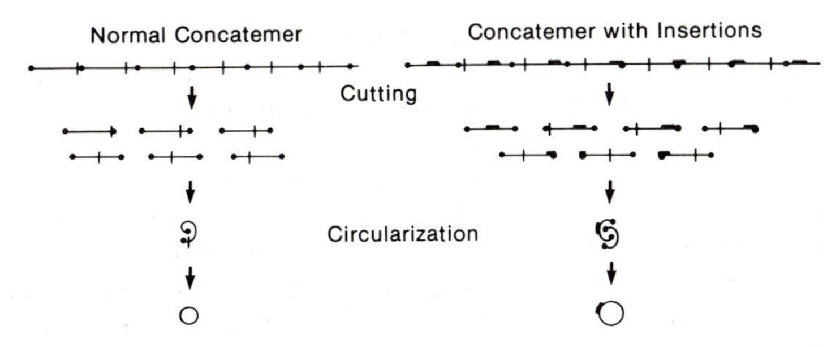

Figure 7-2. Formation of an intact P22 genome from two partial genomes. Each horizontal line represents a portion of a double-strand, concatemeric DNA molecule. The vertical lines indicate genome lengths, and the thick bars represent inserted DNA. At the left of the figure is shown the normal packaging mechanism, which generates circularly permuted terminally redundant molecules by cutting the DNA at the points indicated by the dots. At the right of the figure can be seen the effect of a large insertion on the products of the same cutting mechanisms. The molecules produced are shorter than genome length and must recombine to yield a complete circular phage genome.

Generalized Transduction

Bacteriophage P22

Despite the fact that P22 can integrate itself into the bacterial genome and produce specialized transducing particles, it is able to form generalized transducing particles as well. The formation of these particles occurs not at the time of excision of the prophage but, rather, during the packaging of the DNA into capsids. The production of the generalized transducing particles can occur during a wholly lytic infection and in the absence of any *rec* functions. Therefore P22 DNA apparently is not involved at all in the aberrant packaging.

The headful mechanism of packaging phage DNA generates circularly permuted DNA molecules, which implies that the cutting of the P22 DNA can occur at many sites. Ozeki proposed that generalized transduction occurs by a **wrapping choice** mechanism in which sites on the bacterial chromosome are recognized by the P22 enzymes and used to package bacterial DNA. Because more P22 DNA is packaged than host DNA, the P22 must have a greater number of sites, more efficient sites, or both for the cutting enzyme to act on.

The site on the DNA molecule where the first cut occurs is called a *pac* site. The cutting action is a property of the products of cistrons 2 and 3. The *pac* sites for P22 tend to be clustered in one region of the genome, as heteroduplex analysis shows that the end sequences of the phage DNA molecules are fundamentally similar. This situation occurs in cases where there is an initial cut within a limited region followed by sequential encapsidation of terminally redundant DNA from a concatemer of finite length (Fig. 7-3). Competition between the process of continued encapsidation and the process of reinitiation of packaging at a new *pac* site limits the amount of DNA packaged in any one series of reactions. Available data suggest that the average series length is about 5 phage equivalents, and that the initial cut is somewhat variable in position, as all four bases can be found as the last base at the *pac* terminus. It may be that the nuclease has only limited precision.

If there were only one *pac* site on the bacterial chromosome, all generalized transducing DNA fragments would be homogeneous in the sense that a given marker would appear on one and only one kind of fragment (Fig. 7-4). On the other hand, if fragments can be generated from a variety of sites, many of the transducing DNA molecules would be heterogeneous, meaning that a given bacterial marker might appear on more than one kind of fragment. Genetic analysis of cotransduction can be used to distinguish between the two possibilities. If one cistron can be cotransduced

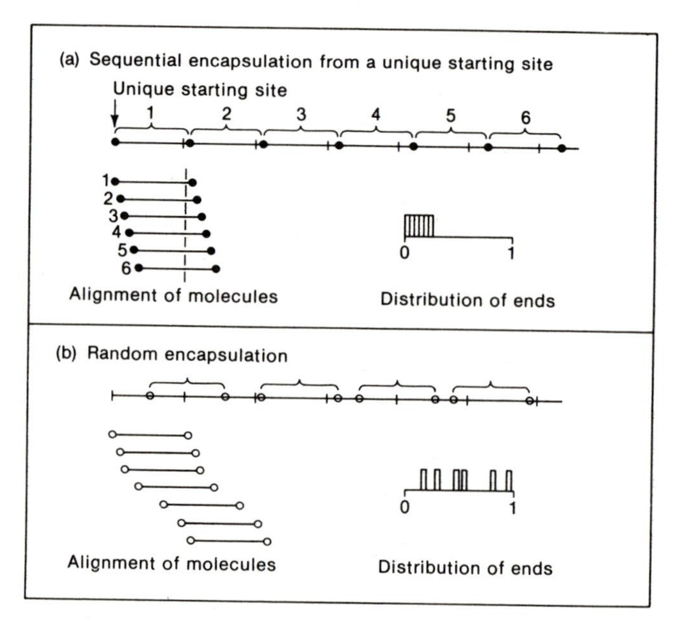

Figure 7-3. Comparison of unique site-sequential encapsulation and random encapsulation. **(a)** If the concatemer is long enough for only a small number of headfuls, sequential encapsulation at a unique starting site results in a restricted distribution of ends (i.e., restricted permutation). **(b)** Random encapsulation results in a random distribution of ends (i.e., random permutation). From Susskind, M.M., Botstein, D. (1978). Molecular genetics of bacteriophage P22. Microbiological Reviews 42:385–413.

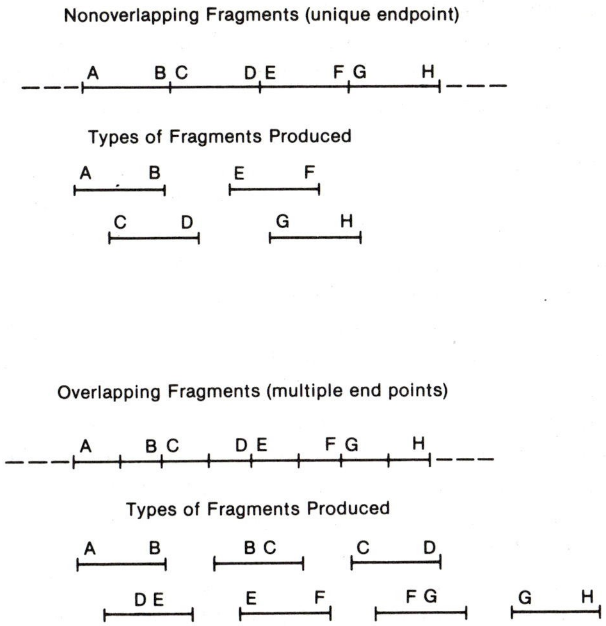

Nonoverlapping Fragments (unique endpoint)

A B C D E F G H

Types of Fragments Produced

A B E F

C D G H

Overlapping Fragments (multiple end points)

A B C D E F G H

Types of Fragments Produced

A B BC C D

DE E F FG G H

Figure 7-4. Production of transducing DNA by the headful method. The horizontal lines represent double-strand DNA, and the vertical lines indicate potential starting points for DNA packaging. Note that if nonoverlapping fragments are produced cistrons B,C or D,E or F,G lie close to one another but can never cotransduce because they will not be on the same DNA fragment. However, if overlapping fragments are produced, all nearby markers are cotransducible.

with either of two other cistrons, but all three cistrons cannot be cotransduced, at least two types of transducing DNA fragment are necessary. This situation was shown by Roth and Hartman to be the case for the *Salmonella ilv* region. Independent estimates suggest that *Salmonella* carries at least five or six *pac* sites.

The existence of multiple cutting sites for the production of transducing DNA fragments also serves to explain another set of observations. It has been known for a long time that deletion mutations can affect the **cotransduction frequency** of nearby markers (the frequency with which two markers are coinherited). If the deletion occurs between the two markers, the case obviously is trivial and the frequency of cotransduction increases. However, instances have been found in which the deletion occurs to one side of the pair of markers and yet changes the cotransduction frequency. In one instance, the deletion occurred so far from the pair of markers that it was not included in the same transducing DNA fragment, yet the cotransduction frequency was still altered. If the deletion mutation either removes a potential site for the initiation of the headful packaging mechanism or else brings the markers closer to such a site, the kinds of DNA transducing fragment produced are altered. The appearance of a new type of fragment carrying only one of a pair of markers would result in a lower cotransduction frequency, whereas the disappearance of such a fragment would result in an increased frequency.

One might assume that, after formation of the transducing particles, the balance of the transduction process would be straightforward, but for P22 it is not the case. At least 90% of all the generalized transducing DNA injected into the host cell fails to recombine and remains as a persistent, nonreplicating DNA fragment. The cistrons in these fragments may be expressed, resulting in the production of **abortive transductants.** These cells are phenotypically recombinant but fail to produce daughter cells with identical phenotypes. Instead, one daughter cell has the DNA fragment (and the recombinant phenotype), whereas the other daughter cell does not. If a nutritional selection is applied to the culture, abortive transductants produce a microcolony (very small colony) composed of the nonrecombinant daughter cells and a single dividing cell, which carries the transducing DNA.

Roughly 2 to 5% of all generalized transducing particles do give rise to complete transductants (true recombinants). The recombination process involves replacement of both DNA strands in the recipient duplex (see Chapter 13). The amount of DNA replaced ranges from 2×10^6 to 2×10^7 daltons (1.3 to 6.6 kb).

Bacteriophage P1

Phage P1 is similar to P22 in many respects, forming both specialized and generalized transducing particles. It is not usually convenient, however, to exploit P1 as a specialized transducing phage, as any recombinants that became P1 lysogens would also acquire the P1 restriction and modification system. The consequence of this acquisition would be that any further genetic manipulation using the recombinant cell as a recipient would require that the donated DNA come from another P1 lysogen. In order to alleviate this problem, the P1 stocks used for transductions are usually mutants in the *virB* cistron that causes overproduction of a repressor bypass protein *(reb),* which allows lytic replication at all times. Such phages cannot produce lysogens even though they inject and circularize their DNA in the same fashion as wild-type P1. By sacrificing specialized transducing ability, an extremely valuable generalized transducing phage can be obtained.

There is evidence to indicate that the mechanism by which generalized P1 transducing particles are produced is the same as that used by P22 (i.e., a mistake by the headful packaging system). However, because the size of the P1 phage head is larger, P1 can carry 2.5 times as much DNA as P22, and therefore transduction experiments in *E. coli* generally involve segments that are more than 2% of the genome, whereas similar experiments in *Salmonella* involve about 1% of the genome. The efficiency of transduction of selected markers varies over a 25-fold range, depending on the marker used. This variability is attributed to fluctuations in the efficiency of the recombination system, but it is assumed that the same phenomenon does not affect cotransduction frequencies for the unselected markers. Despite the differences in efficiency, all regions of the *E. coli* chromosome have been linked by cotransduction to yield the genetic map shown inside the front cover.

One technical note ought to be inserted at this point. Although the transductant survives the original phage infection because, in fact, it receives no viral DNA, it would not survive a subsequent P1 infection. Because at least 99.9% of the P1 virions in any given lysate are functional virulent phages, as soon as the infected cells lyse, large quantities of P1 are released in the culture. These P1 particles normally would infect the transductants and lyse them. However, if calcium ions, which are required for phage adsorption, are removed from the medium before any cells can lyse but after the transducing particles have attached, the transductants are protected against superinfection by P1 virions.

Other Phages

Phage T1 serves as an example of how, under the appropriate conditions, nearly all phages are capable of transduction. Drexler has shown that lethal amber mutants of T1 grown on a host that suppresses the mutation (permissive) occasionally package host DNA instead of viral DNA. If the phage lysate containing such transducing particles is used to infect cells that do not suppress the mutation (nonpermissive), no virus progeny are produced, but certain cells form recombinants and are detectable in the usual fashion. The amount of DNA carried by the transducing particles is approximately 0.5% of the genome, which is roughly consistent with its known head size.

Even phage λ can be shown to carry out generalized transduction, albeit under unusual conditions. Sternberg has shown that if the λ exonuclease function is inactivated and if host cell lysis is prevented by a mutation in the *S* cistron, generalized transducing particles do accumulate in the host cell after 60 to 90 minutes. The length of time required for production of generalized transducing particles is longer than a normal infection, and therefore cells that lyse normally release only specialized transducing phages.

Analysis of Transductional Data

Generalized Transduction

Examples could be taken from a large number of organisms, but for convenience *E. coli* is used because its genetic map is the most detailed and therefore the most amenable to transductional analysis. The phage used is P1, but the analytic principles hold true for any generalized transducing phage.

For any given transduction, four parameters can be measured: number of phage particles added per milliliter; number of recipient cells per milliliter; number of transductants per milliliter; and the number of cotransductants. The cotransduction frequency is determined by replica plating the selected transductants to check for the presence of unselected markers from the donor strain. The other parameters can be determined by the usual colony or plaque counts. An example of the data obtained is shown in Table 7-1.

The relative map positions of the three genetic markers used in the experiment can be determined from the unselected marker analysis. The *pdxJ20* allele is coinherited with *purI* 46% of the time, whereas the *nadB*$^+$ allele is coinherited only 26% of the time. Because the multiplicity of infection was roughly 0.17, the probability of multiply-infected cells as calculated by the Poisson approximation (see Appendix) is only 1%. It is therefore reasonable to assume that coinheritance of two markers means that they were on the same transducing fragment. It implies, however, that the *pdxJ* locus lies closer to *purI* than does *nadB* because there were more DNA molecules carrying *purI* and *pdxJ20* than molecules carrying *purI* and *nadB*. In other words, a higher cotransduction frequency means less recombination and implies that the markers are closer together.

The foregoing analysis is consistent with either of two cistron arrangements: *purI-pdxJ-nadB* or *pdxJ-purI—nadB*. Note that, because there is no way to determine which end of the transducing fragment contains which cistron, it is impossible to distinguish between mirror images of the arrangements (i.e., *nadB-purI-pdxJ* is indistinguishable from *pdxJ-purI-*

Table 7-1. Data from a typical P1 transduction

Donor genotype:	*purI*$^+$	*nadB*$^+$	*pdxJ20*	
Recipient genotype:	*purI66*	*nadB4*	*pdxJ*$^+$	
Selected marker:	*purI*$^+$			
Results of unselected marker analysis:	*nadB*$^+$	*pdxJ*$^+$		3 colonies
	nadB$^+$	*pdxJ20*		10 colonies
	nadB4	*pdxJ*$^+$		24 colonies
	nadB4	*pdxJ20*		13 colonies
	Total			50
Cotransduction frequencies:				
	purI$^+$	*nadB*$^+$		13/50 (0.26)
	purI$^+$	*pdxJ20*		23/50 (0.46)

The multiplicity of infection was 0.17.

Data from Apostolakos, D., Birge, E.A. (1979). A thermosensitive *pdxJ* mutation affecting vitamin B_6 biosynthesis in *E. coli* K-12. Current Microbiology 2:39–42.

nadB). All that remains, then, is to decide whether the unselected markers are on the same side of *purI* or on opposite sides.

Figure 7-5 presents two diagrams showing how crossovers would have to occur for each possible map order to generate the observed recombinants. Note that the left-hand diagram includes one class of recombinants that requires four crossovers whereas all of the recombinants in the right-hand diagram require only two crossovers. Assuming a random distribution of genetic exchanges, four events in a given space should occur less often than two events, and therefore that class of recombinant should be recovered infrequently. Because the $purI^+$ $nadB^+$ $pdxJ^+$ recombinants were indeed the rarest type (Table 7-1), the correct map order must be *purI-pdxJ-nadB*.

For purposes of comparison, Table 7-2 presents data obtained when the unselected markers lie on opposite sides of the selected marker. In this case two genetic exchanges suffice to produce any type of recombinant (Fig. 7-5), and no one class of recombinants can be considered to be significantly rarer than any of the others.

Confirmation of the map orders can be obtained by inspecting the data. If the two unselected markers lie on the same side of the selected marker, inheritance of the distal marker should frequently include inheritance of the proximal marker. It is the case for the data in Table 7-1, as 77% of the $nadB^+$ transductants were also *pdxJ20*. However, for the data in Table 7-2, the opposite is found. Of the $pdxJ^+$ transductants, only 22% were also *glyA8*, confirming that the unselected markers are arrayed on opposite sides of the selected marker and are not frequently coinherited.

The *E. coli* genetic map has as its unit of distance the minute, where 1 minute represents the average amount of DNA transferred from a typical

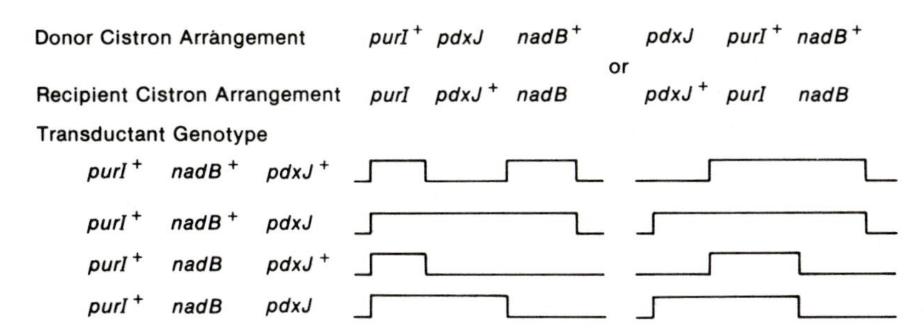

Figure 7-5. Recombination analysis of the data in Table 7-1. The lines indicate how the recombinant DNA molecule would have to be constructed to give the observed results. Each step represents one genetic exchange. Because the donor DNA is only a short fragment, an even number of exchanges is always necessary to regenerate a viable genome. Note that only one arrangement of the cistrons would require a quadruple crossover in order to produce some of the observed recombinants.

Table 7-2. Data from a P1 transduction in which the unselected markers flank the selected marker

Donor genotype:	*purI*$^+$ *pdxJ*$^+$ *glyA8*		
Recipient genotype:	*purI66* *pdxJ20* *glyA*$^+$		
Selected marker:	*purI*$^+$		
Results of unselected marker analysis:	*pdxJ*$^+$	*glyA8*	5 colonies
	pdxJ$^+$	*glyA*$^+$	18 colonies
	pdxJ20	*glyA8*	19 colonies
	pdxJ20	*glyA*$^+$	8 colonies
	Total		50
Cotransduction frequencies:			
	purI$^+$	*pdxJ*$^+$	23/50 (0.46)
	purI$^+$	*glyA8*	24/50 (0.48)

The multiplicity of infection was 0.17.

Data from Apostolakos, D., Birge, E.A. (1979). A thermosensitive *pdxJ* mutation affecting vitamin B_6 biosynthesis in *E. coli* K-12. Current Microbiology 2:39–42.

Hfr cell to an F^- cell in 1 minute. It is possible to convert cotransduction frequencies into minutes by use of an appropriate formula. For *E. coli*, the formula of Wu was adapted by Bachmann and her collaborators to establish all of the genetic map (see inside the front cover):

$$\text{Cotransduction frequency} = \left(1 - \frac{d}{L}\right)^3 \qquad [7\text{-}1]$$

where d represents the distance between the selected and unselected markers in minutes, and L represents the size of the transducing DNA fragment in minutes. Although P1 can theoretically carry 2.2 minutes worth of DNA, in practice only a value of 2.0 is used because genetic exchanges are difficult near the ends of DNA molecules. Using the data of Tables 7-1 and 7-2, it is possible to estimate the distance between *purI* and *pdxJ* as 0.46 minute, the distance between *purI* and *nadB* as 0.72 minute, and the distance between *purI* and *glyA* as 0.43 minute.

Specialized Transduction

Nonlysogenic transductants obtained via specialized transduction can be analyzed in the same manner as transductants obtained via a generalized process. However, lysogenic transductants present difficulties for analysis that preclude their use in genetic mapping. As noted earlier in the chapter, lysogenic transductants actually represent a gene duplication rather than recombination. As such they do not readily yield information regarding the normal cistron arrangement.

Summary

Transduction is a genetic process in which DNA is removed from a cell by a virus, carried through the culture medium inside a virion, and injected into a recipient cell in the same manner as viral DNA. Bacteriophages that do not integrate themselves into the bacterial DNA and do not degrade the host DNA are responsible for generalized transduction, a process by which any bacterial cistron may be transported. Phages that do integrate themselves are responsible for specialized transduction, a process by which only DNA located near the ends of the prophage is transported.

Both types of transducing phage particle seem to result from mistakes by the enzymatic systems responsible for excising and packaging viral DNA. Generalized transducing phages seem to package their DNA by the headful mechanism. If the enzymes responsible for packaging accidentally use bacterial DNA instead of viral DNA, generalized transducing particles result. Each particle contains only bacterial DNA. Specialized transducing phages result when the enzymes that are excising the prophage cut out a proper-sized piece of DNA that is not wholly prophage. The result is a DNA molecule that contains both viral and bacterial DNA. As might be expected, specialized transducing particles are usually defective in one or more functions.

Examples of phages that can carry out generalized transduction are P22, P1, and T1. Analysis of the cotransduction frequencies of unselected markers from generalized transductants permits ordering of cistrons on a genetic map. It is even possible to convert cotransduction frequencies into normal map units by formulas specific for the bacterium and the phage.

An example of a specialized transducing phage is λ. The range of markers it can transduce can be increased by forcing λ to integrate at abnormal sites around the genome. Although not generally used to construct genetic maps, specialized transductants that are also lysogens can be useful biochemically. Induction of such a lysogen in the presence of a helper phage results in production of large numbers of virus particles carrying a particular segment of the bacterial genome. The DNA molecules of such phages frequently are used for some types of sequence analysis.

References

General

Campbell, A. (1977). Defective bacteriophages and incomplete prophages, pp. 259–328. In: Fraenkel-Conrat, H., Wagner, R.R., (eds.) Comprehensive Virology, vol. 8. New York: Plenum Press.

Nordeen, R.O., Currier, T.C. (1983). Generalized transduction in the phytopathogen *Pseudomonas syringae*. Applied and Environmental Microbiology 45:1884–1889.

Susskind, M.M., Botstein, D. (1978). Molecular genetics of bacteriophage P22. Microbiological Reviews 42:385–413.

Specialized

Adams, M.B., Hayden, M., Casjens, S. (1983). On the sequential packaging of bacteriophage P22 DNA. Journal of Virology 46:673–677.

Backhaus, H. (1985). DNA packaging initiation of *Salmonella* bacteriophage P22: determination of cut sites within the DNA sequence coding for gene 3. Journal of Virology 55:458–465.

Roberts, M.D., Drexler, H. (1981). Isolation and genetic characterization of T1-transducing mutants with increased transduction frequency. Virology 112:662–669.

Smith, H.W., Lovell, M.A. (1985). Transduction complicates the detection of conjugative ability in lysogenic *Salmonella* strains. Journal of General Microbiology 131:2087–2089.

Sternberg, N. (1987). The production of generalized transducing phage by bacteriophage lambda. Gene 50:69–85.

Teifel-Greding, J. (1984). Transduction of multi-copy plasmid pBR322 by bacteriophage Mu. Molecular and General Genetics 197:169–174.

Yamamoto, N., Droffner, M.L., Yamamoto, S., Gemski, P., Baron, L.S. (1985). High frequency transduction by phage hybrids between coliphage ϕ80 and *Salmonella* phage P22. Journal of General Virology 66:1661–1667.

Young, K.K.Y., Edlin, G. (1983). Physical and genetical analysis of bacteriophage T4 generalized transduction. Molecular and General Genetics 192:241–246.

Chapter 8
Genetic Transformation

This chapter introduces the first major bacterial genetic exchange process to be discovered, **genetic transformation**. At first glance, the mechanism appears to be improbable. Large DNA fragments (as heavy as several million daltons) are released from donor cells and diffuse through the culture medium to recipient cells. The molecules are then transported across the cell wall and cell membrane into the cytoplasm where recombination occurs. The process is distinct from another biologic phenomenon also denoted transformation, the conversion of normal mammalian cells into tumor cells. In order to emphasize this difference, in this book the bacterial process is always described as genetic transformation.

The natural susceptibility of DNA to degradation by nucleases has prompted many workers to question whether genetic transformation is only a laboratory artifact. However, a wide variety of genera, including *Achromobacter, Azotobacter, Bacillus, Hemophilus, Micrococcus, Mycobacterium, Neisseria, Pseudomonas, Streptococcus, Streptomyces,* and *Synechococcus,* have been reported to have natural genetic transformation systems, and it is reasonable to suppose that so general a phenomenon has some genetic significance. Moreover, studies have shown that DNA bound to soil particles is relatively resistant to nuclease degradation and therefore more likely to remain available for genetic transformation. It is thus probable that genetic transformation has a significance beyond the boundaries of the research laboratory, even if the extent of the naturally occurring process is difficult to assess.

In this chapter, the basic process of genetic transformation and some of its variations are discussed. Only a few of the many transformable genera are considered, primarily *Bacillus, Streptococcus,* and *Hemophilus,* which are the best studied.

Pneumococcus—*Bacillus* Transformation System

Discovery of Genetic Transformation

The original observations on transformation were made by Griffith in 1928 during the course of a series of investigations into the mode of infection of pneumococcus. Griffith noted that there were two colony morphologies present in isolates of these bacteria. One colonial type was smooth and glistening owing to the presence of a polysaccharide capsule surrounding each cell. The other colonial type was much rougher in appearance owing to the absence of a capsule. Both types of colonial morphology were stably inherited and therefore genetically determined.

The presence or absence of a capsule had a profound influence on the virulence of the bacteria. Only the encapsulated bacteria could produce a fatal septicemia (infection of the blood) within a mouse. In the absence of a capsule, the immune system of the mouse soon destroyed the invading bacteria. As might be expected, Griffith showed that encapsulated cells could be heat-killed prior to infection and no septicemia would result.

However, Griffith also showed that if live, nonencapsulated bacteria were mixed with heat-killed encapsulated ones and injected into a mouse, the mouse soon died from the resulting septicemia. When the bacteria that had killed the mouse were cultured, they were found to have capsules, even though the only encapsulated bacteria that had been injected were already dead. Griffith said that the bacteria without capsules had been transformed into bacteria with capsules, but he was never able to show how it was accomplished.

That discovery was left to microbial chemists who painstakingly purified the "transforming principle" and identified it. Finally, Avery, MacLeod, and McCarty showed that transformation was caused by an obscure chemical of unknown function found in all cells and named deoxyribonucleic acid. Their results were the first concrete evidence that DNA was in fact the genetic material of the cell, and so revolutionary was the idea that it did not take hold until 8 years later when the experiments of Hershey and Chase (see Chapter 4) led to the same conclusion. Soon after that, Watson and Crick proposed a molecular structure for DNA, and the molecular biologic era was launched.

Pneumococcus is still extensively studied. However, over the years it has undergone several name changes at the hands of the bacterial systematists. At the moment, it is classified as *Streptococcus pneumoniae*, but other names still in use are pneumococcus or *Diplococcus pneumoniae*. For the balance of this book it is referred to as *S. pneumoniae*.

The other bacterium discussed in this section is *Bacillus subtilis*, but it should be pointed out that not all members of *B. subtilis* are transformable. Indeed, nearly all work has been done with certain special strains, variously denoted "168" or "Marburg," which are derived from that originally used by Spizzizen and co-workers. The differences between these strains and other members of the species have not been elucidated.

Competent Cells

Among the bacteria that are naturally transformable (in contrast to those that can undergo genetic transformation only after special laboratory treatments), nearly all vary in the timing and duration of their **competence**, their ability to bind DNA and protect it from external nucleases. Only *Neisseria gonorrhoeae* seems to lack the ability to regulate its competence. For the remainder, competence is a definite physiologic state inducible by certain growth patterns, usually a "shift-down" (mimicking the transition from the log phase of growth into the stationary phase) in which cells are transferred from a relatively nutrient-rich medium to a nutrient-poor one. Alternatively, *Staphylococcus aureus* can have competence induced by a bacteriophage. In the case of *B. subtilis*, even a shift from 42° to 37°C can induce competence. The number of competent cells in a culture can be variable, ranging from 15% for *B. subtilis* to about 100% for *S. pneumoniae*.

The development of competence may require only a few minutes, but, at least in *B. subtilis*, cultures with competent cells can be maintained for some time even though the competent cells do not divide or replicate their DNA. Chemostat experiments, in which cells are grown in a culture vessel to which fresh medium is constantly added and depleted medium removed, show that competence occurs at doubling times of 150 and 390 minutes. Because the chemostat cultures do not sporulate, these experiments also demonstrate that competence is not necessarily linked to sporulation. The latter conclusion is reinforced by experiments demonstrating that mutations blocking the earliest steps in sporulation have no effect on genetic transformation ability.

Immunologic tests have shown that competent *S. pneumoniae* cells have a new protein antigen on their surfaces. This competence factor is released into the medium and can be used to make noncompetent cells competent by binding to a second specific antigen on the cell surface and triggering the response. The slow release of competence factor is used to account for the relatively high cell density ($\sim 10^8$/ml) required for the production of competent cells. The changes induced by competence constitute a sort of differentiation process in which the cell wall becomes more porous; there is an increase in autolytic enzyme activity; and the average length of chains of cells increases as much as eightfold. Most of the changes occur along the cell's equator, the area where new growth normally is found and where DNA uptake occurs. There often is an association between competence and the synthesis of poly-β-hydroxybutyric acid, which is produced as a storage product. However, the possible role played by the polymer in genetic transformation is unclear, as many nontransformable genera also accumulate it.

DNA Uptake and Entry

Competent and noncompetent cells bind some double-strand DNA to their surfaces; however, only competent cells bind it so that the binding is not readily reversible by simple washing procedures. The addition of ethylenediaminetetraacetic acid (EDTA), a chelator or complexer of divalent cations, allows binding of DNA but prevents any further processing. Measurements of the amount of DNA bound indicate that there are something on the order of 50 DNA binding sites per competent cell. Bound DNA is still sensitive to exogenous nucleases or hydrodynamic shear (as in a blender). Any type of DNA can bind, and foreign DNA, e.g., salmon sperm DNA, is often used in competition experiments to prevent further binding of specific bacterial DNA. The competition experiments have shown that the minimum size for effective DNA binding is about 500 bp and that single-strand DNA or RNA:DNA hybrids rarely if ever bind. Such binding as is seen has been attributed to occasional renatured or persistent double-strand regions.

After binding, the transforming DNA must enter the cell (Fig. 8-1). Before it can happen, though, the DNA must be cut to size by a major endonuclease located near the DNA binding site. This enzyme makes double-strand cuts to generate DNA fragments no larger than about 15 kb, smaller than the size of DNA fragments carried by most transducing phages. It requires magnesium or calcium for function and therefore accounts for the sensitivity of DNA uptake to EDTA.

If the cell is disrupted at this point, the double-strand donor DNA fragments that are released can still be used to transform another cell. However, the next stage is for the DNA to form an **eclipse complex** with a 19.5 kilodalton (kd) competence-specific protein. In *S. pneumoniae* about 68% of the donor DNA enters the eclipse complex, and about 25% of that DNA actually becomes part of the recipient DNA. When the eclipse complex forms, the DNA becomes single-stranded and completely protected from external nuclease activity. The conversion occurs in a linear fashion beginning at one or several points. There does not appear to be any specificity as to which strand is degraded, and the bases from the degraded strand are released into the medium for possible reuse. It has been suggested that the energy derived from the strand degradation might be harnessed to drive other steps in the process. Because single-strand DNA is virtually nonfunctional in transformation, disruption of the cell at this point does not release any donor DNA that can be used to transform a new recipient. In this sense the eclipse phase for transformation closely parallels the eclipse phase for bacteriophage infection.

The single-strand DNA passes through the cell membrane and into the cytoplasm. Electron microscopic studies have demonstrated the close proximity of cell wall and cell membrane in competent cells in a manner reminiscent of the zones of adhesion used during phage T4 infection. The

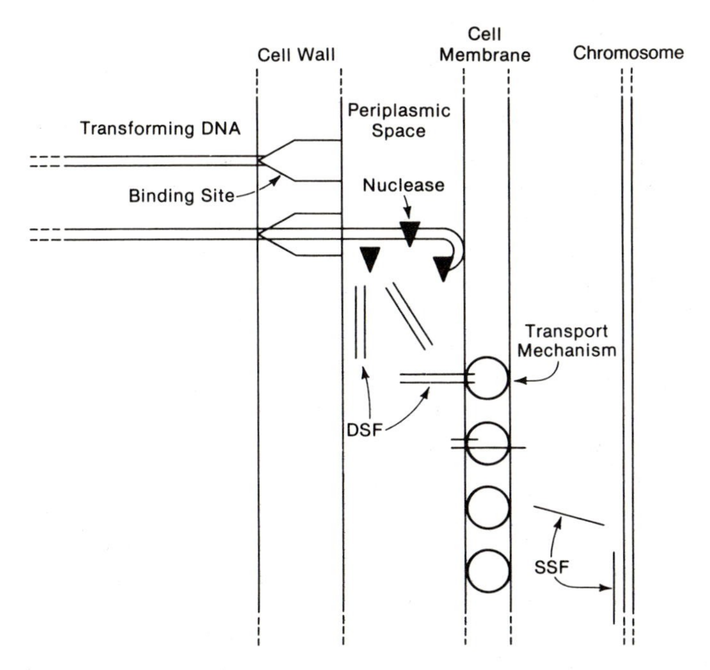

Figure 8-1. Entrance of transforming DNA into *Bacillus subtilis*. Movement of the DNA is from left to right in the figure. Each horizontal line represents a single strand of DNA. Pentagonal structures in the cell wall represent the specific receptors on the surface of the cell. Two sites of nuclease activity are shown: One nuclease (triangles) is found in the periplasmic space and produces double-strand fragments (DSF); the other nuclease occurs in the cell membrane (circles) and produces single-strand fragments (SSF). The single-strand molecules at the bottom right are preparing to recombine with the resident DNA. Figure design courtesy of Dr. William F. Burke, Jr.

bacterial chromosome is also membrane-attached, and therefore the transforming DNA can be delivered right to its target. Venema's laboratory has devoted considerable effort to studying the role of cell membranes in transformation, especially the so-called mesosome. Mesosomes were originally identified as invaginations of the cell membrane seen in fixed preparations. More recent work using frozen sections has suggested that the mesosomes are an artifact of fixation. Nevertheless, specific membrane fractions isolated from competent cells do display differential DNA binding properties and may still play a role in genetic transformation.

Establishment of Donor DNA in the Recipient Cell

Competent cells bind both homologous and heterologous DNA without discrimination. They also process both types of DNA to single-strand fragments. Because a fundamental principle of genetics is that genetic exchange occurs only between similar organisms, there must be some sort of discriminator mechanism that is active to prevent the establishment of heterologous DNA within a cell. In the cases of *Streptococcus* and *Bacillus*, the discrimination occurs at the level of recombination.

Both homologous and heterologous single-strand DNA fragments form an association with the cell membrane and the nucleoid that Venema and his collaborators referred to as the donor–recipient complex. It is possible to demonstrate that there is pairing of the recipient DNA and the single-strand fragments if the donor DNA is radioactive. The donor–recipient complex can be extracted from lysed cells and treated with S1 nuclease. S1 nuclease attacks only single-strand nucleic acid and does not harm correctly hydrogen-bonded strands. When donor–recipient complex from a homologous genetic transformation is tested, it is relatively stable to S1 nuclease, whereas a heterologous complex is sensitive. Failure to correctly hydrogen-bond would prevent the normal recombination process from functioning (see Chapter 13) and thus prevent assimilation of heterologous DNA. This model predicts that recombination deficient mutants would also be defective in transformation, and that is the case. Similarly, treatments that enhance recombination, e.g., introduction of nicks into the recipient DNA by means of x-irradiation, also improve transformation efficiency. Despite the seeming complexity of the genetic transformation process, it is rapid, and the time involved is only about 15 minutes before donor DNA can be recovered as a double-strand molecule.

The single strand of DNA that recombines is literally substituted for one of the two host DNA strands. This substitution leads to production of a transient DNA heteroduplex (Fig. 8-2). To the extent that DNA repair enzymes (discussed in Chapter 13) manage to repair the mismatched bases, potential transformants may be lost. In *S. pneumoniae*, but not in *B. sub-*

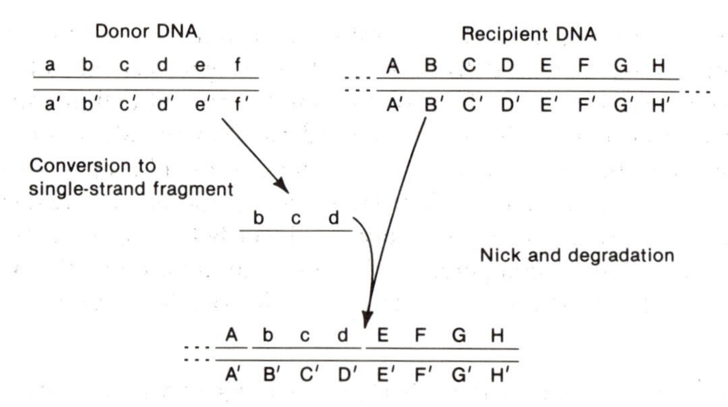

Figure 8-2. Formation of transient heteroduplexes during transformation. Each line represents a single DNA strand, and each letter represents a different genetic marker. Methods for the nick and degradation steps that result in the insertion of the new piece of DNA are discussed in Chapter 13.

tilis, there is an enzyme system coded for or controlled by the *hex* cistron, which seems to bind to various specific sites on the genome and to correct any mismatched bases in the immediate vicinity. As a consequence, in *hex*$^{+}$ strains markers that recombine near one of the Hex enzyme-binding sites tend to be lost owing to the correction process, whereas markers that recombine further away are not lost. The former type of marker is called low efficiency (LE), and the latter type is called high efficiency (HE). The Hex system is limited by the number of enzyme molecules available, and therefore one method of bypassing its effects is to overload the cell with transforming DNA.

The requirement for recombination as an integral part of genetic transformation might be thought to be applicable only to chromosomal and not plasmid DNA. After all, by definition plasmids are self-replicating entities, and therefore simple arrival in the cytoplasm ought to be sufficient to yield a transformant carrying a plasmid. However, that analysis overlooks the mechanism of transformation. The plasmid DNA, even if it is a small molecule, is linearized by the nucleases at the cell surface. Moreover, it arrives in the cytoplasm as a single-strand entity, whereas it must be a double-strand circle for successful replication. In light of this analysis, it is not surprising that genetic transformation of a cell by a single plasmid molecule is rare, and that when successful plasmid transformation is observed extra DNA has often been acquired.

The extra DNA may be a second copy of the plasmid (i.e., a concatemer) or DNA that is homologous either to the bacterial chromosome or to another plasmid that is already resident in the cell. The process is also recombination-dependent. The model based on these observations is that if two plasmid copies are available there are regions of homology that can be used for circularization (Fig. 8-3). Similarly, regions of homology with another replicon can anchor the DNA and allow circularization to proceed. In general, the larger the region of homology, the greater is the number of transformants obtained. Thus when homologous DNA is spliced into a vector and transformation attempted, the result is a high frequency of genetic transformation. However, if eukaryotic DNA is spliced into the same vector, the transformation frequency may be low unless the vector itself provides homology or there is some accidental homology between the DNA insert and the host cell. Experiments in which a resident plasmid is used to improve the recovery of transformants are often denoted marker rescue experiments.

Other Transformation Systems

Hemophilus influenzae

Hemophilus influenzae is a gram-negative organism that is naturally transformable. Stewart and Carlson argued that although there are some specific differences in detail between the mechanism used by *Hemophilus*

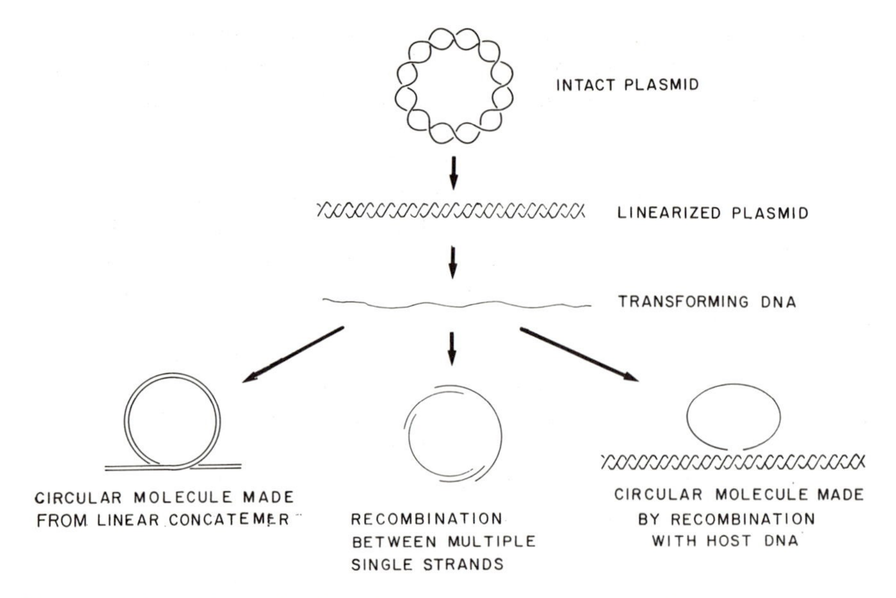

Figure 8-3. Possible mechanisms for successful genetic transformation by plasmid DNA. A plasmid is linearized by cutting at one site. During transformation the strands are separated so that only single-strand DNA arrives in the cytoplasm. Circular DNA can be regenerated by recombination with the host cell chromosome, pairing of multiple single strands of DNA followed by gap-filling DNA synthesis, or synthesis of a complementary DNA strand followed by self-recombination.

and that discussed in the preceding section, the fundamental mechanism of genetic transformation is the same in both gram-positive and gram-negative organisms. For example, a *Hemophilus* culture generally requires a nutritional shift-down to produce competent cells. Moreover, an appropriate treatment that blocks nucleic acid synthesis but allows continued protein synthesis leads to essentially 100% of the cells in a culture becoming competent. Unlike *S. pneumoniae,* there does not seem to be a competence protein that is released from the cells. The competent state can persist for a reasonable period of time so long as the cells are not allowed to grow. Transfer into rich medium rapidly eliminates competence.

The competent cells exhibit more selectivity in the DNA they bind compared with gram-positive organisms. Although *H. influenzae* takes up DNA from other *Hemophilus* species, it rejects DNA from disparate organisms such as *E. coli* or *Xenopus laevis* (South African clawed toad). The basis for the specificity can be determined by competing total unlabeled *Hemophilus* DNA with pieces of radioactive *Hemophilus* DNA for cell surface binding sites. The minimum binding sequence is apparently 11 bp and can be found roughly every 4000 bp (about 600 per genome). Barouki and Smith noted that some DNA fragments lacking the 11-bp recognition sequence are nonetheless taken up by competent cells. They suggested that there may be more than one such sequence or another unique structure within the DNA.

Entry of the transforming DNA into the cell is also different from the previously described mechanism. In competent *H. influenzae* cells a series of membranous extrusions appear on the cell surface. The double-strand transforming DNA is cut on the surface of the cell and then transported not into the cytoplasm but into the membranous extrusions that subsequently bud off to form **transformasomes**, independent double-strand DNA-containing vacuoles within the cytoplasm (Fig. 8-4). The binding data fit a model in which there is a single DNA receptor site per transformasome.

A comparison of the efficiency of genetic transformation by chromosomal DNA and plasmid DNA indicates that the latter is inefficient. Apparently only linear DNA exits the transformasome normally, and a linear plasmid has difficulty recircularizing, as was the case with the gram-positive cells (each transformasome would normally have only a single plasmid molecule). Plasmid establishment by double-strand DNA escaping from the transformasome has been reported by Pifer, but the frequency is low: 10^{-5} to 10^{-7}. The same considerations shown in Fig. 8-3 should apply to plasmid transformation of *Hemophilus*.

Despite the double-strand nature of the transformasome DNA, the recombinant DNA molecules are still heteroduplexes. Therefore only a single DNA strand is exchanged, as is the case with the gram-positive cells. Electron microscopic experiments have shown that during the acquisition of competence single-strand gaps appear in the chromosomal DNA. A

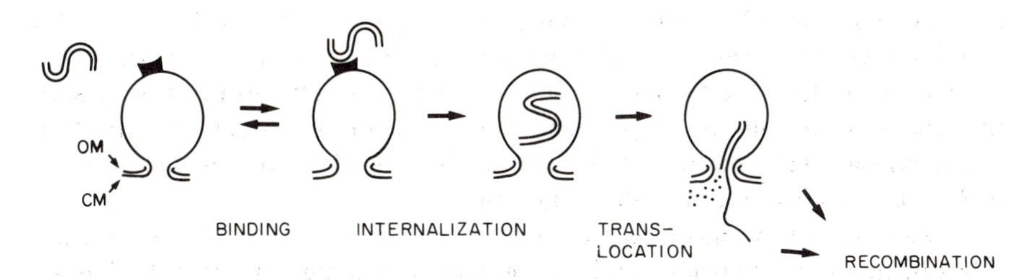

Figure 8-4. Hypothetical model for DNA entry in *H. influenzae* cells. DNA entry in *H. influenzae* constitutes the initial step leading to the genetic transformation of these bacteria by DNA present in the medium. The model assumes that small vesicles of cell membrane (CM) protrude through the outer membrane (OM) of the gram-negative cell wall. DNA first binds reversibly to specific receptors present at the surface of the transformasome and is then irreversibly internalized into the vesicle. Once inside the vesicle, the DNA is translocated into the cytoplasm as a single-strand entity with concomitant degradation of the complementary strand. Recombination is assumed to occur elsewhere. DNA uptake is defined as the summation of the binding and internalization step. Redrawn from Barouki and Smith (1986).

recombination deficiency mutation that prevents transformation also prevents gap formation. The transforming DNA is processed to yield short, single-strand tails that are presumed to serve as initiators of the recombination process beginning at the gaps (see Chapter 13).

Escherichia coli

Although also a gram-negative organism, *E. coli* is unlike *Hemophilus* because it does not undergo transformation spontaneously without exposure to conditions drastically different from its normal environment. For example, **spheroplasts** (cells that have lost most of their cell wall) can take up DNA and be genetically transformed by it. At the present time most genetic transformation in *E. coli* is based on the original observation of Mandel and Higa that intact cells shocked by exposure to high concentrations of calcium ions become competent. The most elaborate study on the conditions leading to competence is that of Hanahan, who found that, in addition to calcium ion, treatment should include cobalt chloride and dimethylsulfoxide. All cell manipulations must be done on ice, and the transforming DNA is added as a chilled solution. Even so, as many as 95% of all cells may be killed by the treatment. After allowing time for adsorption, a temperature shift from ice to 37° to 40°C is used to stimulate uptake and entry of the DNA.

Reusch and co-workers have shown that competent *E. coli* cells accumulate poly-β-hydroxybutyrate (PHB) in their membranes, and that the amount of PHB correlates well with the transformability of the cell. The addition of PHB (and possibly other substances as well) results in a change in the lipid structure of the cell membrane. All lipids have a characteristic phase transition temperature at which they go from being relatively solid to relatively more fluid. As the intensity of the *E. coli* phase transition increases, so does the transformability of the cells. The authors suggested that the role of calcium in genetic transformation is to activate the biochemical pathway leading to the synthesis of PHB, making competence in *E. coli* also a regulated phenomenon.

A wide variety of plasmids readily transform *E. coli*, which is one reason it has been so widely used as a host for genetic cloning experiments. Unlike the naturally transformable organisms, *E. coli* does not seem to linearize plasmids during transformation. If the plasmids are linearized by the experimenter prior to the transformation, low levels of transformants are obtained. Conley and co-workers have shown that the rare successfully transformed linear plasmids usually contained deletions that included the site of linearization. Sequence analysis indicated that recombination occurred between 4 to 10 bp directly repeated sequences.

The enzyme responsible for failure of linearized plasmid transformation is exonuclease V, the same enzyme that must be inactivated to allow rolling

Table 8-1. Representative transformation efficiencies in various bacteria using homologous donor DNA

Organism	Transformants/ viable cell
B. subtilis	0.001
S. pneumoniae	0.05
H. influenzae	0.01
E. coli	0.05

circle replication. Because it attacks only linear duplexes, circular plasmid DNA, even if nicked, is not a substrate, and transformation can proceed normally. Many strains of *E. coli* that are used for transformation are *recB recC* (to inactivate exonuclease V) and *sbcB* (to suppress the recombination-deficient phenotype). *E. coli* thus possesses two natural barriers to transformation—lack of competent cells and presence of exonuclease V—both of which can be overcome in the laboratory. A comparison of the transformation frequencies for the various organisms that have been discussed is given in Table 8-1.

Transfection

An unusual variant of transformation is transfection, a process in which the source of the donor DNA is not another bacterial cell but a bacteriophage. The result of successful uptake and entry of the DNA into the recipient cell is an infected cell (infectious center), and the success of the procedure can be measured by a standard plaque assay. By varying the amount of phage DNA added, it can be shown that the number of infectious centers is proportional to the amount of DNA added to the cells.

The process of transfection has proved to be a powerful tool for investigating the transformation process itself as well as other genetic problems. This situation is due to the fact that, unlike the DNA extracted from bacterial donors, phage DNA, when carefully extracted, is a homogeneous population of molecules. Therefore it is much easier to detect the action of various bacterial enzymes on the DNA, as all molecules were originally identical. An important use of transfection is to study the DNA recombination and repair process by adding artificially created heteroduplex phage DNA molecules to cells and observing the pattern of inheritance of the heteroduplex region.

Escherichia coli as a Recipient

Oddly enough, the initial observation of transfection did not take place in the *Bacillus* or *Streptococcus* systems but, rather, in *E. coli*. In 1960 Kaiser and Hogness observed that purified DNA from λ*dgal* high-frequency transducing lysates could cause the production of gal^+ recombinants, provided the recipient cell was simultaneously infected with a nontransducing λ helper phage to provide certain unspecified functions.

Further examination of this **"helped transfection"** system showed that it had several interesting properties. The helper phage must not be ultraviolet (UV)-inactivated, and it must contain functional DNA. In general, the best multiplicity of infection for the helper phage was high, around 5 to 15 per cell. Under these conditions, the efficiency of transfection was increased from 10^{-7} to 10^{-3}, making the λ transfection system one of the most efficient known.

Certain strictures also applied to the λ DNA molecules used in the experiment. If less-than-intact genomes were used, each transfecting fragment must contain the *cos* site. Additionally, the *cos* sites must be identical, or nearly so, to the *cos* sites of the helper phage. Benzinger interpreted these results to mean that, as the helper DNA is injected into the bacterial cell in the usual manner, the *cos* site of the transfecting fragment attaches to the helper DNA, allowing the transfecting fragment to be pulled into the cell along with the infecting DNA. Fragments without *cos*, according to this model, fail to transfect because they cannot enter the cell.

Helped transfection is of more general use than might be expected because markers other than *gal* can be used. In fact, if the helper phage carries a mutation that prevents its replication in the recipient cell, the results of phage recombination can be studied. Phages other than λ, such as ϕ186 or P2, also participate in transfection. Their *cos* DNA sequences are distinct from those of λ, so they cannot act as helpers for λ transfection. However, they do help each other, as 17 of the 19 bases comprising their *cos* sites are identical.

Nonhelped transfection is also found in *E. coli*. In this case the phage DNA enters the cell by a more usual uptake and entry mechanism. However, before the DNA can do so, the cell must be made competent. One way to do it is by calcium shock, as described above. An alternative method is removal of most of the cell wall to give a spheroplast. The most commonly used method for preparing spheroplasts for transformation is EDTA–lysozyme treatment. EDTA removes magnesium ions from the cell wall, resulting in functional destruction of the outer membrane, thereby allowing the enzyme lysozyme access to the underlying peptidoglycan layer. Lysozyme hydrolyzes the glycosidic linkages of the peptidoglycan layer of the cell wall, resulting in the loss of large chunks of it. The cells, however, remain viable, provided the osmotic strength of the medium is adjusted to prevent cell lysis, and phage multiplication is unaffected. New phage infections due to released progeny virions cannot occur because the attachment sites for the phage have been destroyed.

The mechanism of transfecting DNA uptake and entry seems to depend on both the method used to prepare the cell and the DNA used. Certain phage DNA molecules carry out transfection only when a particular pretreatment is used (Table 8-2). A few spheroplasts seem to account for all of the DNA binding and may have as many as five to ten DNA molecules attached to their surface. This relatively large number of molecules per cell is consistent with studies on *Bacillus* (see below) that indicate cooperativity effects.

In addition to the phages already discussed, a number of others can be used as sources of transfecting DNA (Table 8-2). In fact, many *E. coli* phages transfect a wide range of the enterobacteria. It is interesting to note that even RNA phages produce transfectants under the proper conditions, which usually include a treatment to remove ribonuclease from the culture.

Table 8-2. Some typical bacterial transfection systems

Genus and species	Genetically transformable	Transfecting bacteriophage nucleic acid	Best efficiency of transfection	Transfection assay system
Pseudomonas aeruginosa	+	φX174 DNA	10^{-9}	Lysozyme-EDTA spheroplasts
		PP7 RNA	10^{-7}	Lysozyme-EDTA spheroplasts
Agrobacterium tumefaciens	+	LR-4 DNA	10^{-7}	Competent cells
Escherichia coli	+	T1 (T3, T7) DNA	10^{-5}	Lysozyme-EDTA *recB*$^-$ spheroplasts Ca^{2+}-shocked *recB*$^-$ r_k^- m_k^- cells
		T2 (T4, T6) DNA	10^{-4}	Lysozyme-EDTA spheroplasts
		φX174 (S13) DNA	10^{-2}	Lysozyme-EDTA spheroplasts
		Filamentous phage DNA	10^{-5}	Lysozyme-EDTA spheroplasts
		R17, MS2, M13, Qβ RNA	10^{-5}	Washed lysozyme-EDTA spheroplasts

Table 8-2. (Continued)

Genus and species	Genetically transformable	Transfecting bacteriophage nucleic acid	Best efficiency of transfection	Transfection assay system
		λ (434, 21, 186) DNA	10^{-3}	Helped transfection Lysozyme-EDTA spheroplasts
		P1 DNA	2×10^{-6}	Lysozyme-EDTA spheroplasts
		P2 DNA	10^{-5}	Helped transfection
		P22 DNA	10^{-6}	*recB*⁻ lysozyme-EDTA spheroplasts Ca^{2+}-shocked cells
		Mu DNA	10^{-7}	Ca^{2+}-shocked *recBC endoI sbcA cells*
Enterobacter aerogenes	+	T4 DNA	10^{-8}	Helped transfection
		φX174 DNA	10^{-4}	Lysozyme-EDTA spheroplasts
Salmonella typhimurium	+	P22 DNA	10^{-7}	Ca^{2+}-shocked cells
		φX174 DNA	2×10^{-6}	Lysozyme-EDTA spheroplasts
Shigella paradysenteriae		φX174 DNA	10^{-6}	Spheroplasts
Klebsiella pneumoniae	+	φX174 DNA	5×10^{-8}	Spheroplasts
Proteus vulgaris		φX174 DNA	10^{-9}	Spheroplasts
Serratia marcescens		φX174 DNA	2×10^{-8}	Spheroplasts
Bacillus subtilis	+	φ29, SP02, φ105, SPP1, H1, SP01, 2C, SP50, φ1, φ25, SP3, SP82G DNA	10^{-2}–10^{-8}	Competent cells, helped transfection

Adapted from Benzinger, R. (1978). Transfection of Enterobacteriaceae and its applications. Microbiological Reviews 42: 194–236.

Bacillus subtilis as a Recipient

Transfection in this organism is much simpler than in *E. coli* because the normally competent cells serve as recipients without any special treatment. The rate of entry of the phage DNA seems to be much slower than that of chromosomal DNA. Conversion of the phage DNA to a nuclease-insensitive state may require an hour or more, depending on the temperature. DNA uptake and entry by these cells has been clearly shown to be a cooperative phenomenon. Trautner and co-workers demonstrated that there are various levels of cooperativity, with smaller phages such as φ29, whose DNA tends to aggregate, showing first-order kinetics and larger phages tending toward second- and third-order kinetics (Fig. 8-5). The kinetic order is an indication of the number of individual molecules or aggregates that must come together to give the observed result, in this case an infected cell. The presumed explanation for these observations is fragmentation of the phage genome as it attempts to enter the cell and the necessity for two single strands of opposite polarity to produce a replicating DNA molecule. This situation results in several fragments being necessary to provide the equivalent of an intact phage DNA. After transformation, the infectious process proceeds normally, although the yield of phage is lower.

There are some indications that transfection as a process differs from transformation. A strain has been isolated that carries a mutation that makes it nontransformable owing to an inability to bind DNA (less than 1% of normal activity). Nevertheless, the mutant strain undergoes transfection with DNA from φ29 but not with DNA from SP01. The difference between these two phages is that SP01 requires recombination for DNA circularization, whereas φ29 is circularized by a protein linker. Apparently phage DNA can attach to competent mutant cells even though bacterial DNA cannot, which suggests the existence of at least two mechanisms for DNA uptake. The lack of transfection by SP01 suggests that the strain in question is also recombination-deficient, which seems to be borne out by other evidence.

Genetic Mapping Using Transformation

Analysis of Transformational Data

Conceptually the analysis of transformational data is similar to that of transductional data. In both cases one marker is selected, and other unselected markers are then checked to determine cotransfer frequencies. Also, in both cases the deduced map order is only relative, as there is no way to unambiguously distinguish left from right on the donor DNA molecule.

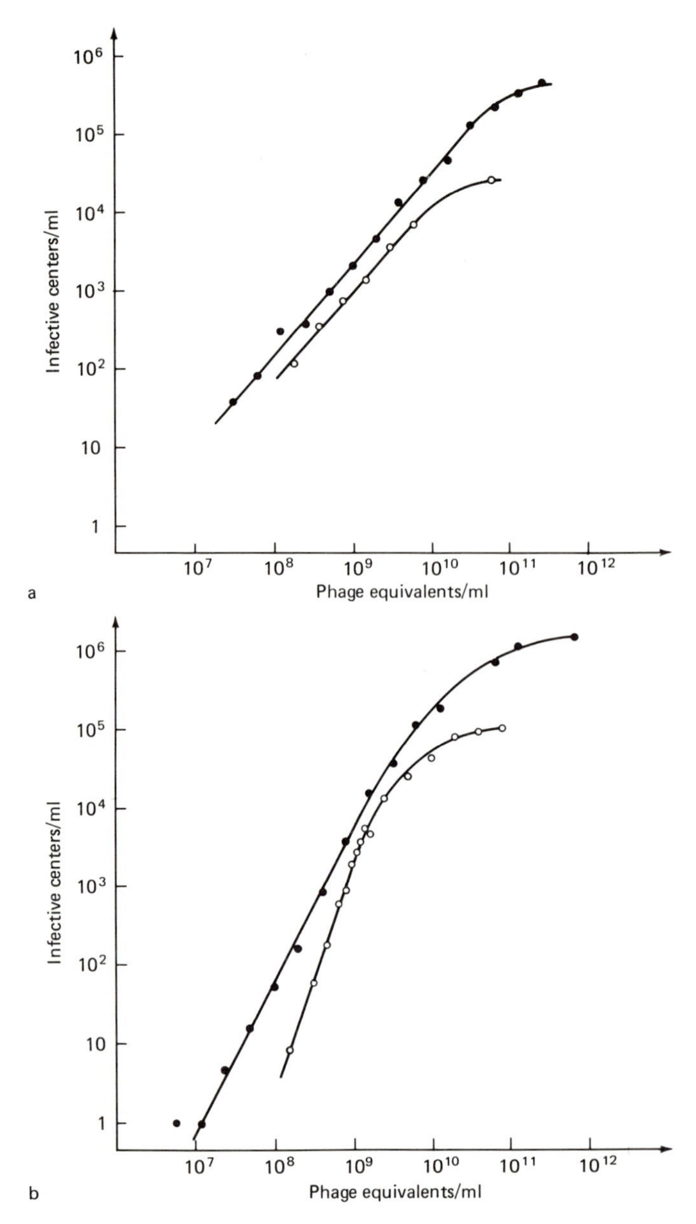

Figure 8-5. Cooperativity effects during transfection. The number of infectious centers observed is plotted as a function of the number of genome equivalents of phage DNA added to the cells. For first-order kinetics, a tenfold increase in phage equivalents would result in a tenfold increase in infectious centers. For second-order kinetics, a tenfold increase in phage equivalents would result in a 100-fold increase in infectious centers, etc. The curves in **a** approximate first-order kinetics, whereas the curves in **b** approximate second- and third-order kinetics. The phage DNA molecules used were ϕ29 (closed circles) and SPO2 (open circles) **(a)**; and SPP1 (closed circles) and SP50 (open circles) **(b)**. From Trautner, T.A., and Spatz, H.C. (1973). Transfection in *B. subtilis*. Current Topics in Microbiology and Immunology 62:61–88.

Table 8-3. Typical data from a *Bacillus subtilis* transformation experiment

Donor strain:	*gua*$^+$ *pac-4*	*dnaH*$^+$	
Recipient strain:	*gua-1* *pac*$^+$	*dnaH151*	
Selected marker = Gua$^+$			
Distribution of unselected markers:	*pac-4*	*dnaH*$^+$	6
	pac-4	*dnaH151*	1
	pac$^+$	*dnaH*$^+$	44
	pac$^+$	*dnaH151*	56
	Total transformants tested		107
Cotransformation frequencies:	*gua*$^+$	*pac-4*	6.5%
	gua$^+$	*dnaH*$^+$	47.0%
Selected marker = DnaH$^+$			
Distribution of unselected markers:			
	gua$^+$	*pac-4*	6
	gua$^+$	*pac*$^+$	30
	gua-1	*pac-4*	5
	gua-1	*pac*$^+$	55
	Total transformants tested		96
Cotransformation frequencies:	*dnaH*$^+$	*pac-4*	11%
	dnaH$^+$	*gua*$^+$	38%
Deduced map order:	*gua-dnaH---pac*		

Abbreviations are listed in Table 2-2, with the exception of *pac*, which refers to resistance to the antibiotic pactamycin.

The data are from Trowsdale, J., Chen, S.M.H., Hoch, J.A. (1979). Genetic analysis of a class of polymyxin resistant partial revertants of stage 0 sporulation mutants of *B. subtilis*. Molecular and General Genetics 173:61–70.

Parameters that can be measured during a transformation experiment include the total number of viable recipient cells, the amount of donor DNA added, the number of transformants obtained, and the frequency with which the unselected markers were obtained. A sample of such data is given in Table 8-3. As with transduction, the cotransfer frequencies are inversely proportional to the genetic distance between the selected and unselected markers. By diagramming the possible genetic exchanges as in Figure 7-5, it is possible to deduce the indicated cistron order.

Genetic Map for *Bacillus subtilis*

The genetic map for *B. subtilis* has been developed piecemeal using both transformation and transduction. The current map is presented in Figure 8-6, and a comparison with the *E. coli* map (based on conjugation and transduction and shown inside the front cover) demonstrates that it is much less detailed. This lack of detail can be attributed to two problems:

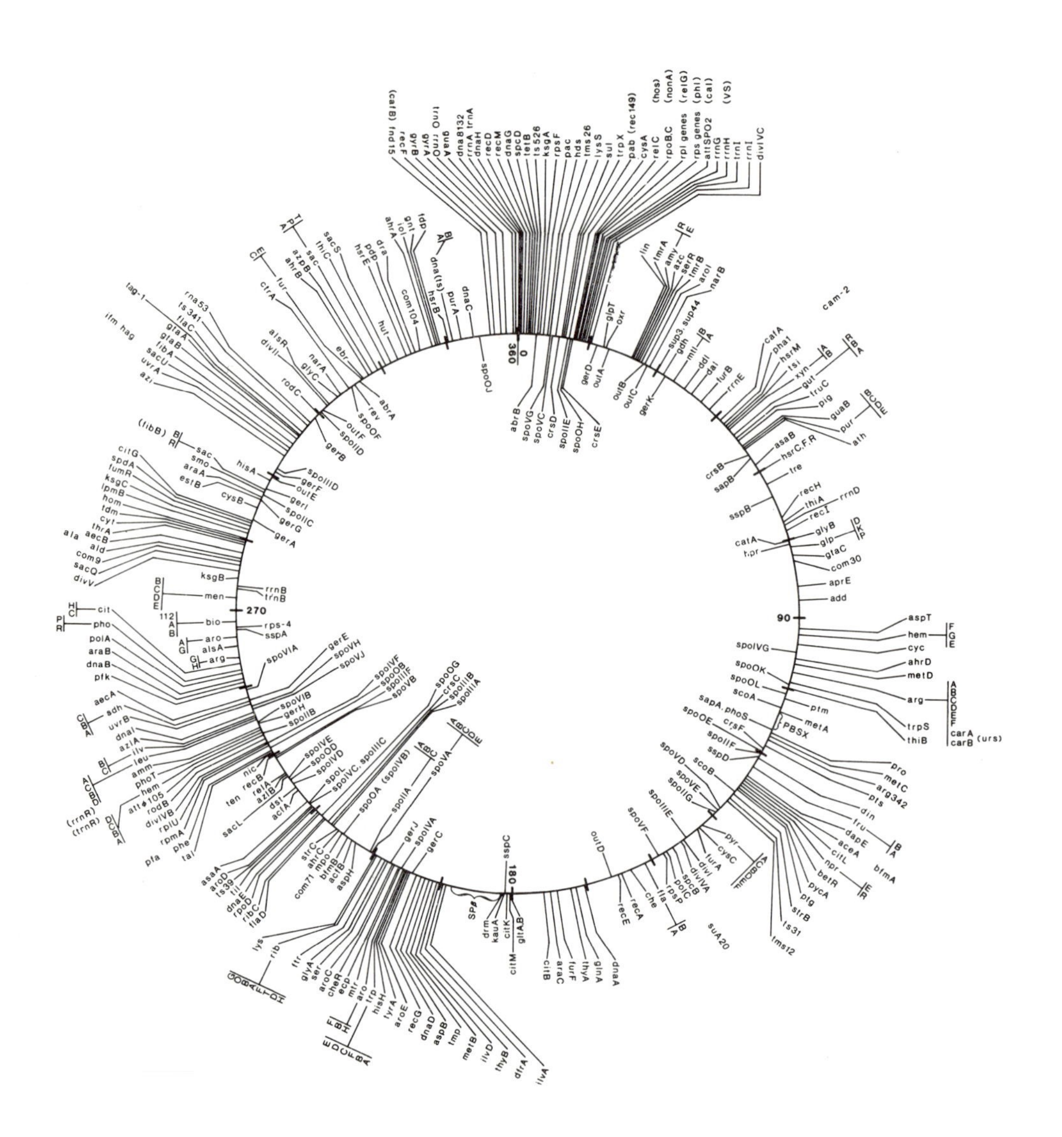

(1) fewer mutations have been isolated in the *B. subtilis* genetic system; and (2) only small regions of the genome can be transferred at one time by transformation or transduction, especially as compared to the average amount of DNA transferred by conjugation.

The largest *Bacillus* transducing phage, SP01, can transfer only some 10% of the genome at one time and using gently lysed protoplasts allows for transfer of only 2.3% of the genome. On the other hand, during conjugation *E. coli* can easily transfer 30% or even 50%. Therefore basic relations between widely separated markers are more difficult to establish in *Bacillus*. An indication of the difficulty is that until 1978 the total number of markers was not sufficient to permit every known marker to be cotransformed or cotransduced with at least two others. It was therefore impossible to establish the circularity of the genome prior to that time.

The *Bacillus* genetic map does not show any obvious similarities to genetic maps for other bacteria (compare with Figures 9-3, 10-1, 10-2, 10-3). It is interesting to note that the sporulation cistrons seem to be scattered about the genome, rather than being clustered in one area. The problem inherent in regulating such a system is considered in Chapter 12.

Summary

Genetic transformation was the first genetic exchange process to be observed in bacteria; despite its comparative antiquity, however, the molecular details are only now being clarified. The process requires that the recipient cells be in a state of competence for the binding and transport of double-strand DNA. In most organisms competence is regulated in some fashion, usually by growth rate, so that stationary phase cells tend to be competent and logarithmically growing cells are incompetent. Competence is often associated with the appearance of poly-β-hydroxybutyrate in the cell membrane. Competence-specific proteins also appear on the cell surface. In some cases these proteins can be purified and used to induce competence in other cells.

DNA bound to competent cells is processed by nucleolytic attack to yield the shorter double-strand fragments that penetrate the cell. Entry of the DNA seems to be linear and may involve either a single strand (*Bacillus, Streptococcus*) or a double strand (*Hemophilus*). The DNA appears directly in the cytoplasm in the case of gram-positive bacteria, but

Figure 8-6. Genetic map of the *Bacillus subtilis* chromosome. The outer circle presents the complete genome, and the inner circle includes only those functions affecting sporulation. Many of the genotype abbreviations used in this map are presented as part of Table 2-2. Additional information can be found in the original reference. From Piggott and Hoch (1985).

in vesicular transformasomes in the case of *Hemophilus*. Normal exit of DNA from a transformasome is as a linear molecule. In all cases recombination is necessary for completion of genetic transformation, and only a single donor strand is inserted into the recipient chromosome.

Genetic transformation by plasmid DNA presents some difficulties. *E. coli* is unusual in allowing high-efficiency plasmid transformation, for in most organisms the processing of the incoming DNA results in formation of a linear molecule that must be recircularized before it can self-replicate. The recircularization is fundamentally a recombination process and is facilitated by the presence of plasmid concatemers, multiple transforming DNA molecules, or homology to DNA sequences already present in the host.

In most cases the recipient cells exhibit little selectivity with respect to the origin of the DNA they take up. *E. coli*, for example, can be transformed by a wide variety of plasmids as well as by its own DNA. The exception is *Hemophilus influenzae* whose DNA contains about 600 copies of an 11-bp sequence that is required for normal uptake of DNA. Those organisms lacking specificity of uptake compensate for it by a stringent recombination process once the DNA arrives in the cytoplasm. Nonhomologous DNA is rapidly degraded.

All genetically transformable organisms also undergo transfection in which purified phage DNA molecules are transported into the cell and normal infection is established. Transfection systems have proved to be particularly advantageous for analyzing the biochemistry of various transformation, recombination, or repair processes, as they use such a homogeneous population of molecules.

References

General

McCarty, M. (1985). The Transforming Principle: Discovering that Genes Are Made of DNA. New York: Norton.

Piggott, P.J., Hoch, J.A. (1985). Revised genetic linkage map of *Bacillus subtilis*. Microbiological Reviews 49:158–179.

Stewart, G.J., Carlson, C.A. (1986). The biology of natural transformation. Annual Review of Microbiology 40:211–235.

Specialized

Aardema, B.W., Lorenz, M.G., Krumbein, W.E. (1983). Protection of sediment-adsorbed transforming DNA against enzymatic inactivation. Applied and Environmental Microbiology 46:417–420.

Akamatsu, T., Sekiguchi, J. (1987). Characterization of chromosome and plasmid transformation in *Bacillus subtilis* using gently lysed protoplasts. Archives of Microbiology 146:353–357.

Barouki, R., Smith, H.O. (1986). Initial steps in *Haemophilus influenzae* transformation: donor DNA binding in the *com*10 mutant. The Journal of Biological Chemistry 261:8617–8623.

Biswas, G.D., Burnstein, K.L., Sparling, P.F. (1986). Linearization of donor DNA during plasmid transformation in *Neisseria gonorrhoeae*. Journal of Bacteriology 168:756–761.

Conley, E.C., Saunders, V.A., Jackson, V., Saunders, J.R. (1986). Mechanism of intramolecular recyclization and deletion formation following transformation of *Escherichia coli* with linearized plasmid DNA. Nucleic Acids Research 14:8919–8931.

Conley, E.C., Saunders, V.A., Saunders, J.R. (1986). Deletion and rearrangement of plasmid DNA during transformation of *Escherichia coli* with linear plasmid molecules. Nucleic Acids Research 14:8905–8917.

Hanahan, D. (1983). Studies on transformation of *Escherichia coli* with plasmids. Journal of Molecular Biology 166:557–580.

McCarthy, D., Kupfer, D.M. (1987). Electron microscopy of single-stranded structures in the DNA of competent *Haemophilus influenzae* cells. Journal of Bacteriology 169:565–571.

Pifer, M.L. (1986). Plasmid establishment in competent *Haemophilus influenzae* occurs by illegitimate transformation. Journal of Bacteriology 168:683–687.

Reusch, R.N., Hiske, T.W., Sadoff, H.L. (1986). Poly-β-hydroxybutyrate membrane structure and its relationship to genetic transformability in *Escherichia coli*. Journal of Bacteriology 168:553–562.

Te Riele, H.P.J., Venema, G. (1984). Molecular fate of heterologous bacterial DNA in competent *Bacillus subtilis:* further characterization of unstable association between donor and recipient DNA and the involvement of the cellular membrane. Molecular and General Genetics 195:200–208.

Weinrauch, Y., Dubnau, D. (1987). Plasmid marker rescue transformation proceeds by breakage-reunion in *Bacillus subtilis*. Journal of Bacteriology 169:1205–1211.

Chapter 9
Conjugation in *Escherichia coli*

The discovery of the process of conjugation in prokaryotes was due to one of the most fortuitous experimental designs in recent scientific history. Many other scientists had tried to demonstrate conjugation and had failed. Nevertheless, Lederberg and Tatum, working at Yale University, chose to make another attempt using a common laboratory strain of bacteria, *Escherichia coli* K-12, as their experimental system. They also decided to test only a few isolates of the strain. By chance, they chose not only one of the few organisms that readily undergoes conjugation but also a fertile stain of that organism. In so doing they started an extremely fruitful branch of bacterial genetic research. *E. coli* remains the standard conjugation system to which all others are initially compared. In this chapter the *E. coli* system is presented. Several alternative conjugation systems are discussed in Chapter 10.

Basic Properties of the *E. coli* Conjugative System

Discovery of Conjugation

The experimental design used by Lederberg and Tatum was both simple and elegant. They realized that to observe a rare event it is necessary to have a powerful method of selection. Yet most biochemical mutations revert to wild type with a frequency of 10^{-6} to 10^{-8}, a rate that may well be greater than that of the desired event. In order to circumvent this problem, they decided to utilize strains that were multiply auxotrophic and to look for complete conversion to prototrophy. This design permitted the system to be as sensitive as necessary, if it was assumed that simultaneous transfer of all prototrophic markers was possible. For example, if a cell were doubly auxotrophic and each mutation reverted at a frequency of 10^{-7}, the probability of a double revertant, assuming independence of reversion events, would be 10^{-14}.

Pairwise tests of various *E. coli* strains gave results indicating that mixtures of some strains would produce prototrophs at a rate of 10^{-6} instead of the expected 10^{-14}. It soon became apparent that there were really two types of *E. coli* in the culture collection: those that were fertile when mixed with any other *E. coli* strain and those that were fertile only with certain strains. The former type was designated F^+ (possessing fertility) and the latter F^-. The various pairings gave the following results:

$F^+ \times F^+$ = prototrophs
$F^+ \times F^-$ = prototrophs
$F^- \times F^-$ = no prototrophs

The observed genetic transfer appeared to be a generalized phenomenon, as all types of auxotrophs tested tended to give rise to prototrophs at the same low frequency. Interestingly, the prototrophs were invariably F^+, but their frequency (10^{-6}) was substantially lower than the conversion of F^- cells to F^+ cells in the culture as a whole (70% in a 1-hour mating).

It was quickly realized that the F^+ quality was unique. Not only was it infectious, it was independent of the other genetic properties of the cell. It became customary to speak of the "F factor" as a discrete entity. In fact, the F factor was a plasmid and is now known as the F plasmid.

In order to demonstrate that the genetic transfer process involving F^+ cells was unlike any other, Davis performed an experiment using a U-shaped tube. He observed that if one auxotrophic strain of *E. coli* (either F^+ or F^-) was placed in one arm of the U and an auxotrophic F^+ strain was placed in the other, prototrophs were produced. However, if the bottom of the U was closed off by a microporous fritted glass filter that would allow soluble factors such as DNA molecules or viruses to pass through but that blocked the passage of intact cells, no prototrophs were observed. From these results, it was concluded that cell-to-cell contact was necessary for transfer. Lederberg and Tatum were inclined to interpret these results as demonstrating homothallic cell fusion (in contrast to the heterothallic fusion described for *Saccharomyces* in Figure 1-1), but their opinion was not universally accepted.

Discovery of Efficient Donor Strains and Partial Transfer

Hayes was one scientist who disagreed with the idea of homothallic cell fusion, and he decided to inquire about obvious differences in the roles played by the F^+ and F^- strains. He used two pairs of strains with each pair carrying identical auxotrophic markers, but one member of the pair was F^- and the other was F^+. Hayes mated the F^+ strain from one pair with the F^- strain from the other pair and selected for prototrophic recombinants. He then examined the unselected markers of the prototrophs for vitamin production, sugar utilization, and streptomycin resistance. The data presented in Table 9-1 indicate that 87 to 99% of the transconjugants carried *mal* and *rpsL* markers appropriate to the F^- cell. The conclusion to be drawn from the Hayes and Davis experiments is that the process is not one of cell fusion but of conjugation (direct transfer of DNA from cell to cell) and that one cell plays the role of a **donor** of genetic material and the other the role of a **recipient.** Transfer is therefore one way rather than reciprocal. By obvious analogy, donor cells are sometimes referred to as male cells and recipient cells as female cells.

Not much information could be gained about a system in which genetic transfer took place so inefficiently. Therefore a more thorough understanding of the conjugation process had to await the discovery of some method that would increase the frequency with which genetic exchange took place. Two donor strains of *E. coli,* representing a new sort of fertility, were independently isolated from F^+ cultures by Cavalli-Sforza and Hayes. These strains gave rise to prototropic transconjugants at a much higher frequency than did F^+ cultures, i.e., 10^{-1} instead of 10^{-6}. For this reason they were designated high-frequency-of-recombination strains, or **Hfr strains.** In honor of their discoverers, they are still referred to as HfrC and HFrH.

The type of genetic transfer carried out by Hfr cells appeared to be different from that carried out by F^+ cells. Although it was true that certain types of prototrophs arose with a high frequency, other types of prototrophs were produced rarely if at all. Moreover, the two Hfr strains differed with respect to the types of transconjugant that appeared at high frequency after direct selection (Table 9-2), and the transconjugants that did arise were nearly always F^-. These data provided further evidence that something more complex than simple cell fusion was occurring.

Table 9-1. Effect of reversal of F polarity on the genetic constitution of recombinants from otherwise similar crosses

Strain A = thr^+, leu^+, met^-, mal^+, $rpsL^-$
B = thr^-, leu^-, met^+, mal^-, $rpsL^+$
Select: Met^+, Thr^+, Leu^+
Examine: 300 colonies for unselected markers (*rpsL* and *mal*)

Strains mated		Percent of transconjugants		
F^+	F^-	Mal^+ $RpsL^-$	Mal^- $RpsL^+$	Other combinations
A	B	0	99	1
B	A	86.8	8.3	4.9

Adapted from Hayes, W. (1953). The mechanism of genetic recombination in *E. coli.* Cold Spring Harbor Symposia on Quantitative Biology 18:75–93.

Table 9-2. Results of Hfr matings using two donors[a]

Donor used	Transconjugant cells/ml		
	Arg^+ $RpsL^-$	Leu^+ $RpsL^-$	Lac^+ $RpsL^-$
Similar to HfrH	0	1.9×10^6	8.7×10^5
Similar to HfrC	3.2×10^3	2.8×10^5	2.4×10^6

[a]Recipient strain genotype = *leu, arg, lac, rpsL.*

Nature of the Transfer Process

The Hayes Hfr was soon utilized for an important series of experiments by Jacob and Wollman at the Pasteur Institute in Paris. They conjugated a multiply marked F^- strain to HfrH and then observed the distribution of the unselected markers. Data from a similar series of experiments are presented in Table 9-3. In cross 1 it can be seen that there is apparently a gradient of inheritance, or polarity gradient, of the unselected markers among the Leu^+ transconjugants. Some are coinherited with varying frequency, whereas others are never inherited. The data are arranged to emphasize this gradient.

However, the data from cross 2 show a strikingly different pattern. There are basically two classes of unselected markers: those that are coinherited with the selected marker frequently and those that are rarely or never coinherited. There seems to be little middle ground. The explanation for such results is not obvious.

One possible explanation, which occurred to Jacob and Wollman, was that transfer might be not only unidirectional but also linear. In such a case, if the DNA were transferred from the donor such that the selected marker arrived early **(proximal marker),** the unselected markers need not be inherited by the recipient cell. However, if the selected marker were transferred late **(distal marker),** then of necessary the recipient cell had to receive all of the proximal markers as well (Fig. 9-1).

Such an interpretation also suggests, but does not prove, the possibility

Table 9-3. Gradient of transfer in Hfr × F^- cross

Cross no.	Selected marker	Transconjugants/ ml	Percent of transconjugants inheriting the Hfr allele for					
			Leu	Pro	Lac	Trp	His	Thy
1	Leu^+	3.5×10^6	100	42	32	1	0	0
2	Trp^+	4.1×10^4	71	60	52	100	7	0

Matings were carried out at 37°C using a 10:1 ratio of F^- to Hfr cells. After 40 minutes, aliquots of the culture were placed on agar selective for the indicated markers. When the transconjugant colonies had grown, they were tested for the unselected markers by replica plating.

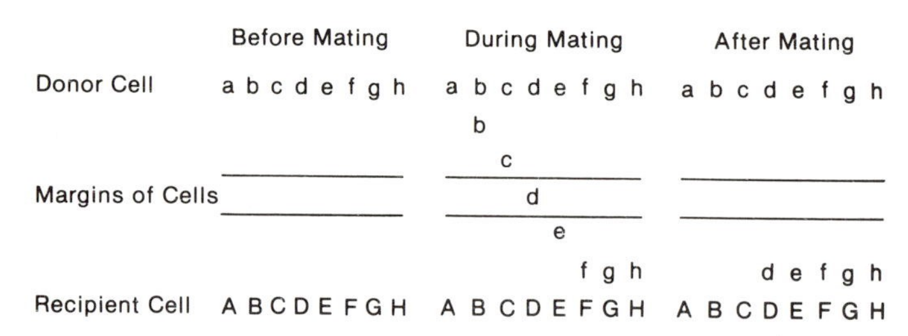

Figure 9-1. Hypothesis of unidirectional transfer. The double horizontal lines represent the boundaries between the donor and recipient cells and are the points at which the cells separate. Each genome is represented in linear form. It is assumed that transfer is preceded by replication so the donor cell is not depleted of DNA. Note that transfer of *e* is impossible unless *f, g,* and *h* precede it. After mating, the recipient cell is shown as a merodiploid. It is assumed that any untransferred DNA in the donor cell is lost.

that only part of the donor chromosome is transferred to the recipient. One way to ascertain if partial transfer does occur is to take advantage of the phenomenon of zygotic induction (see Chapter 6). Because induction of the transferred prophage was postulated to be due to the lack of the λ repressor in the F^- cytoplasm, any transfer of the prophage should result in a phage infection. According to the hypothesis of partial transfer, zygotic induction should be observed only when the transferred fragment includes the prophage. Data from an experiment demonstrating the proof of this proposition are given in Table 9-4.

Crosses 1 and 2 represent positive controls showing that the presence of a λ prophage in both conjugating strains does not influence the distribution of the unselected markers regardless of marker selection. Crosses 3 and 4 are additional controls designed to show that two DNA molecules can recombine essentially normally even though one molecule lacks a segment of DNA that is present in the other. In this case the deficiency is the lack of prophage in the Hfr strain, and inheritance of the Hfr DNA

Table 9-4. Effect of prophage transfer on unselected marker inheritance

Cross	λ Prophage present in		Selected	Transcon-	Percent of transconjugants inheriting the Hfr allele for			
no.	Hfr	F^-	marker	jugants/ml	Leu	Lac	Gal	Bio
1	Yes	Yes	Leu^+	7.1×10^5	100	29	18	13
2	Yes	Yes	Gal^+	8.4×10^4	25	27	100	93
3	No	Yes	Leu^+	7.3×10^5	100	19	11	9
4	No	Yes	Gal^+	9.3×10^4	9	13	100	99
5	Yes	No	Leu^+	1.1×10^4	100	20	5	1
6	Yes	No	Gal^+	1.2×10^3	36	30	100	75

Mating conditions were as described for Table 9-3.

means loss of the prophage DNA. Induction does not occur in this case, as the λ repressor is present at all times until the prophage DNA is degraded or lost by cell division.

The crucial test of the hypothesis of partial transfer is made in crosses 5 and 6 in which zygotic induction is possible. The data show that the total number of transconjugants recovered is dramatically reduced. When selection is made for Leu^+, there is a specific loss of Gal^+ recombinants. This result is expected from the hypothesis because of the location of the λ prophage next to *gal*. Cells that would have become Gal^+ transconjugants usually also acquire the λ prophage and lyse. Note that some Gal^+ recombinants are recovered and that they are generally Bio^-, thereby confirming that the λ prophage is located between *gal* and *bio*. The occasional inheritance of *bio* can happen if the transferred prophage manages to produce sufficient repressor to turn itself off prior to cell lysis. The same phenomenon, of course, is observed during normal phage infection of a cell. Note also that direct selection for Gal^+ requires that the prophage be successfully turned off in order to obtain the transconjugants in the first place. Therefore the unselected marker frequencies are essentially normal, although the number of transconjugants recovered is low.

The partial transfer of donor DNA might result from mechanical forces in the culture disrupting the mating aggregates. However, if the mixture of cells to be mated is immobilized on membrane filters, the polarity gradient seen in Table 9-3 is still observed. Therefore the property of partial genetic transfer is inherent in the conjugation system itself and is not a function of culture agitation, etc.

The conclusions drawn from the data of Tables 9-3 and 9-4 lead to a model for conjugation in which transfer of DNA begins at a specific point on the genome of the Hfr cell and proceeds in a linear fashion into the F^- cell. Apparently, this transfer is spontaneously and randomly stopped, or interrupted, in some fashion during the normal course of events. Jacob and Wollman reasoned that if this model were correct what could be done by nature could also be done artificially. They developed an experimental design for an **interrupted mating,** in which the conjugating bacterial cells would be broken apart and further transfer prevented by dilution or immobilization of the cells in an agar layer. By taking samples from a mating culture at various times and disrupting the cells, it should be possible to observe a specific **time of entry** for any particular genetic marker.

A variety of methods for interruption were tried by Jacob and Wollman, as well as by subsequent workers. The most successful have been the following: (1) lysis from without using T6 phage and sensitive Hfr cells but resistant F^- cells in a manner analogous to premature lysis experiments; (2) a blending device that applies intense hydrodynamic shear to the mating bacteria causing them to separate and breaking any DNA connecting them; (3) the chemical nalidixic acid, which instantaneously stops transfer by stopping DNA replication. Because DNA replication is required

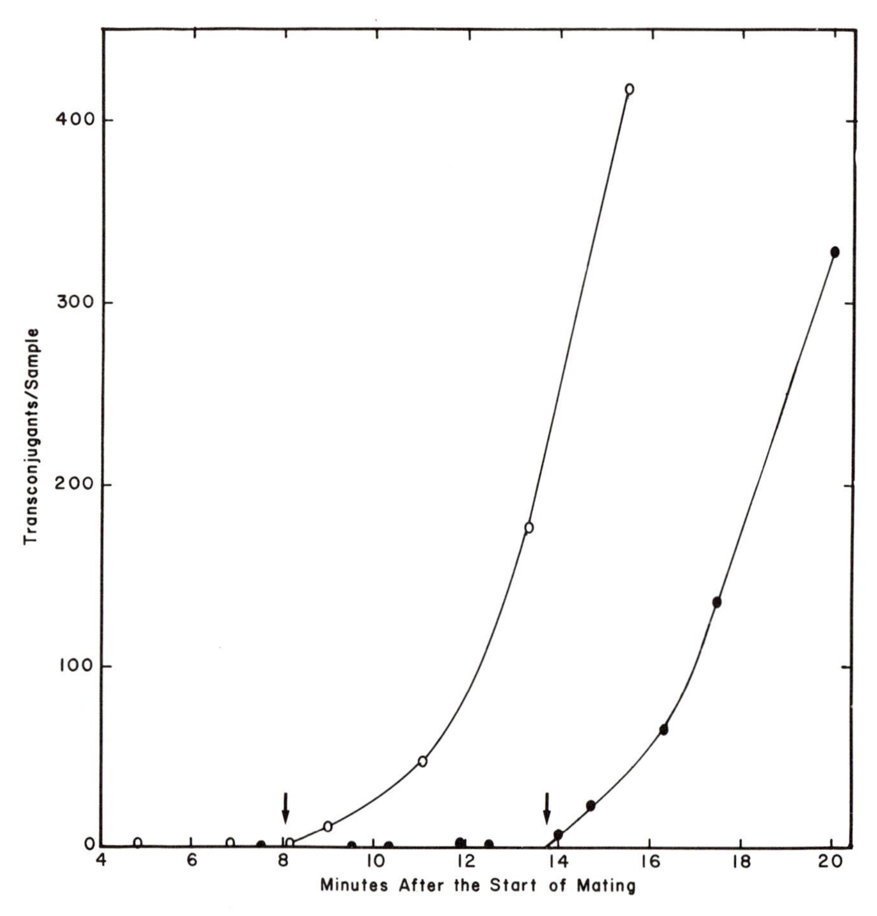

Figure 9-2. Typical interrupted mating curves. The Hfr (donor) and F^- (recipient) cells were mixed at time zero in a ratio of 1:100. The mixed culture was incubated at 37°C. Samples of 0.05 ml were removed at various times, blended to break apart the mating aggregates, and then plated on selective medium. The open circles are Leu^+ transconjugants, and the closed circles are $ProA^+$ transconjugants. The arrows indicate the extrapolated times of entry.

for transfer (see below), when replication halts, so does DNA transfer. Regardless of the method of interruption used, the same type of graph is obtained; a typical one is shown in Figure 9-2. It can easily be seen that each type of transconjugant has a specific time of entry, which is obtained by extrapolation of the early time points back to zero and that can be expressed in terms of minutes elapsed since the Hfr and F^- cells were mixed together. This confirmation of the prediction of the model was the final piece of evidence in the development of a basic theory of conjugation. By means of such an experiment, it is also possible to verify that *gal* transfers before λ prophage, the time differential being about 0.5 minutes.

Construction of the First Genetic Map

Interrupted mating is a powerful tool for the unambiguous determination of gene order, so an obvious next step was to use the data obtained from such experiments to construct a genetic map. Most genetic maps have as their unit of distance some function of the number of recombinants observed in various crosses. However, the interrupted mating experiments permitted easy determination of times of entry for different markers. Therefore the basic unit of the *E. coli* genetic map became the minute, the amount of DNA transferred by an average Hfr cell in 1 minute. For example, Figure 9-2 indicates that *proA* and *leu* are about 5 minutes apart on the genetic map. Note, however, that there is no inherent directionality to the genetic map; i.e., there is no structure equivalent to the centromere in eukaryotic cells, only the Hfr origin itself.

However, a particular Hfr strain does not transfer all markers efficiently (Table 9-2), and therefore other Hfr strains would have to be found in order to complete the genetic map. F^+ cultures were systematically tested for the presence of Hfr cells and even treated with ultraviolet radiation (UV) (which enhances DNA recombination) to see if new types of Hfr strains could be obtained. It soon became apparent that the number of types of Hfr strains was not only finite but actually small. Approximately 22 discrete types of naturally occurring Hfr strains have been observed.

After adoption of a few conventions, the bulk of the data obtained could be presented in a simplified genetic map such as that shown in Figure 9-3. The necessary conventions are the following:

1. After taking into consideration the overlapping regions of DNA transferred by various Hfr strains, the length of the genetic map is taken to be 100 minutes (early estimates were 90 minutes).
2. An arbitrary zero point is picked to lie at *thr*.
3. The direction of transfer of HfrH is considered to be clockwise and is represented by an arrowhead pointing in a counterclockwise direction.

Figure 9-3 presents a simplified *E. coli* K-12 map together with the points of origin and directions of transfer of some commonly used Hfr strains. Note that, in confirmation of the physical studies described in Chapter 1, the genetic map is circular rather than linear. This finding was a surprise at the time, although circular maps have since been found to be the rule rather than the exception among prokaryotes. In order to enter the data obtained from P1 transduction it is necessary to use a conversion factor. Bachmann assumed that Pl carries a piece of DNA equivalent to 2 minutes on the genetic map and used the formula of Wu (Eq. 7-1). The complete *E. coli* genetic map is given inside the front cover.

Note the correlation between the map positions determined by interrupted matings and the gradient of transfer seen in Table 9-3. It is consistent with the model of partial unidirectional transfer. If the distance between the various markers is accurately known and is not too great, it is possible

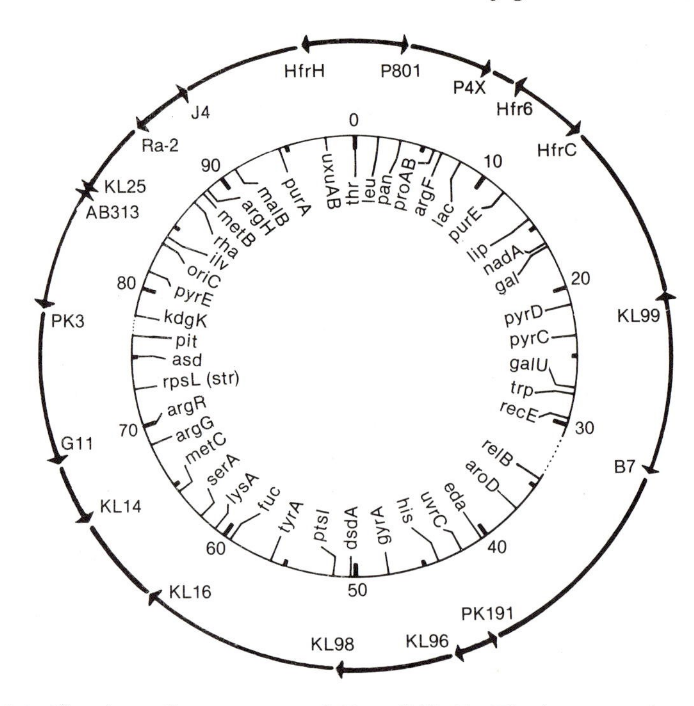

Figure 9-3. Circular reference map of *E. coli* K-12. The large numbers refer to map position in minutes, relative to the *thr* locus. From the complete linkage map presented inside the front cover, 43 loci were chosen on the basis of greatest accuracy of map location, utility in further mapping studies, or familiarity as long-standing landmarks of the *E. coli* K-12 genetic map. Outside the circle, the orientations and leading transfer regions of a number of Hfr strains are indicated. The tip of the arrow marks the first DNA transferred in each case. Adapted from Bachmann, B.J., Low, K.B. (1980). Linkage map of *E. coli* K-12, edition 6. Microbiological Reviews 44:1–56.

to determine the probability of a spontaneous interruption in the DNA transfer per minute of DNA transferred. In this particular case it works out to about 12% of all DNA molecules being interrupted for each minute of DNA passed along to the recipient cell.

Some of the same genetic oddities seen in viral genetic maps have also been detected in bacterial genetic maps. There are several examples of overlapping cistrons known, usually of the type where the overlap is in the same reading frame. An example is the *cheA* cistron that codes for two chemotaxis proteins. The *dnaQ* and RNase H cistrons do not directly overlap but are transcribed from divergent, overlapping promoters.

A genetic map should not be thought of as a static entity. Rather, it is constantly being changed by mutation and evolutionary processes. Sometimes these changes can be dramatic. Two variant strains of *E. coli* K-12 carrying substantial inversions have been identified. W3110 and its descendants carry an inversion of the DNA lying between *rrnD* and *rrnE*, whereas 1485IN carries an inversion of 35% of the chromosome (roughly *gal* to *his*).

Interactions of the F Plasmid with the Bacterial Chromosome

The F plasmid can be isolated as an independent entity from F^+ cells but not from Hfr cells. Nevertheless, blotting and heteroduplex electron microscopy experiments clearly demonstrate the presence of F plasmid sequences in both types of cell. The difference between the two types of cell is that in Hfr cells the plasmid is integrated into the bacterial DNA in a manner analogous to a λ prophage. Any plasmid that can exist in either the autonomous or integrated state is said to be an **episome.**

Even when replicating autonomously, the F plasmid is nonetheless in close spatial association with the host cell chromosome. As was noted in Chapter 1, the bacterial chromosome exists in a series of giant supercoiled loops. The covalently closed, circular F DNA is also supercoiled and resembles one of the chromosomal loops. When F^+ *E. coli* cells are gently lysed with a detergent and the DNA examined, the F plasmid is generally found to be entangled in the fibers of the extracted bacterial chromosome. It has been shown that this association between the F supercoil and the bacterial chromosome is not strand-specific. The association is presumably useful to F because such an arrangement would automatically segregate F DNA to the daughter cells along with the host cell chromosomal DNA.

Integration of the F Plasmid

The interaction of the F DNA with the bacterial DNA to effect integration is similar to the situation that occurs when λ phage integrates itself. In both cases circular molecules are formed subsequent to entry into the cell. Therefore the Campbell model for integration of λ should function equally well for F integration. The subsequent discussion is based on this assumption.

The Campbell model also predicts that F^+ and Hfr cells are in equilibrium and that aberrant excisions of the F plasmid from the bacterial chromosome are occasionally seen. In λ there are size constraints on DNA that is incorporated into specialized transducing particles. However, the F plasmid does not have to be packaged, and so the size of the bacterial DNA excised along with the F plasmid can range from a fraction of a minute up to 10 to 15% of the genome. However, because larger plasmids have a greater retarding effect on cell growth, there is a strong natural selection for shorter F plasmids within a culture carrying a large plasmid. Low found that this shortening was due to recombination events, and that it could be prevented or greatly reduced by using *recA* strains as hosts for F plasmids carrying bacterial DNA.

To distinguish an F plasmid carrying bacterial DNA sequences from a normal F plasmid, the former is designated **F-prime** (F′). F′ molecules are usually stable but can be isolated in various sizes and may carry any portion of the bacterial DNA (Fig. 9-4). They can also exist in the integrated form but are difficult to maintain and distinguish from normal Hfr cells. Each unique F′ molecule is designated either by a number or the cistron symbol for the bacterial markers it carries. Thus F42 is also referred to as F′ *lac*.

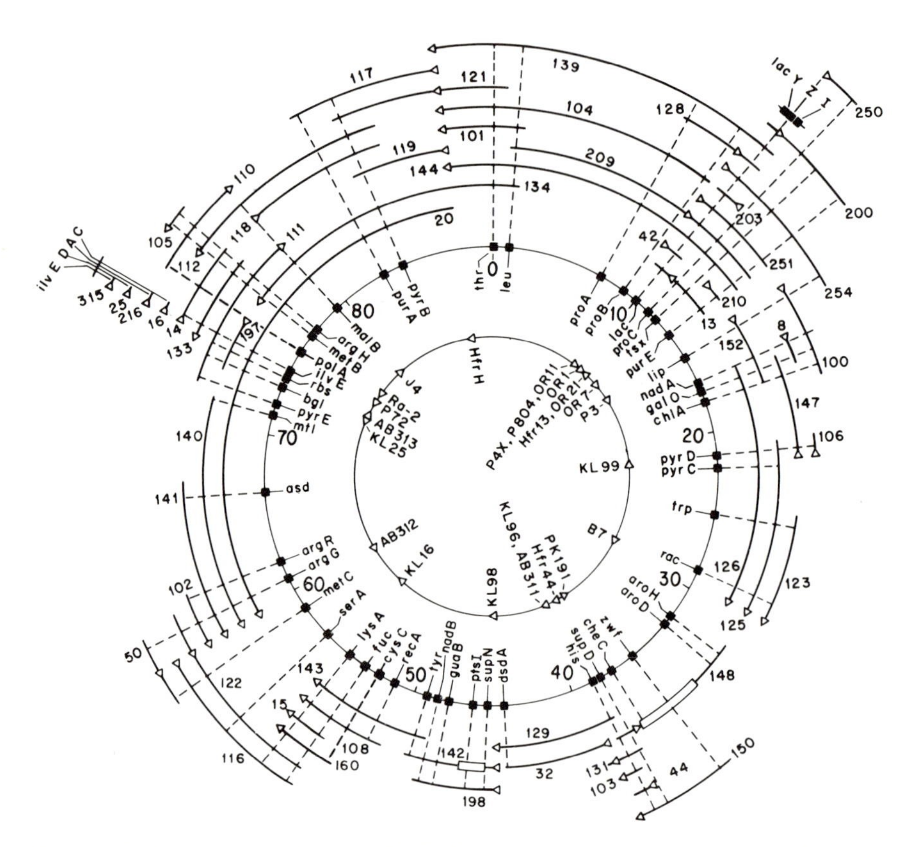

Figure 9-4. Genetic map of *E. coli* K-12 showing approximate chromosomal regions carried by selected F′ plasmids. Each F′ is represented by an arc that has an arrowhead drawn to show the point of origin of the ancestral Hfr strain (see inner circle). The dashed lines, which extend radially from the genetic markers on the outer circle, indicate the approximate termini of the F′ plasmids as far as they are known. Those deletions that are known to be present are indicated by narrow rectangles (e.g., F142 is deleted for *ptsI*). From Low, K.B. (1972). *E. coli* K-12 F-prime factors, old and new. Bacteriological Reviews 36:587–607.

In addition to the naturally occurring Hfr strains, a number of Hfr strains have been produced by essentially the same process of directed transposition as that used by Beckwith for phage ϕ80. The Hfr process was also developed by Beckwith and co-workers and involves taking an F′ plasmid that carries a temperature-sensitive mutation for DNA replication and putting it into an F^- strain that carries a deletion for the genetic material carried on the F′. When the temperature of the culture is raised, the plasmid stops replicating. However, if selection is applied to the culture for the cistron carried on the F′ plasmids, the bacteria need to avoid the loss of the F′ DNA if they are to survive. They can do it by **integrative suppression,** which is integrating the F′ into the bacterial genome where it can be replicated by bacterial enzymes. The integration would normally occur in the bacterial region corresponding to the bacterial DNA carried by the F′. However, because this material has been deleted, there is no extensive region of homology between the F DNA and the bacterial DNA, and the integration occurs more or less at random. If the integration occurs in the middle of a bacterial cistron, the cistron is inactivated. Therefore the site of F integration can be "directed" by an appropriate selection that incidentally results in transposition of the bacterial DNA carried by the F′ from its normal location to a new site on the genome.

Although it seems that the F plasmid should establish the same sort of relationship with the host cell each time it integrates itself, it does not seem to be the case. Some Hfr strains, e.g., HfrC, are extremely stable. Only rarely does the F plasmid manage to excise itself and convert a cell from Hfr to F^+. Most Hfr strains are somewhat less stable, segregating F^+ cells at a perceptible but not inconvenient rate. However, certain Hfr strains are extremely unstable, segregating F^+ cells at a high frequency. In practice, these unstable Hfr strains must be purified by isolation streaking every few weeks, or they come to consist almost entirely of F^+ cells. Therefore when thinking of Campbell's model as it applies to F, the process of integration should be looked on as a dynamic one in which the equilibrium state may lie anywhere along the continuum from Hfr to F^+ cell.

All the observations discussed thus far suggest that, unlike the situation for λ, there are multiple specific sites on the bacterial genome at which F can insert. Originally these sites were called *sfa* (sex factor affinity) and were considered to be unique. However, electron microscopic studies (see below) have indicated that the *sfa* sites are merely examples of special discrete DNA sequences known as insertion sequences (IS), which are found on both the F plasmid and the bacterial DNA. Integration can therefore be considered a normal recombination event between homologous DNA regions. However, no F plasmid enzyme system analogous to the λ *int* protein has been identified. Instead, the normal cellular recombination system seems to catalyze most of the integration events. Thus in a *recA* strain, which lacks normal recombination ability, the amount of F integration is reduced 100- to 10,000-fold.

Excision of the F Plasmid

As noted above, it is possible for an integrated F plasmid to excise itself and convert an Hfr cell to an F^+ cell. Campbell's model also accounts for this process as it did for λ excision. There are, however, major differences between F excision and λ excision. Because F is not a virus, the excision event does not affect the host cell significantly. It does not result in cell lysis, for example, and F pili continue to be produced.

Because there is no size limit to F′ DNA, there is also no restriction as to the location of the aberrant excision event. Scaife proposed that F′ plasmids be categorized according to the original position of the bacterial DNA they carry (Fig. 9-5). Type IA F′ plasmids carry bacterial DNA sequences located near the origin of the ancestral Hfr, whereas type IB plasmids carry bacterial DNA sequences located near the terminus of the ancestral Hfr. Type II plasmids carry both proximal and distal bacterial

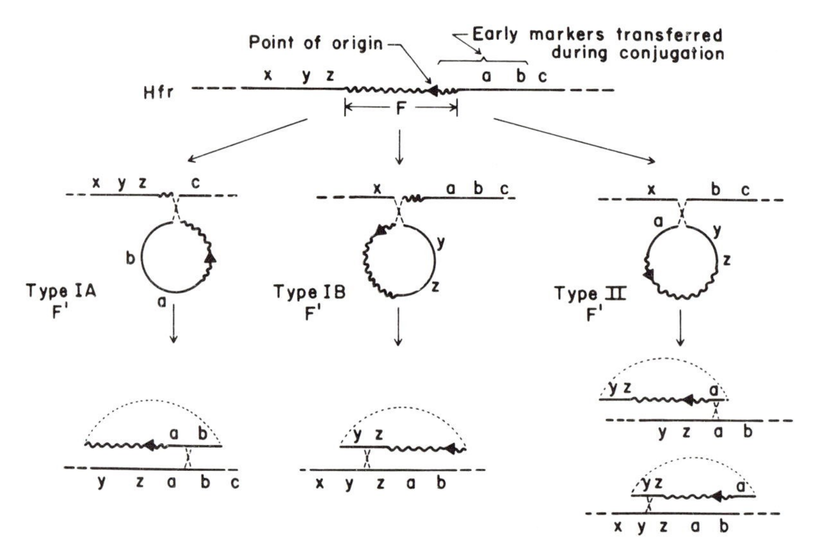

Figure 9-5. Variations in the topology of F′ formation and chromosome mobilization. The line in the top portion of the figure represents part of the chromosome of a hypothetical Hfr strain that transfers the genetic markers *a*, *b*, and *c* early and *x*, *y*, and *z* late in conjugation. The middle part of the figure indicates the relative orientation of the F plasmid and chromosomal markers during the formation of the three types of F′ plasmids shown. The bottom portion of the figure indicates regions of homologous pairing and crossover with the chromosome when the various types of F′ plasmids are in secondary F′ strains (i.e., transferred into new F^- hosts). The dotted lines symbolize the circularity of the F′. From Low (1972).

sequences and are the most prevalent type of F′ isolated, although the relative amounts of proximal versus distal bacterial DNA may be disproportionate.

Physiology of Conjugation

Using the observations outlined above, it is now possible to consider some of the information that has been obtained about the conjugation process. A good place to begin is by considering whether there are donor cells other than F^+, F′, or Hfr. Given that Hfr cells occur spontaneously within a population of F^+ cells, the question is whether such cells can account for all of the observed transfer, as suggested by Jacob and Wollman. Reeves performed a reconstruction experiment in which he added known amounts of Hfr cells to F^+ cultures and measured the amount of genetic transfer that occurred. From these data and making a few assumptions about the frequency of transfer of certain markers by spontaneously arising Hfr strains, he was able to predict how many Hfr cells would have to be found in an F^+ culture to provide the observed level of transfer. His predicted percentage of Hfr cells turned out to be higher than what he observed, from which he concluded that some other type of donor cell was possible. Subsequent studies have never found such a cell, and therefore it is presumed that some of the assumptions made by Reeves were incorrect and that all transfer of the bacterial chromosome results from Hfr or F′ cell types.

Formation of Mating Aggregates

The Hfr cells form aggregates with the F^- cells that are stabilized by the *ompA* outer membrane protein. Walmsley used an electronic particle counter to show that, as mating aggregates form after the Hfr and F^- cultures are mixed, there is a net decrease in the total number of particles counted. Blending the cultures breaks apart the aggregates and restores the original number of particles. Achtman and co-workers have shown that mating aggregates with Hfr cells are more stable than those with F^+ or F′ cells. They proposed that the difference in stability may be due to the amount of DNA that can be transferred.

At one time it was assumed that the mating aggregates were formed on a 1:1 basis, but this assumption is not valid. Achtman and co-workers used a particle counter to demonstrate that mating aggregates usually consist of either 2 to 4 cells or 8 to 13 cells. Such an analysis, however, does not indicate the relative proportion of Hfr and F^- cells within the aggregate. Conjugation experiments are customarily performed with an Hfr/F^- ratio of approximately 1:10. Skurray and Reeves have shown that when the ratio is reversed the F^- cells suffer severe membrane damage and die, a phenomenon they called **lethal zygosis.** An unexplained and extremely puzzling observation is that the phenomenon is seen only with Hfr cells. All other types of donor are harmless, and even Hfr cells cause no damage if the culture is not aerated during the mating. Recombination-deficient cells seem to be particularly sensitive to lethal zygosis, even in nonaerated cultures.

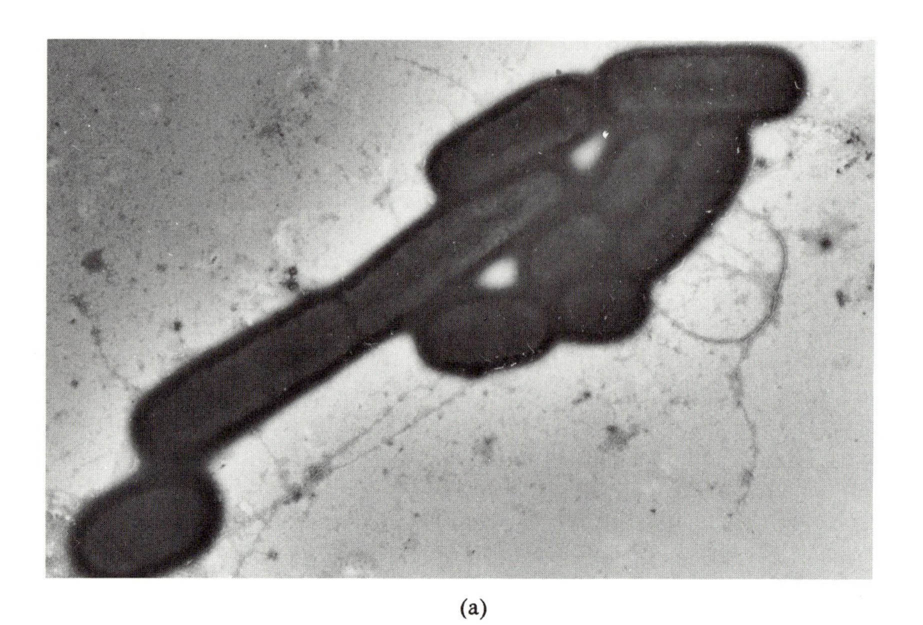

(a)

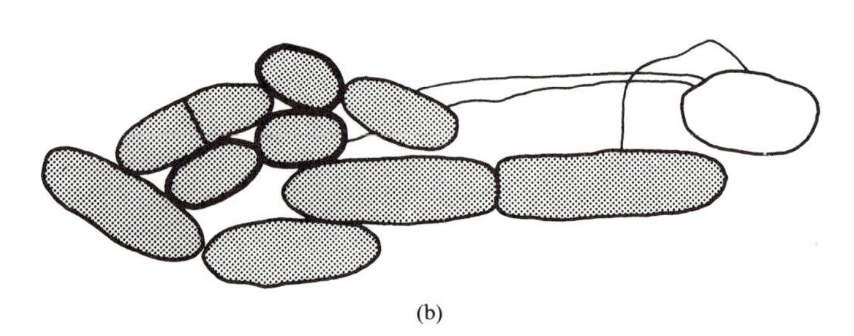

(b)

Figure 9-6. Hfr $\times$ F^- mating aggregates. **(a)** Formaldehyde-fixed cells were negatively stained with phosphotungstic acid and examined in the electron microscope. The cells are differentiated on the basis of size, with the longer cells being Hfr cells. **(b)** Interpretation of the aggregate. The three thick-walled cells were classified as F^-, the unshaded cell was unclassifiable, and the thin-walled cells were classified as Hfr. The long, thin structures connecting various cells are F pili. From Achtman, M., Morelli, G., Schwuchow, S. (1978). Cell–cell interactions in conjugating *E. coli:* role of F pili and fate of mating aggregates. Journal of Bacteriology 135:1053–1061.

Mating aggregate formation depends on the presence of donor-specific pili, or **F pili,** which are long filamentous structures visible on electron micrographs of the cell surface (Fig. 9-6). They have a vague similarity to eukaryotic flagella but are not used for locomotion and have a different biochemical structure. The F pilus protein is encoded on the F plasmid DNA, and the pili are produced in the quantity of one to three per cell whenever F^+, F', or Hfr cells are grown at temperatures above 33°C. The pili disappear when the cells are grown at lower temperatures, enter the stationary phase of growth, or are cured of the F plasmid. Donor cells that lack F pili behave like recipient cells during conjugation experiments and therefore are referred to as **F^- phenocopies.** Other surface structures (fimbriae) may be present at all times. The simplest assay system to detect the presence of F pili involves the use of certain male-specific phages, e.g., f1, MS2, or M13, that use the pili as attachment sites for initiating infections.

It is the tips of the F pili that are important for conjugation. Ou has demonstrated that MS2 phage particles, which can be shown by electron microscopy to bind to the sides of the F pili, have little effect on mating aggregate formation, whereas f1 phage particles, which bind at the tip of the pili, rapidly prevent mating aggregate formation. Electron micrographs such as that shown in Figure 9-6 seem to show a connection between cells via the F pilus, and it has been proposed that the function of the pili may be to overcome the mutual repulsion of the negative charge normally found on the cell surface. Although it has been suggested that DNA might be transferred through the F pilus, Achtman and co-workers demonstrated that using the detergent sodium dodecyl sulfate (SDS) to disaggregate the F pili after mating aggregate formation does not necessarily disrupt DNA transfer. Figure 9-7 presents a model in which the F pilus brings the cells together so membrane fusion can occur, and DNA is directly transferred from one cytoplasm to the other.

Transfer DNA Synthesis

In order for DNA transfer to take place, it is necessary that recipient as well as donor cells be active metabolically. Starvation of recipient cells or uncoupling of their oxidative phosphorylation results in failure of DNA transfer. The metabolic necessity is, among other things, for a signal of unknown nature to be transmitted from the recipient to the donor cell, which initiates a special round of **transfer DNA synthesis.** The only long-term biosynthetic activity in the donor that is essential for transfer is this DNA synthesis, as streptomycin-sensitive donor cells replicated onto streptomycin-containing agar still form mating aggregates and transfer some of their DNA.

Transfer DNA synthesis is a special process that has been analyzed by Sarathy and Siddiqi (Fig. 9-8). They used density shift experiments similar

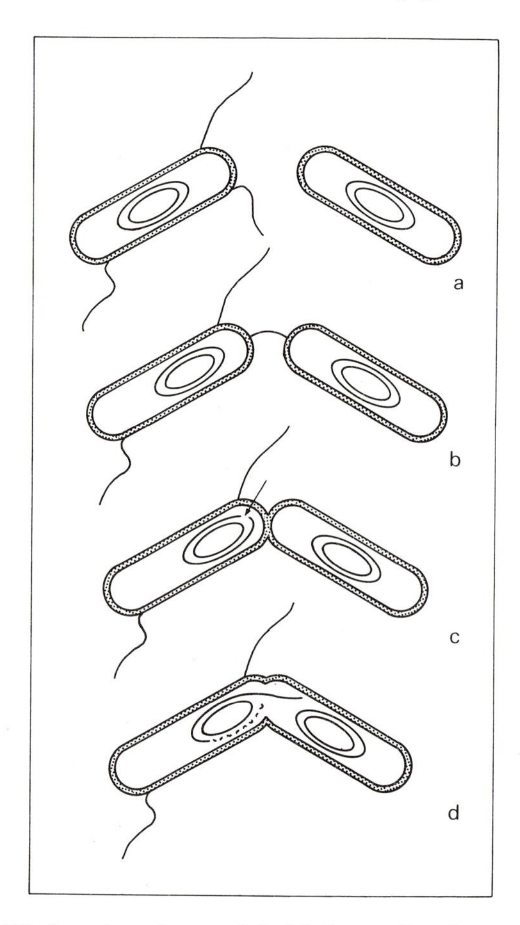

Figure 9-7. Simplified conjugation model. **(a)** Two cells of opposite mating type are shown. The donor cell carries specific sex pili. **(b)** Initial mating contact is shown. The F pilus plays an essential role. **(c)** A stable mating pair. The cells are tightly connected; the F plasmid DNA has been nicked; and conjugative replication is beginning. **(d)** Transfer of donor DNA into the recipient cell is occurring. New DNA is being synthesized (dotted line), and old DNA is being transferred to the recipient cell. Note that although large areas of the cytoplasm are shown as continuous, it cannot be the case because mating interruption by blending does not seem to harm the cells. From Hoekstra, W.P.M., Havekes, A.M. (1979). On the role of the recipient cell during conjugation in *E. coli*. Antonie van Leeuwenhoek Journal of Microbiology 45:13–18.

to those of Meselson and Stahl that established semiconservative DNA replication to demonstrate that when transfer DNA synthesis is initiated the normal replicating forks stop and a new round of replication begins. The new synthesis results in the premature appearance of light-light DNA (DNA that carries no density label in either strand) but at a rate that is one-half normal. No such effect is seen if two donor cell cultures are

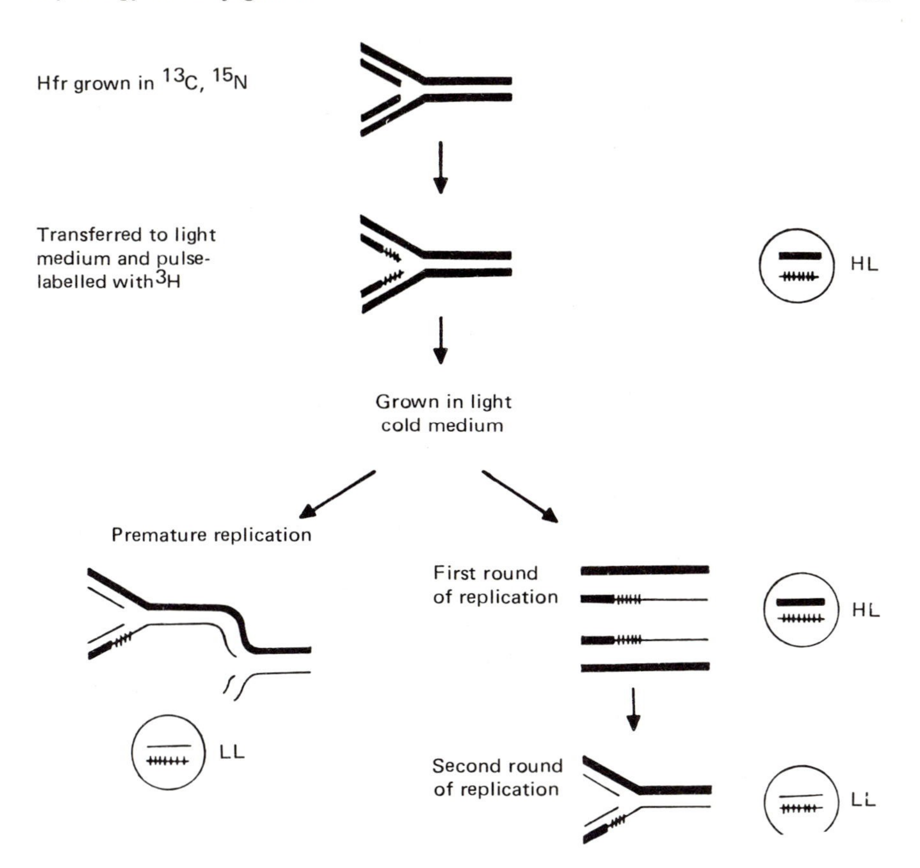

Figure 9-8. Replication of a segment of DNA pulse-labeled with ^{3}H. The experiment is a variation on the classic experiment of Meselson and Stahl. The DNA is initially labeled with heavy isotopes of carbon and nitrogen (thick lines) and then shifted to a medium containing the normal light isotopes (thin lines). Each line represents a single strand of DNA. At the time of the shift to light medium, ^{3}H-thymidine is added to the medium for a brief period and incorporated into the DNA (short vertical lines). The density of the resulting ^{3}H-labeled DNA is shown in the upper inset. The right-hand side of the figure shows the normal course of events, whereas the left-hand side shows the effects of a premature initiation of DNA replication. The density of the ^{3}H-labeled DNA at each step is shown by the insets. Note that premature initiation results in premature conversion of the ^{3}H-labeled DNA to the LL density. From Sarathy, P.V., Siddiqi, O. (1973). DNA synthesis during bacterial conjugation. Journal of Molecular Biology 78:427–441.

mixed. If DNA synthesis is prevented by expression of a conditional *dnaB* mutation, transfer occurs regardless, but the Hfr cells rapidly become nonviable. Transfer DNA synthesis is less accurate than normal replication, as the transferred DNA tends to accumulate mutations at known hotspots.

Transfer DNA synthesis is unique in yet another way: Its product is a single strand of DNA that is transferred without any significant amount of Hfr cytoplasm into the F^- cell. To prove the latter point, an experiment was performed in which an F^+ cell that was synthesizing the enzyme β-galactosidase was mated to an F^- cell that could not produce the enzyme. After DNA transfer had occurred, the F^+ cells were lysed by temperature induction of a λ prophage, and the amount of β-galactosidase associated with the F^- cells was measured (F^+ proteins adsorbed to cell surfaces had been eliminated with the enzyme chymotrypsin). Within the limits of error of the techniques used, no transfer of β-galactosidase was observed. In terms of the conjugation model shown in Figure 9-7, these results imply that the connection between the cells is substantially smaller than is diagrammed. In a different conjugation system, that of ColIb (see Chapter 10), transfer of donor proteins into recipient cells has been demonstrated.

The proof that only a single strand of DNA is transferred during conjugation was more complex and was achieved independently in the laboratories of Rupp and of Tomizawa. One set of experiments took advantage of zygotic induction. An Hfr (λ^+) culture whose DNA was labeled with ^{32}P was mated to an F^- (λ^-) culture in the absence of ^{32}P. After lysis of the zygote, the progeny phage were collected and the DNA extracted. Barring degradation and resynthesis, any ^{32}P counts in this preparation represented DNA inherited directly from the DNA transferred from the Hfr. The strands of λ DNA were separated by poly(UG) binding and checked for radioactivity. Because the direction of transfer of the Hfr and the orientation of the prophage DNA were known, it could be calculated that only a single DNA strand was transferred, beginning with the 5′-phosphate end (Fig. 9-9). It is assumed that the single strand of DNA transferred is produced by a rolling circle mechanism, which allows the Hfr cell to have a complete set of genetic information at all times. The molecular mechanism involved is discussed in Chapter 11.

The problem of what happens to DNA synthesis in an Hfr cell that has its mating aggregate sheared apart was considered by Kusnierz and Lombaert. They mixed an Hfr strain with a recipient strain F_1^- and allowed mating to proceed for a period of time. After interruption, a new recipient strain, F_2^-, was added and the culture allowed to mate. An interrupted mating curve was determined for transconjugants from the second F^- strain. The time and order of appearance of the transconjugant types indicated that the Hfr cells could begin a second mating immediately after interruption of the first, and that transfer began at the usual spot on the genome instead of at the point of interruption.

The process of DNA transfer is not entirely synchronous. For example, two cultures can be mixed and allowed to form mating aggregates for 5 minutes before dilution prevents additional aggregates from developing. If interrupted mating curves are prepared for the culture, entry of the selected marker is spread over a period of about 20 minutes, even though all mating aggregates perforce developed over a 5-minute interval. There-

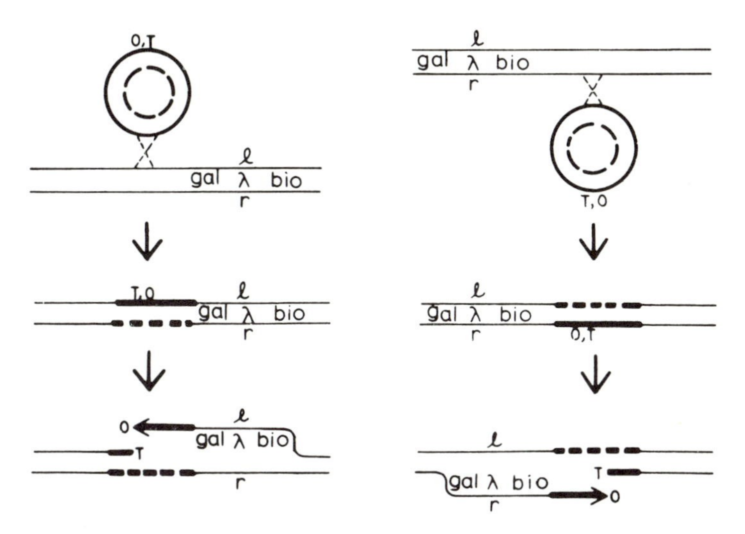

Figure 9-9. Insertion of the F plasmid and transfer of one strand. The thick lines represent the complementary strands of the plasmid DNA and the thin lines a portion of the bacterial DNA. During mating, one strand of the F plasmid is broken specifically between O (the origin) and T (the terminus), with O being transferred first and T last. Attachment of either strand, *l* or *r*, to the same plasmid strand is the consequence of inserting the plasmid in opposite orientations. From Rupp, W.D., Ihler, G. (1968). Strand selection during bacterial mating. Cold Spring Harbor Symposia on Quantitative Biology 33:647–650.

fore some cells experienced a **transfer delay** of 15 minutes. Other experiments have shown that once transfer begins the rate of transfer is essentially constant over the temperature range 30° to 40°C but is slower at all other temperatures. Therefore transfer delay occurs prior to the actual moment at which the first DNA arrives in the recipient.

Analysis of the F Plasmid

General Structure

One of the first questions to arise in the study of the F plasmid was that of **copy number,** the number of plasmid copies per bacterial genome equivalent. There are always some problems in the determination of the copy number for a particular plasmid due primarily to the difficulty of extracting DNA from a cell without having any of it fragment into smaller pieces. Any fragmentation of the DNA being measured gives a copy number that is too high if both fragments react with the detection system. The most satisfactory results have been obtained either by using radioactive probes to make heteroduplex molecules with the F plasmid or by comparing the rate of reversion of a marker carried on an F′ plasmid to the rate of reversion of the same marker carried on the bacterial chromosome. In the latter case, the ratio of the reversion rates is then the copy number. There is general agreement that the copy number for F is low, perhaps on the order of two copies per genome in cells that are growing slowly. In rapidly growing cells, the number must of course be larger to allow for the shorter time between cell divisions.

The first technique of physical analysis to be applied to the F plasmid was that of heteroduplex mapping. The work was begun in the laboratory of Norman Davidson and continued primarily by Deonier and Ohtsubo, his collaborators. When analyzing a circular DNA molecule, the first necessity is to establish reference points for use when measuring distances along the DNA molecule. Unless at least two such points are available, it is impossible to distinguish the clockwise from the counterclockwise direction. The first reference point, an arbitrarily chosen zero point, is obtained by examining a heteroduplex molecule composed of one strand of F DNA and one strand of F100 DNA. The point at which the insertion loop (representing the bacterial DNA of the F′) diverges from the double-strand DNA is taken as the origin of the F physical map. Because the F DNA is 94.5 kb in length and is circular, the zero point can also be called 94.5 and is usually designated 94.5/0 F. In order to define the clockwise direction, an F plasmid [FΔ(33-43)] that has a deletion of some 10 kb was used as a reference, and the deleted segment was arbitrarily assigned to the interval 33F to 43F.

As more F and F′ DNA molecules were examined by the heteroduplex technique, other differences were found. Small insertions or deletions were observed. Also certain base sequences were seemingly present several times in any one F DNA strand and could, under the proper conditions, anneal with one another. There seemed to be several of these sequences, and they were designated by paired Greek letters: αβ, γδ, εζ. In order to differentiate between repeats of the same sequences, they were given subscripts $\alpha_1\beta_1$, $\alpha_2\beta_2$, etc. Complementary sequences were designated $\beta_1'\alpha_1'$, $\beta_2'\alpha_2'$, etc. It soon became obvious that the junction point between F DNA and bacterial DNA in an F′ tended to occur at one of these sequences. The same appeared to be true for F integrated to form an Hfr. The natural conclusion was that these sequences represented the actual points of recombination for F and corresponded functionally to the *att* site on lambda.

Other work has shown that some of the repeated sequences are the same as certain insertion sequences that seem to be able to move themselves about the chromosome and that may form the ends of transposons. These sequences are discussed in greater detail in Chapter 13, but a brief description is necessary here. Each unique sequence has been given a number (e.g., IS*1*, IS*2*), and it can be shown that several copies of the IS DNA may be found on any genome. Hybridization analysis has shown that αβ is the same as IS*3* and εζ is the same as IS*2*, which means that

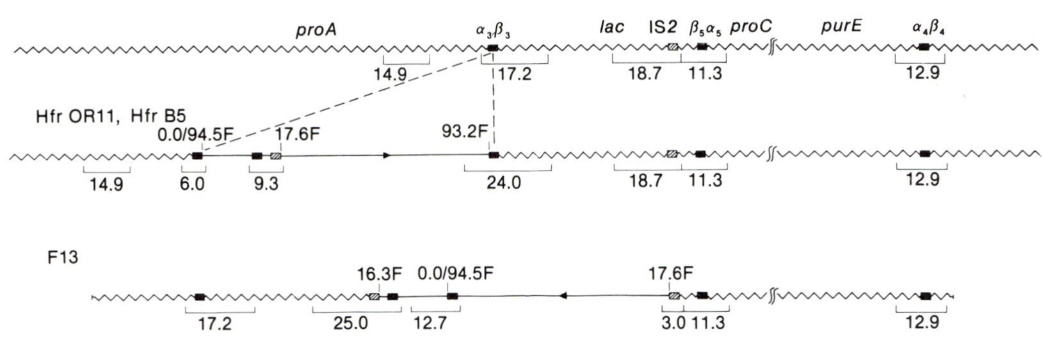

Figure 9-10. Summary of the physical structures of an Hfr and an F′. A physical map of a portion of the bacterial chromosome is shown at the top of the figure with the approximate locations of selected cistrons and IS elements indicated. A portion of the bacterial DNA between *proC* and *purE* has been omitted for clarity. Below this area the sequence arrangements of Hfr strains OR11 and B5 and F′ F13 are shown. Sawtooth lines represent bacterial DNA, and smooth lines represent the integrated F sequences. Selected F coordinates are indicated by the numbers above the lines (see also Fig. 9-11). The small arrowheads indicate the point of origin and direction of transfer. The various IS*3* elements are designated in the αβ notation with subscripts used to identify the individual components of each IS*3* sequence. They are represented by the solid boxes. The IS*2* elements are represented by the hatched boxes. Note that in the Hfr the F plasmid integrated between the two halves of $\alpha_3\beta_3$ in a left-to-right orientation. In F13 the orientation is reversed, and the site of the F plasmid insertion is an IS*2* element. From Hadley, R.G., Deonier, R.C. (1979). Specificity in the formation of type II F′ plasmids. Journal of Bacteriology 139:961–976.

homologous DNA sequences exist on F DNA and bacterial DNA that might be used for integrative recombination. The current model for F integration takes these findings into account and is shown in Figure 9-10. Different Hfr strains seem to use different repeated sequences for integration, which results in the integrated F plasmids being circular permutations of one another. All Hfr strains have *oriT* (transfer DNA origin of replication) located away from the ends of the plasmid, which means that part of the plasmid can transfer only after the entire bacterial chromosome and accounts for the lack of Hfr transconjugants.

F′ plasmids are assumed to result when pairing takes place between either two repeat sequences inverted with respect to each other and flanking the integrated F plasmid or between one such sequence and one of the normal F termini. The possible genetic consequences of recombination events of this sort are discussed in Chapter 13.

Genetic Analysis of the F Plasmid

The major barrier to genetic analysis of the F plasmid was the existence of **incompatibility,** the inability of two plasmids to occupy the same cell at the same time, which precluded the use of conventional *cis-trans* tests to assign mutations to cistrons. Achtman and co-workers overcame this problem by means of an extremely ingenious experimental design.

The basic plan of action was to take advantage of the fact that two strains carrying F′ plasmids can be mated if one of them has been F^- phenocopied, which gives rise to phenocopy cells carrying two different F′ plasmids. These cells would, of course, soon exclude one of the F′ plasmids (such exclusion could be prevented by using *inc* plasmids, defective for incompatibility, but they were not discovered until later). However, prior to its exclusion, the genetic information the plasmid carried would be expressed. The result of that expression during the transient merodiploid state would be a potential phenotypic change that would permit a *cis-trans* test, provided the trait could be assayed before exclusion had time to operate. One easily assayed trait is transfer ability itself, as plasmids are ready to retransfer within 50 minutes of their arrival in a new cell. Therefore, if the two plasmids were both *tra,* and the merodiploid cell were mated to an F^- tester strain before exclusion could become effective, a positive *cis-trans* test would be indicated by transfer of one or the other of the F′ plasmids into the tester strain.

Achtman and co-workers began by isolating a series of *tra* mutations in an F′*lac* plasmid. From among these mutant strains, they selected *tra* mutants that could be suppressed by a nonsense suppressor and moved the appropriate suppressors into strains carrying the plasmids, giving them a set of donor strains that were genotypically *tra* but phenotypically Tra^+. A second set of F′ carrying strains that had no suppressors and therefore were genotypically and phenotypically *tra* were prepared for use as recipients.

The actual experiment consisted of mating a suppressed donor F′ strain (Tra^+) to an F^- phenocopied recipient F′ strain (Tra^-). After about 45 minutes of mating, the donor strain was destroyed by lysis from without by phage T6. The recipient strain, which was resistant to phage T6, was unaffected. The culture was then diluted to prevent any retransfer of the plasmid between cells, and samples were removed and tested for donor ability. The tester F^- strain was resistant to T6 *(tsx)*, resistant to the antibiotic spectinomycin *(rpsE)*, and unable to utilize lactose as a sole carbon source *(lac)*. If resistant cells appeared that could ferment lactose, they must have received an F′*lac* from the initial recipient strain, which would imply that either complementation or recombination had occurred between the two F′ plasmids. The two alternatives were easily distinguished, because complementation would cause transfer of a *tra* plasmid that would be incapable of further transfer, whereas recombination would cause transfer of a tra^+ plasmid that could be retransferred indefinitely.

The experiments were so successful that the same workers devised a way to test *tra* mutations that could not be suppressed. Instead of mating the first donor strain to a phenocopied F′ strain, they grew Pl phage on it. They then used transduction to carry portions of the F′ plasmid into other plasmid cells. Once again the donor DNA was expressed, and complementation was possible and could be detected in the usual way. However, the efficiency of transfer was, of course, considerably lower, and therefore the transductional method lacked the sensitivity of the direct mating.

Broda and co-workers devised an interesting variation of the same experimental scheme in which the primary donor was an Hfr strain rather than an F′ strain. They were investigating the prediction that only part of the F plasmid transfers at the beginning of the Hfr DNA, a prediction based on the observation that the transconjugants from an Hfr × F^- mating are generally F^- and not F^+. If the prediction were true, a given Hfr should be able to complement only certain *tra* mutations during a mating in which only short pieces of DNA are transferred. In this instance, experiment exactly corresponded to theory, and complementation was obtained. As indicated in Figure 9-10, subsequent work confirmed that not all Hfr strains transfer equal amounts of the F DNA during the initiation of conjugation owing to the manner of the integration of the F plasmid.

It is obvious that the experimental method discussed above does not suffice to study the genetics of F traits that do not affect transfer ability. The electron microscopic technique of physical mapping via heteroduplex DNA molecules has been applied to F with considerable success, as described above. This technique has been augmented by the use of restriction enzymes to cut the F DNA into fragments that can be individually analyzed or spliced into other DNA molecules for analysis, as discussed in Chapter 11. In fact, the *tra* complementation experiments can be done much more easily if cloned DNA fragments are used in vectors that can coexist with F.

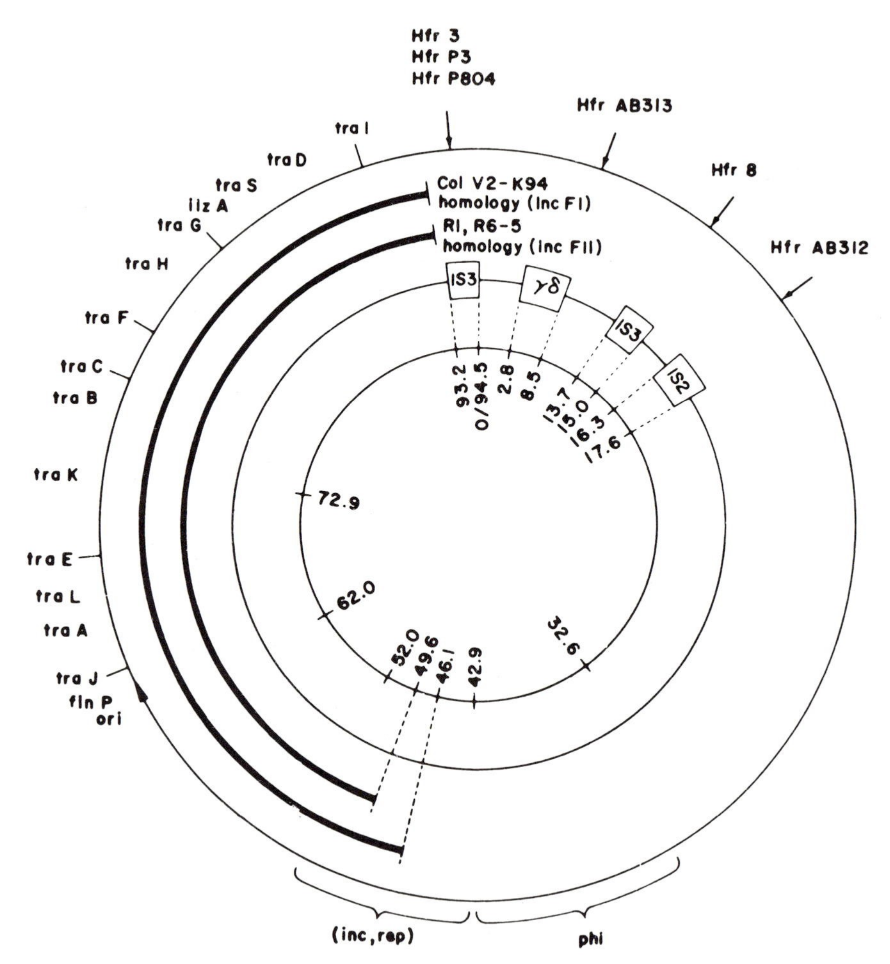

Figure 9-11. Simplified genetic map of the F plasmid. The inner circle gives some physical coordinates in kilobase units. The next circle indicates the locations of identified insertion elements. The two arcs show regions of extensive homology with ColV2-K94, R1, and R6-5 plasmids. The outer circle indicates the locations of certain genetic loci involved in phage inhibition (*phi*), incompatibility (*inc*), replication (*rep*), transfer (*tra*), fertility inhibition (*fin*), and immunity to lethal zygosis (*ilz*). The origin of transfer replication (*ori*) is also indicated. The positions where insertion elements on F recombine with the bacterial chromosome to form Hfr's are also indicated. A detailed map of the *inc* and *rep* region can be found in Figure 11-2 and a detailed map of the *tra* region in Figure 11-4. From Shapiro, J.A. (1977). F, the *E. coli* sex factor, p. 671. In: Bukhari, A.I., Shapiro, J.A., Adhya, S.L. (eds.) DNA Insertion Elements, Plasmids and Episomes. Cold Spring Harbor, N.Y.: Cold Spring Harbor Laboratory.

The result of these investigations is a map for F, shown in Figure 9-11. Although not as detailed as some genetic maps, it nonetheless represents a large amount of work using a difficult experimental system. Indicated on the map are both the known cistrons and the known repeated sequences.

Summary

The conjugation process was first observed in *E. coli* and has proved to be a powerful tool for rapid elucidation of its genetic map. The basic process is one of linear, ordered transfer of DNA from a donor to a recipient. The length of DNA transferred is variable. Donor ability is conferred by the presence of the F plasmid, and cell-to-cell contact is required for DNA transfer to occur. Mating aggregates form as a result of the presence of a specific F pilus on the cell surface. DNA is transferred from either Hfr or F′ cells. The former carry an integrated F plasmid, and the latter carry an autonomous plasmid that has incorporated some bacterial DNA in an aberrant excision process. DNA transfer in Hfr cells has been observed from specific points of origin on the bacterial chromosome and requires that a special transfer DNA synthesis occur. The DNA is transferred as a single-strand entity beginning with the 5′-end.

The genetic map of *E. coli* is based on conjugative data and has the minute as its unit of linear measure. This unit represents the average amount of DNA transferred by a donor cell in 1 minute under standard physiologic conditions. The quickest method of determining map order involves the use of interrupted matings.

The F plasmid is 94.5 kb in size and includes several insertion sequences within its DNA. These sequences serve as regions of homology for integration of the F plasmid during Hfr formation. Excision of the F plasmid may result in accurate regeneration of the F plasmid or in production of an F′ plasmid carrying all of the F DNA as well as some bacterial sequences. F plasmid genetics have been investigated using transfer-defective mutants and various types of suppressor.

References

General

Bachmann, B.J. (1983). Linkage map of *Escherichia coli* K12, edition 7. Microbiological Reviews 47:180–230.

Jacob, F., Wollman, E.L. (1961). Sexuality and the Genetics of Bacteria. New York: Academic Press.

Lederberg, J. (1986). Forty years of genetic recombination in bacteria: a fortieth anniversary reminiscence. Nature 324:627–628.

Willetts, N., Wilkins, B. (1984). Processing of plasmid DNA during bacterial conjugation. Microbiological Reviews 48:24–41.

Wollman, E.L., Jacob, F., Hayes, W. (1956). Conjugation and genetic recombination in *Escherichia coli* K-12. Cold Spring Harbor Symposia on Quantitative Biology 21:141–162.

Specialized

Achtman, M., Morelli, G., Schwuchow, S. (1978). Cell-cell interactions in conjugating *E. coli:* role of F pili and fate of mating aggregates. Journal of Bacteriology 135:1053–1061.

Achtman, M., Willetts, N.S., Clark, A.J. (1972). Conjugational complementation analysis of transfer-deficient mutants of F*lac* in *E. coli:* Journal of Bacteriology 110:831–842.

Hadley, R.G., Deonier, R.C. (1979). Specificity in formation of type II F′ plasmids. Journal of Bacteriology 139:961–976.

Kunz, B.A., Glickman, B.W. (1983). The infidelity of conjugal DNA transfer in *Escherichia coli*. Genetics 105:489–500.

Kusnierz, J.P., Lombaert, M.A. (1971). Etude de l'initiation du transfert chromosomique chez *E. coli* K-12. Comptes Rendues Hebdomadaires des Séances de l'Academie des Sciences, Série D 272:2844–2847.

Ou, J.T., Reim, R.L. (1978). F^- mating materials able to generate a mating signal in mating with HfrH *dnaB* (ts) cells. Journal of Bacteriology 133:442–445.

Singleton, P. (1983). Zeta potential: a determinative factor in F-type mating. FEMS Microbiology Letters 20:151–153.

Skurray, R.A., Reeves, P. (1973). Characterization of lethal zygosis associated with conjugation in *Escherichia coli* K-12. Journal of Bacteriology 113:58-70.

Xia, X-M., Enomoto, M. (1986). A naturally occurring large chromosomal inversion in *Escherichia coli* K-12. Molecular and General Genetics 205:376–379.

Chapter 10
Other Plasmids and Other Conjugation Systems

Not all plasmids are conjugative, and, even among those that are, they do not necessarily have chromosome-mobilizing ability (capability of promoting transfer of chromosomal markers to a recipient cell). Nevertheless, these plasmids can have major economic and genetic importance. The diversity of plasmids that have been identified is staggering, and they are ubiquitous. In one study, 34 of 87 hospital isolates of enteric bacteria or *Pseudomonas* carried at least one plasmid. *Bacillus megaterium* routinely has eight or more plasmids in its cytoplasm. All of these plasmids have only two things in common. They can be identified in cell lysates as autonomous DNA molecules, and they are capable of self-replication. If the presence or absence of a plasmid has no observable effect on the cell phenotype, it is said to be a **cryptic plasmid.** Plasmids are often subdivided according to whether they are capable of self-transfer from one host cell to another. Many, but not all, of those plasmids that are self-transmissible are also capable of mobilizing the bacterial chromosome of the host cell and causing it to be transferred into another cell in a manner analogous to that of the F plasmid.

This chapter is intended to serve as an introduction to a variety of commonly encountered plasmids. It is important to note, however, that *E. coli* alone is reported to have more than 270 naturally occurring plasmids. The number of plasmids that can be discussed in a text such as this one is obviously highly restricted, and therefore the choices are frequently arbitrary. As in Chapter 9, the emphasis here is on the physiology and classic genetics of these systems. The molecular biologic aspects are deferred until Chapter 11. The first section here deals with chromosome-mobilizing plasmids, the second with a particular set of *E. coli* plasmids called colicinogenic plasmids, and the last with a general group of plasmids, the R plasmids.

Major Chromosome-Mobilizing Plasmids

Escherichia coli may get most of the publicity, but there are well documented conjugation systems in many other bacterial genera. Their number is such that it is not possible to even list them all, and therefore this chapter considers only selected examples from both the gram-negative and gram-positive bacteria (Table 10-1).

Salmonella

The enteric bacteria, of which *E. coli* is a member, are closely related, and therefore it is not surprising that conjugation systems have been found among several of them. The best-studied genus other than *Escherichia* is *Salmonella*, whose genetic apparatus seems to have many of the attributes of the *E. coli* system. Hfr-type cells can be produced by transfer of the F plasmid from *E. coli* to *S. typhimurium*. Using such Hfr strains, it is possible to transfer DNA from *Salmonella* to *Escherichia*, or vice versa, and to obtain transconjugants within the limits of DNA homology.

One unique feature of *Salmonella typhimurium* is that the length of the genetic map is longer than that for *E. coli*. A map constructed in the same manner as that for *E. coli* has a length of 135 minutes instead of 100 minutes, even though the relative order and spacing of most markers is the same. After a conference between the coordinators of the respective maps, it was decided that this stretching of the map is due to slower transfer of DNA by *Salmonella* Hfr strains, and the *Salmonella* map has been recalibrated to 100 minutes. After recalibration, extensive areas of homology appeared in the two genetic maps (compare the inside front and back covers).

Nevertheless, there is one striking difference between the two maps. In addition to small insertions/deletions, which account for generic differences, there is a large inversion of about 10% of the genome that is roughly bounded by *purB* and *aroD* (25 to 37 minutes). As far as has been ascertained, there is no phenotypic effect due to this inversion, and the

Table 10-1. Some plasmids with chromosome-mobilizing ability

Organism	Plasmid
Salmonella typhimurium	F
Pseudomonas aeruginosa PAO	FP2
	FP39
	FP110
	R91-5 (transposon-mediated)
Streptococcus faecalis	pAD1
Streptomyces coelicolor A3(2)	SCP1
	SCP2*

reason for its stabilization within the population remains unclear. Some effort has been made to map the endpoints of the inversion, but there is interstrain heterogeneity present that prevents an accurate determination.

Riley and Anilionis proposed that the small insertions/deletions observed when the maps of *Escherichia* and *Salmonella* are compared indicate one of the ways in which evolution of the bacterial genome occurs. They postulated that various insertion elements such as plasmids or transposons may cause the addition or deletion of genetic material in the course of their normal insertion/excision processes. Possible mechanisms by which such events might occur are discussed in Chapter 13, and some evolutionary considerations are presented in Chapter 14.

Pseudomonas

The genus *Pseudomonas* is one of the most nutritionally diverse of all the prokaryotes, a fact due in large measure to the presence of an extensive number of plasmids. Among this large body of plasmids are some that have chromosome-mobilizing ability superficially similar to that of F in *E. coli*, and conjugation has been reported in *Pseudomonas aeruginosa*, *P. putida*, *P. fluorescens*, *P. glycinea*, *P. syringae*, and *P. morsprunorum*. Three such plasmids have been described in *P. aeruginosa*: FP2, FP5, and FP39. Transducing phages are also known in four of the species.

The map developed for *P. aeruginosa* strain PAO is shown in Fig. 10-1. Like *E. coli*, the genetic map is circular, but the total genome size of 3600 kb is somewhat smaller than that for *E. coli* (4500 kb). There are no obvious genetic homologies with the genetic map for *E. coli*. One particularly interesting observation is that the clustering of similar biosynthetic functions in *E. coli* and *S. typhimurium* (e.g., the *trp* or *his* cistrons) is not seen in *P. aeruginosa*, where *trp* cistrons are located at four discrete sites. Another difference between the Enterobacteriaceae and *Pseudomonas* is that Holloway and Morgan have estimated that all *P. aeruginosa* isolates carry at least one prophage.

Most conjugation in *Pseudomonas* takes place by an unknown mechanism that is obviously different from that discussed in connection with *E. coli*. Only recently has any sort of Hfr strain been observed, and that only with special plasmids. The FP plasmids and others commonly used for conjugation experiments do not integrate themselves into the bacterial chromosome, although they do transfer from specific points of origin, as indicated in Figure 10-1. Other plasmids can be forced to integrate by preparing transposon-carrying derivatives whose integration is then mediated by the transposon itself. The FP plasmids do not code for pili, and therefore the process for mating aggregate formation must also be different. Finally, some 30 to 80% of the transconjugants inherit the conjugative plasmid, whereas it is rare for an *E. coli* Hfr donor to give a plasmid-carrying transconjugant.

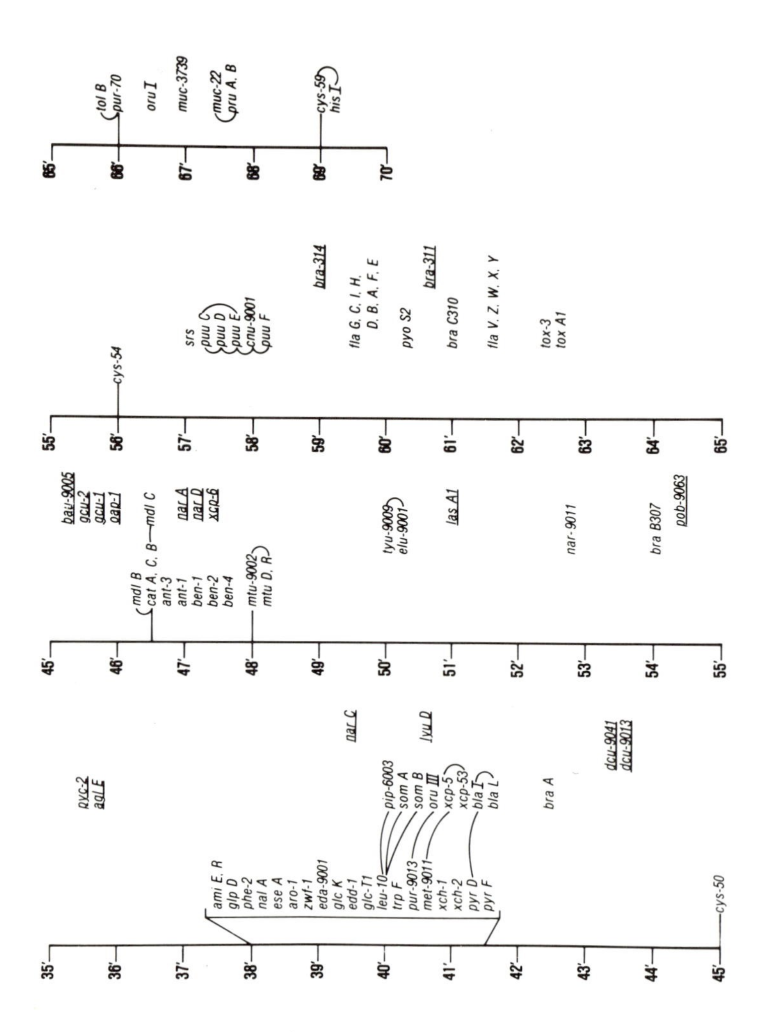

35'
36'
37'
38'
39'
40'
41'
42'
43'
44'
45'
ami E, R
glp D
phe-2
nal A
ese A
aro-1
zwf-1
eda-9001
glc K
edd-1
glc-T1
leu-10
trp F
pur-9013
met-9011
xch-1
xch-2
pyr D
pyr F
pip-6003
som A
som B
oru III
xcp-5
xcp-53
bla I
bla L
pyc-2
agl F
nar C
lyu D
bra A
dcu-9041
dcu-9013
cys-50
46'
47'
48'
49'
50'
51'
52'
53'
54'
55'
mdl B
cat A, C, B
mdl C
ant-3
ant-1
ben-1
ben-2
ben-4
mtu-9002
mtu D, R
bau-9005
gcu-2
gcu-1
gap-1
nar A
nar D
xcp-6
tyu-9009
elu-9001
las A1
nar-9011
bra B307
pob-9063
56'
57'
58'
59'
60'
61'
62'
63'
64'
65'
cys-54
srs
puu C
puu D
puu E
cnu-9001
puu F
bra-314
fla G, C, I, H,
D, B, A, F, E
pyo S2
bra-311
bra C310
fla V, Z, W, X, Y
tox-3
tox A1
66'
67'
68'
69'
70'
tol B
pur-70
oru I
muc-3739
muc-22
pru A, B
cys-59
his I

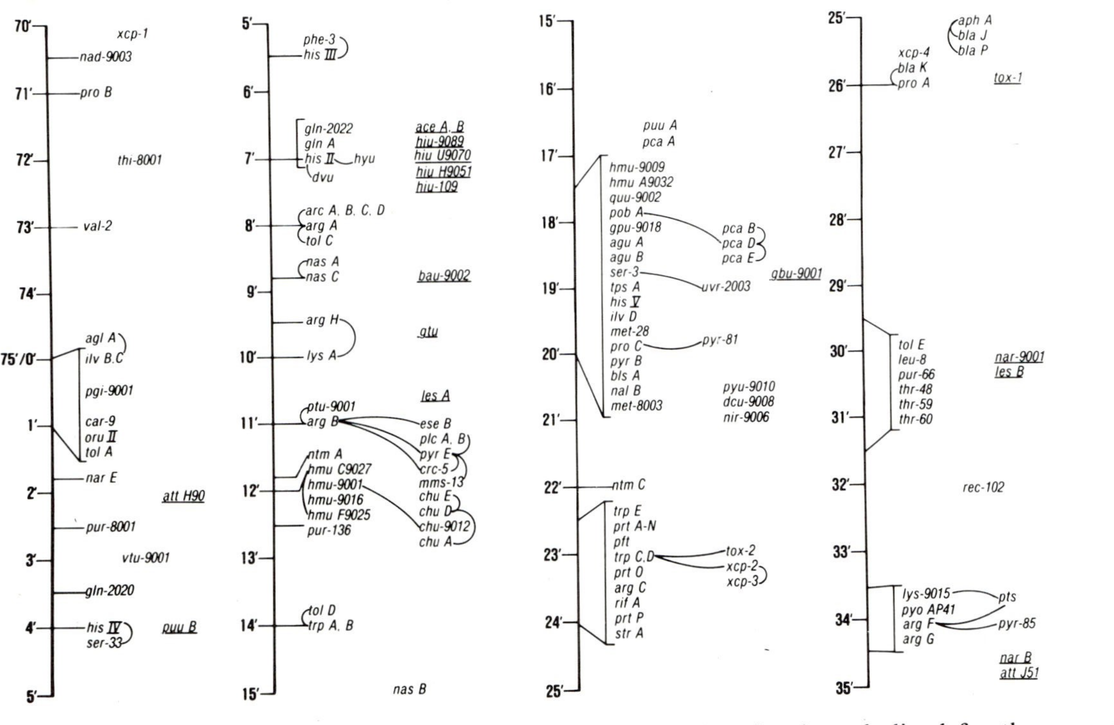

Figure 10-1. Chromosome map of *Pseudomonas aeruginosa* PAO. The origin (0 minutes) has been arbitrarily designated at the *ilvB/C* locus. The following notation has been used. (1) Markers whose location is indicated by a bar joining the locus designation to the map have been located by interrupted matings using FP2, pMO514, or IncP-1 insert donors. (2) Curved lines indicate that the markers so joined are cotransducible using one or more of the bacteriophages F116, F116L, G101, and E79*tv-2*. (3) The marker designation is underlined for those cases where there is evidence to locate the marker in the area in which the symbol is placed, but the relations to flanking markers have not been determined. Unusual abbreviations are listed in the original reference. From Holloway, R.W., O'Hoy, K., Matsumoto, H. (1987). Chromosome map of *Pseudomonas aeruginosa* PAO, p. 223. In: O'Brien, S.J. (ed.) Genetic Maps 1986. Cold Spring Harbor, N.Y.: Cold Spring Harbor Laboratory.

Streptococcus

The study of conjugation in the members of the genus *Streptococcus* has uncovered several unusual phenomena. In *Streptococcus faecalis* there are conjugative plasmids known with chromosome-mobilizing ability, one of which is pAD1. Cells lacking pAD1 produce a specific **pheromone**, a compound that elicits enhanced mating aggregate formation from the plasmid-carrying cell. The pheromone is designated cAD1 and is a short polypeptide with the sequence

$$HOOC\text{-}Leu\text{-}Phe\text{-}Ser\text{-}Leu\text{-}Val\text{-}Leu\text{-}Ala\text{-}Gly\text{-}NH_2$$

Cells carrying pAD1 respond to cAD1 by synthesizing an aggregation substance that results in mating aggregate formation to an extent that the cells of the mixed culture can be seen to visibly clump. Not surprisingly, the presence of the pheromone improves plasmid transfer 200- to 300-fold.

Plasmid-carrying cells do not clump with each other because the plasmid chemically modifies the cAD1 molecule so that it is no longer functional. They also produce a protein inhibitor, iAD1, whose structure is

$$HOOC\text{-}Leu\text{-}Phe\text{-}Val\text{-}Val\text{-}Thr\text{-}Leu\text{-}Val\text{-}Gly\text{-}NH_2$$

The marked similarity of sequence to that of cAD1 has led to the assumption that iAD1 presumably acts as a competitive inhibitor of the binding of cAD1 to the cell surface. Similar results have been obtained for another pheromone and its inhibitor, cPD1 and iPD1.

There is another, strikingly different mechanism of conjugation observed in some streptococci, i.e., **conjugative transposition**.The process occurs during **filter matings** in which the donor and recipient cells are not mixed in liquid culture but, rather, trapped on the surface of a membrane filter. After an appropriate period of time has elapsed, the filter is transferred to selective medium, and the transconjugants are allowed to grow. The process meets all the criteria for conjugation. It is insensitive to the presence of DNase, and it requires cell to cell contact. A culture filtrate is not sufficient. Nevertheless, it is not possible to identify any sort of plasmid molecule in the donor cells. The markers transferred in this manner are generally antibiotic markers, and it can be shown that within their original cell they behave like transposons. They can move from bacterial chromosome to plasmid or from one site to a different site on the same DNA molecule.

Perhaps the best studied member of the conjugative transposons is the one designated Tn*916* and originally observed in *S. faecalis*. It transposes with a frequency of 10^{-5} to 10^{-8}/cell/generation. The transconjugants have Tn*916* inserted into a variety of sites on their chromosomes and may receive more than one copy of the transposon. It has been suggested that the latter phenomenon may be the result of an effect similar to that of zygotic induction where the transposition machinery is normally kept turned off by a repressor that must accumulate to appropriate levels within the cell. If the repressor is slow to form, additional transpositions can occur within the recipient. No one has yet formulated a satisfactory model to explain the mechanism of conjugative transposition.

The two conjugation methods are not mutually exclusive, for the *S. faecalis* plasmid pCF10 seems to embody them both. It is a 58-kb plasmid conferring tetracycline resistance that carries a 25-kb TRA region coding for the usual pheromone responses and a separate 16-kb region that includes the antibiotic resistance. The 16-kb region behaves as a conjugative transposon and has been designated Tn*925*. In addition, conjugative transposition is not restricted to *S. faecalis*. Transposon Tn*1545* has been shown to transfer into a variety of gram-positive bacteria including *Bacillus subtilis*.

Streptomyces

Most of the genetic work with *Streptomyces*, an industrially important genus, has been with *Streptomyces coelicolor* A3(2) and *Streptomyces lividans*. Hopwood and his collaborators have identified two principal conjugative plasmids in *S. coelicolor* A3(2). The first is SCP1, a large plasmid at least 150 kb in size that has never been isolated intact, although portions of it have been cloned. It codes for the antibiotic methylenomycin. The second is SCP2, a 31-kb cryptic plasmid. Both of these plasmids appear to have low copy numbers, but there are other plasmids known in *Streptomyces* that may have copy numbers as high as 800. Both of these former plasmids as well as at least seven others can mobilize the chromosome of *S. lividans*. Most *Streptomyces* plasmids cause a **pock formation** in the cultures that carry them which is the result of an inhibition or delay in the production of aerial mycelia and spores. In a lawn resulting from a mixture of cells, some of which are carrying a plasmid, the plasmid-containing clones can be readily identified by their lower profile.

Numerous crosses have been made with SCP1 in *S. coelicolor* A3(2). Unlike the case for *E. coli*, SCP1 promotes recombination between two plasmid-carrying strains (NF strains) as well as between a plasmid-free strain and an NF strain. The efficiency of plasmid transfer itself approaches 100%, but marker exchange is a more conventional 3×10^{-6} to 5×10^{-6}. What is particularly unusual about the recombinants is that donor markers are inherited bidirectionally from a point centered on the 9 o'clock position of the genetic map (Fig. 10-2). The diffusible agarase cistron located at this site is inactivated in all NF strains. Another bidirectional donor has been obtained from an SCP1-prime strain, and this time the bidirectional gradient is centered on the region corresponding to the bacterial DNA carried by the plasmid. From these observations it is concluded that SCP1 is integrated into the bacterial chromosome. Some donors have been isolated that give only a unidirectional gradient of donor marker inheritance,

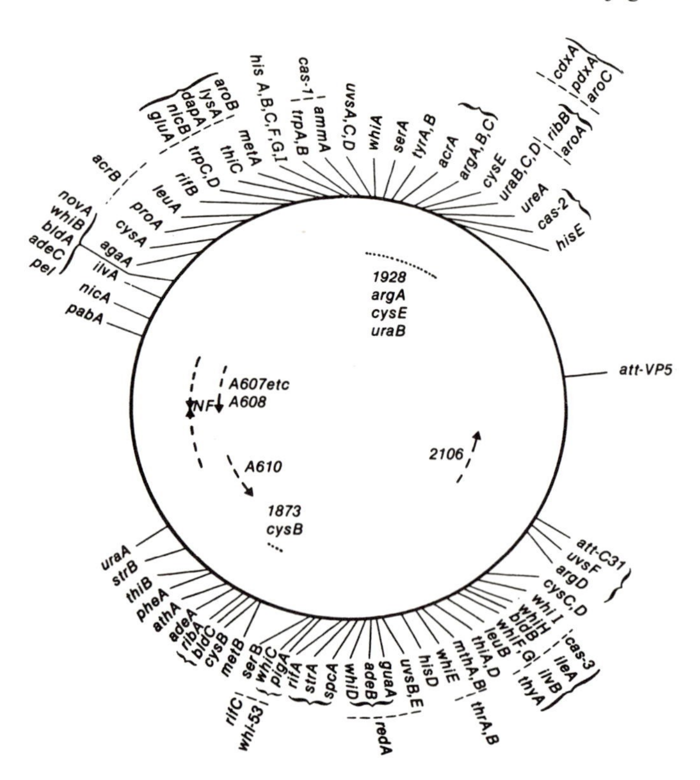

Figure 10-2. Circular linkage map of *Streptomyces coelicolor* A3(2). The orders of markers within groups of bracketed loci are unknown. Loci outside the dashed lines have not been ordered relative to loci inside these lines. Single arrows indicate unidirectional donors, and a double arrow indicates a bidirectional donor. The *dagA* locus is the diffusible agarase that is disrupted by insertion of the SCP1 plasmid. Many of the other marker abbreviations are listed in Table 2-2. Adapted from Hopwood, D.A. (1976). Linkage map and list of markers of *Streptomyces coelicolor,* pp. 723–728. In: Fasman, G.D. (ed.) Handbook of Biochemistry and Molecular Biology, 3rd ed. Boca Raton: CRC Press.

but the difference between these donors and the conventional NF type is not known.

SCP2 plasmids do not normally promote genetic exchange efficiently. Spontaneous variants denoted SCP2* have been obtained whose transfer is somewhat more efficient. However, artificially prepared SCP2*-primes transfer their cloned DNA efficiently and have been used for complementation studies. The added homology of the SCP2*-primes apparently promotes their chromosomal integration because strains carrying these plasmids become bidirectional donors in a manner similar to SCP1-carrying strains.

There are many other plasmids within the genus *Streptomyces*, far too many to discuss here. However, there is one particularly unusual plasmid that deserves mention. It is pSRM and is found in *S. rimosus*. Its claim to fame is that it is a linear rather than a circular plasmid. The suggestion is that the DNA is circularized by proteins in the same manner as that of phage ϕ29.

Agrobacterium

The conjugation system found in members of the genus *Agrobacterium* is unique. It promotes transfer not from one bacterial cell to another but from a bacterial cell to a plant cell. The best studied case is that of *Agrobacterium tumefaciens*, and the observation is that if any one of a variety of dicotyledonous plants is infected by the bacterium near the crown of the plant a gall or tumor develops. If there is no infection, a tumor is not produced. If the bacterium does not carry the Ti (tumor-inducing) plasmid, there is no tumor. Blotting experiments using DNA extracted from tumor cells have shown that a portion of the DNA from the tumor cell binds to a Ti plasmid probe. By using various restriction fragments of the Ti DNA, it is possible to demonstrate that only a relatively small region of the Ti plasmid, the T-DNA, is actually incorporated into a plant chromosome.

Transfer of the T-DNA requires the interaction of bacterial, plant, and plasmid proteins. The bacterial cistrons are *chvA* and *chvB*, and they code for plant attachment functions. The plant produces molecules that act as specific inducers of *vir* functions located on the plasmid. Seven *vir* cistrons have been identified, although their individual contributions are not completely known. The *virA* protein is an inner membrane protein that seems to act as a regulator and may recognize chemical signals from the plant. The *virD* protein promotes excision of the T-DNA from the Ti plasmid in the form of a circle. DNA sequencing experiments have shown that the T-DNA is bounded by two nearly perfect 25-bp repeats. If the right-hand border segment is deleted, transfer of T-DNA is abolished, but the left-hand border does not appear to be necessary. The process of circle formation therefore is assumed to begin with the right-hand border sequence. Confirmation of this model came from Peralta and co-workers, who identified a 24-bp sequence called *overdrive* that is required for right-hand border function and is located adjacent to the T-DNA. The DNA that is transferred into the plant cell is single-stranded, as in the case of the *E. coli* conjugation system. The amount of DNA to be transferred is determined only by the positions of the boundary sequences. Insertions of transposons into the T-DNA have no effect on DNA transfer unless they accidentally inactivate the *vir* functions. For example, integration of the transferred DNA into the plant chromosome requires *virE* function.

The T-DNA system is the only confirmed example of the transfer of prokaryotic DNA into eukaryotic cells. As such, it is an exception to the genetic principle that genetic exchange occurs only between closely related individuals. From the point of view of the bacterium, the relationship is an unusual form of parasitism. The transformed plant cells produce a variety of amino acid derivatives called opines that can be used as carbon sources by *Agrobacterium* cells carrying the Ti plasmid. From the point of view of the geneticist, the Ti plasmid system represents a remarkable opportunity to carry out genetic engineering experiments with plants. For further discussion on this point, see Chapter 14.

Bacteriocins

General Properties

Microbiologists have long known that mixtures of certain bacteria are incompatible because only one of the organisms so combined survives for more than a few hours. Gratia first observed that one type of *E. coli* released diffusible substances into the medium that were lethal to some other bacteria, including *E. coli,* but did not produce the virulence (V) factor. Fredericq, in collaboration with Gratia, showed that the V factor was proteinaceous in nature (i.e., sensitive to the proteolytic enzyme trypsin) and was merely one example of a large number of antibiotics produced by various types of bacteria.

The general name given to substances of this type is **bacteriocin**, and individual bacteriocins are named according to the species of organism that originally produced them. Thus there are colicins from *E. coli,* subtilisins from *Bacillus subtilis,* influenzacins from *Hemophilus influenzae,* and pyocins from *P. aeruginosa* (formerly *P. pyocyanea*). Different protein molecules within each type of bacteriocin are identified by letters and/or numbers. In all cases, cells of the producing culture are immune to their own bacteriocin but are sensitive to different bacteriocins from the same species. However, the same bacteriocin may be produced in slightly different forms by different strains or species of bacteria. Therefore each bacteriocin designation also includes the strain number of the producing culture. In the case of Gratia's colicin V, for example, a typical designation might be colicin V-K357.

Many bacteria produce bacteriocins, and genetic analysis of these bacteriocins has shown that each type is encoded by certain specific DNA sequences. However, all of the bacteriocins can be subdivided into two general categories. One type, exemplified by colicin V, occurs either as a pure protein molecule or as a protein complexed to a portion of the outer membrane of the producer cell. The colicin V protein is generally found complexed to the O antigen from the *E. coli* host. The other type of bacteriocin, best exemplified by pyocin R from *P. aeruginosa,* closely resembles all or part of a bacteriophage, except that there is no DNA associated with it (Fig. 10-3). The latter type of bacteriocin has always

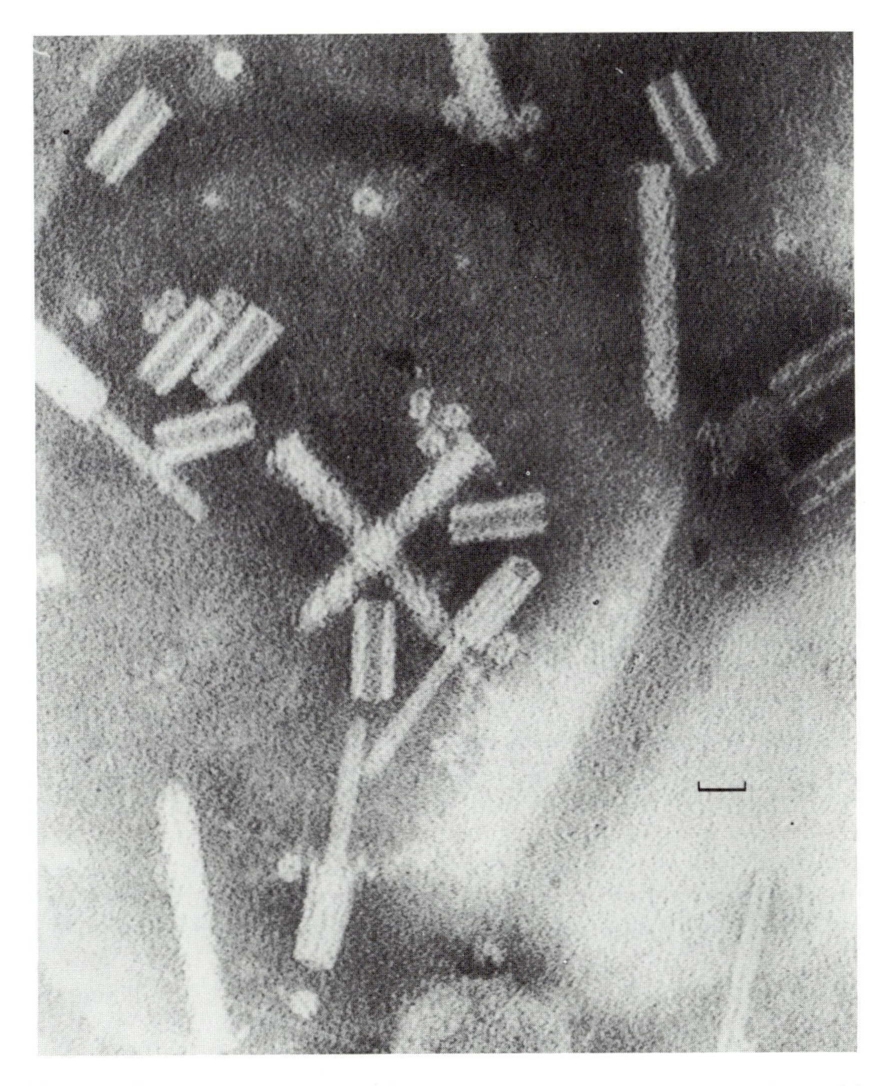

Figure 10-3. Electron micrograph of pyocin R. These structures were obtained by treating a culture of *Pseudomonas aeruginosa* with mitomycin C. Structures resembling both contracted and noncontracted phage tails can be seen. The bar indicates a length of 25 nm. From Bradley, D.E. 1967. Ultrastructure of bacteriophages and bacteriocins. Bacterological Reviews 31:230-314. Print courtesy of Dr. B.W. Holloway.

been reported to be produced by DNA located within the bacterial chromosome and is assumed to be a defective prophage. The former type of bacteriocin is associated with the presence of a plasmid within the cell, and it is this type that is further discussed.

Although bacteriocins are present in producing cultures at all times, the amount of bacteriocin present can be greatly increased by the same treatments that cause induction of the λ prophage. In recognition of this similarity, the process is called bacteriocin induction. In most cases bacteriocin production parallels induction of a prophage in that a producing cell is killed as a result of the induction whereas nonproducing cells survive. The producing cell is thus sacrificed in order to kill neighboring sensitive bacteria to the presumed betterment of the remainder of the culture of colicinogenic cells.

The number of producing cells in a culture can be measured using a variation of the plaque assay technique. A diluted cell suspension of a bacteriocin producer is mixed with an indicator bacterium in soft agar. When spread on a plate, the bacteriocin producers result in the development of holes, or **lacunae**, in the lawn of indicator bacteria. Each lacuna represents a single producer cell and has in its center a small colony resulting from that cell.

The bacteriocin released by the producing cells can be shown to bind to specific receptors on the surface of the cell. Elimination of the receptors by mutation results in a bacteriocin-resistant cell. However, the protoplasts or spheroplasts made from these resistant cells are found to be once again sensitive to the bacteriocin, which means that the cell wall must protect the cytoplasmic membrane from the colicin. The inference is that the receptor facilitates penetration of the bacteriocin through the cell wall. In contrast to the resistant type of mutant is the tolerant mutant in which the receptors are still present on the surface of the bacterial cell, but changes have occurred in the cytoplasmic membrane so that the bacteriocin no longer has any effect. Operationally it is frequently difficult to distinguish between resistance and tolerance mutations.

Colicins, the Best Studied Bacteriocins

As might be expected, the *E. coli* bacteriocins have had the most intensive study. One colicin, produced by strain 15, has been identified as a defective prophage, but the remainder are plasmid-derived. A colicin-producing plasmid is frequently denoted as Col followed by the appropriate letter. A short list of the nonphage colicins is presented in Table 10-2, along with the modes of action and receptors for the protein molecules, if known. Note that in every case examined the colicin receptor functions in the transport of some other substance into the cell. The relation between this substance and the colicin, if any, is not known. The cross resistance grouping refers to the fact that certain *E. coli* mutations confer resistance

Table 10-2. Properties of some colicins produced by plasmids[a]

Colicin	Cross-resistance group	Mode of action	Normal function of receptor	*E. coli* cistron coding for receptor
A	A	Inhibits RNA synthesis	Vitamin B_{12} transport	*btuB*
B/D	B	Inhibits macromolecular synthesis	Enterochelin transport	*fepA*
E1	A	Depolarizes cytoplasmic membrane	Vitamin B_{12} transport	*btuB*
E2	A	Inhibits DNA replication; induces DNA breakdown	Vitamin B_{12} transport	*btuB*
E3	A	Cleaves 16S rRNA	Vitamin B_{12} transport	*btuB*
I	B	Depolarizes cytoplasmic membrane	Chelated iron transport?	*cir*
K	A	Depolarizes cytoplasmic membrane	Nucleoside transport	*tsx*
M	B	Cell lysis	Ferrichrome transport	*fhuA*
V	B	—	—	—

[a]The colicins are assigned to one of two cross-resistance groups, and any cell that becomes resistant to the action of one colicin is also resistant to all other members of the group.

Adapted from Reeves, P. (1979). The concept of bacteriocins. Zentralblatt für Bakteriologie, Parasitenkunde, Infections krankheiten und Hygiene, Abt. 1 Orig. Reihe A 244: 78–90.

to all members of a particular group of colicins. Cells that are *tolA* are tolerant to all group A members, whereas cells that are either *fhuA* (phage T1 resistant) or *exbB* (enterochelin excretion) are tolerant to group B.

The plasmids that code for colicin production may be either conjugative or nonconjugative. As Table 10-3 shows, they are rather heterogeneous in size and copy number. Plasmids with a copy number near one are said to be stringently regulated, whereas those with copy numbers near ten or more are considered relaxedly regulated.

Conjugative Colicin Plasmids

One important conjugative plasmid is that which produces colicin I or, more correctly, the two closely related plasmids that produce colicins Ia and Ib. The proteins they produce have molecular weights of about 80,000

Table 10-3. Some properties of plasmids that produce colicins

Plasmid	Fertility	MW (daltons)	Copy number
ColB-K77	Conjugative	70×10^6	—
ColD-CA23	Nonconjugative	3.3×10^6	15–30
ColE1-K30	Nonconjugative	4.2×10^6	10–15
ColE1-16	Nonconjugative	6×10^6	—
ColE2-P9	Nonconjugative	5.0×10^6	10–15
ColE3-CA38	Nonconjugative	5.0×10^6	10–15
ColIb-P9	Conjugative	62×10^6	1–2
ColK-K235	Nonconjugative	6×10^6	—
ColV2	Conjugative	94×10^6	10–15
ColV-K94	Conjugative	85×10^6	—
ColVB-K260	Conjugative	107×10^6	1–2

Modified from Konisky, J. (1978). The bacteriocins, pp. 71–136. In: Ornston, L.N., Sokatch, J.R. (eds.) The Bacteria, A Treatise on Structure and Function, vol. 6. New York: Academic Press.

daltons and similar modes of action. However, the producing strains are not cross immune, and the colicins are therefore given separate designations.

The colicin I plasmids produce their own pili as part of the conjugal apparatus, and these pili are thinner and more flexible than F pili. They confer sensitivity to a different group of male-specific phages than do the F pili, and they are immunologically distinct from the F pili. In cells having compatible plasmids coding for F and I pili, each plasmid seems to use its own pili preferentially. The colicin I plasmid promotes transfer of bacterial DNA but does not give rise to stable Hfr strains.

The receptor for colicin I (Table 10-2) passes the protein through the outer membrane (lipoprotein–lipopolysaccharide layer of the cell wall) and into the periplasmic space. Once the colicin molecule has arrived at the cytoplasmic membrane, it creates channels within the membrane that allow potassium ions to leak out, resulting in membrane depolarization. The depolarization prevents further ATP synthesis, which results in cessation of all macromolecular synthesis and cell death. Cells mutant at the *tolI* locus are specifically immune to I colicins but also suffer from many membrane defects that block active transport and electron transport.

Colicin B and D plasmids code for similar proteins, but the producing strains are not cross immune. The plasmids also code for F pili but do not show any ability to instigate the transfer of bacterial DNA. The protein products are roughly the same size (about 89,000 daltons) and seem to affect the cell membrane in some fashion that causes the cessation of macromolecular synthesis. Strains that are *tolB* tolerate colicins B, A, and K. No tolerance mutations are known that are specific for only colicins B or D.

The role of colicin V is even more mysterious than that of colicin B/D, as neither its receptor nor its mode of action are known. No specific *tol* mutations have been identified either. On the other hand, colicin V plasmids form excellent Hfr strains possessing F pili, and several of these Hfr strains have been widely used for genetic crosses, after appropriate mutations were introduced to prevent colicin production.

Nonconjugative Colicin Plasmids

An important group of nonconjugative plasmids is the colicin E group. Colicin A (originally found in *Citrobacter freundii*) probably also belongs in this group. The original colicin E designation has been subdivided into three distinct groups, E1, E2, and E3, each having its own unique killing mechanism. Colicin E1 is a 56,000-dalton protein whose mode of action is similar to that of colicin I. No specific tolerance mutations are known. Colicin E2 is slightly larger at 62,000 daltons and inhibits DNA synthesis. The apparent direct effect is to activate exonuclease I, which causes DNA degradation. Colicin E2-specific tolerance loci include *tolD* or *tolE*. Colicin E3 is intermediate in size (60,000 daltons) and has a well defined mode of action. It causes the specific cleavage of about 50 bases from the 3′-end of the 16S ribosomal RNA molecule in either 30S ribosomal subunits or 70S ribosomes. The *tolD* and *tolE* loci also confer tolerance to colicin E3.

One unusual aspect of the colicin E group is that the mechanism of immunity within a producing cell has been identified. Colicin E2- and E3-producing cells also produce a small protein of about 10,000 daltons that acts as an immunity substance. Complexes of colicin and immunity protein have been identified in which the stoichiometric ratio is 1:1. Removal of the protein causes an increase in the in vitro lethality of the colicin. Just as the modes of action of colicins E2 and E3 are different, so also are the immunity proteins.

Although the colicin E plasmids are not self-transmissible, they and many other nonconjugative plasmids can transfer into recipient cells if they are in the presence of a compatible self-transmissible plasmid. The mechanism of this mobilization is not certain, but at least in the case of ColE3 the mobilization does not require F functions other than those necessary to form mating aggregates. Research has shown that a colicin E1 derivative, pBR322 (which is often used as a cloning vector), can also be mobilized by F. Those pBR322 DNA molecules that do transfer have acquired the γδ insertion sequence from F at the same time. The significance of this result is uncertain, but γδ may represent a region of homology for transient formation of a composite plasmid or may act as an origin of transfer.

Two other nonconjugative colicins have been tested. Colicin K has a mode of action like that of colicin I and a size of about 43,000 daltons. It is sometimes found complexed to the O antigen. Colicin M has been obtained as a complex of protein and phosphatidylethanolamine with a molecular weight of 27,000 daltons. Although its mode of action is unknown, the result is cell lysis.

Resistance Plasmids

General Properties

Resistance plasmids were originally identified in Japanese clinical isolates of *Shigella* strains during the 1950s. Researchers noted that the *Shigella* strains were simultaneously resistant to several antibiotics and that the combination of resistances tended to be the same. In other words, it did not appear that random mutations to individual antibiotic resistance were occurring. Further testing showed that the drug resistances were inherited by an infectious process just as in the case of the F plasmid. By 1971, 70 to 80% of all *Shigella* clinical isolates in Japan were found to carry multiple drug resistances. Moreover, the same phenomenon was being observed in other bacteria of clinical importance. Bacteria growing in areas polluted with heavy metals were also found to have acquired a kind of resistance that was inherited in the same manner as the antibiotic resistance.

Soon after the original observations, investigators showed that the DNA coding for the antibiotic resistances was plasmid DNA. The plasmids were called **R plasmids** (originally R factors), and many types of R plasmids were identified. Various combinations of resistance were found, and the nature of the resistance itself was shown to be unusual. Consider, for example, the case of resistance to the antibiotic streptomycin. Mutations on the bacterial genome can alter the 30S ribosomal subunit so that the cell becomes resistant to a high level of streptomycin (as much as 1 mg/ml), but these mutations occur infrequently (10^{-9}). By comparison, R plasmids may code for enzymes that chemically modify the streptomycin by adding adenylyl, phosphate, or acetyl groups to the molecule. This type of resistance can be overcome if the external concentration of the antibiotic is sufficiently high.

Classification of R plasmids is done using a system of incompatibility tests with other R plasmids as well as the F plasmid and various colicin plasmids. The name of an incompatibility group is assigned according to the predominant plasmid member. Thus F is the primary member of the IncF group of plasmids. Sometimes it is later necessary to subdivide an incompatibility group, which is done using Roman numerals appended to the group designation. In the case of the IncF group, for example, the F plasmid is now assigned to group IncFI. and the R100 plasmid is a member of IncFII. At present more than 30 incompatibility groups are known for *E. coli*, and *Pseudomonas* has eight more, five of which represent plasmids that do not transfer to *E. coli*. *Staphylococcus aureus* plasmids have been assigned to seven incompatibility groups. *Bacillus subtilis* has no naturally occurring R plasmids but propagates many *Staphylococcal* plasmids transferred to it by transformation or transduction. There are indications that certain *Bacillus* strains can carry out conjugation using the filter mating technique.

All of the originally observed R plasmids were self-transmissible, but more extensive investigation has shown that nonconjugative R plasmids also exist. Conjugative R plasmids using the F and I transfer systems are known, but incompatibility groups N and P have also been found to code for their own unique conjugation systems, complete with pili. IncN plasmids can mate only on solid surfaces (filter matings) or in liquids that are foaming. If more than one kind of conjugative plasmid is present in a cell, each system preferentially uses its own pili. The various conjugation systems also recognize different cell surface structures as part of mating aggregate formation. For example, F′*lac* and an R1 derivative (both IncF) transfer well to *lps* mutants, but an R64 derivative (IncIα) does not.

Despite the large number of incompatibility groups, it is rare to find bacteria carrying more than one or two R plasmids, probably because of the retarding effect on cell growth that results from the extra DNA. In a book of this size it is impossible to discuss even a representative number of R plasmids, and therefore only one such plasmid is considered as an introduction to this large group.

Plasmid R100

Plasmid R100 is a comparatively simple R plasmid that was isolated from *Shigella flexneri* in Tokyo. As originally isolated, it conferred resistance to streptomycin 100 μg/ml, chloramphenicol 200 μg/ml, tetracycline 12.5 μg/ml, and sulfonamide 200 μg/ml. The tetracycline resistance has been shown to be due to changes in cell permeability that reduce influx of the antibiotic and promote its efflux, as well as to the presence of an intracellular inhibitor. In contrast, the chloramphenicol resistance has been shown to be due to the presence of a chloramphenicol acetylating enzyme, and the streptomycin resistance has been shown to be due to the presence of a streptomycin adenylating enzyme. The molecular basis for the sulfonamide resistance has not been determined. Although the plasmid is self-transmissible and produces F pili, it belongs to the IncFII incompatibility group and therefore can coexist with the F plasmid, which is IncFI. R100 can transfer between *Escherichia, Klebsiella, Proteus, Salmonella,* and *Shigella,* but not into *Pseudomonas*. Two synonyms for the plasmid are NR1 and 222.

In an attempt to understand the way in which this plasmid behaves in a cell, Rownd and co-workers transferred it into a *Proteus mirabilis* strain. There were two principal advantages to working with this particular plasmid–host combination. First, the plasmid DNA was observed to be less associated with the bacterial chromosome than it was in *E. coli,* with about 20 to 25% of all plasmid molecules completely autonomous. This situation greatly facilitated the preparation of pure plasmid DNA molecules. The second advantage was that the densities of the plasmid and bacterial DNA molecules were different, which made separation of the two types of DNA by means of density gradients easy.

Density gradients are conveniently prepared by adding a dense chemical, usually cesium chloride or cesium sulfate, to a tube containing DNA. The powdered chemical is added until the density of the solution is approximately the same as that of the DNA, in this case about 1.7 g/cm^3. The tube is then placed in an ultracentrifuge rotor and spun at high speeds for periods ranging from 6 hours to several days until an equilibrium is reached. At equilibrium, some of the heavy metal ions have migrated to the bottom of the tube as a result of the forces generated by the spinning rotor. The shift in the position of the metal ions results in generation of a concentration gradient for the metal ions (and therefore a gradient of density as well) that is maintained so long as the rotor is spinning. Once the centrifuge has stopped, the process of diffusion gradually destroys the gradient, but it occurs so slowly it does not influence the normal experiment. The concentration difference between the top and bottom of the tube (the steepness of the gradient) is a function of the rate of rotation in the centrifuge. The faster the rotor spins, the steeper is the gradient.

Density gradients can be analyzed in one of two ways. The centrifuge can be stopped and the tube removed. The liquid in the tube can be carefully removed in small aliquots and each aliquot assayed for the presence of DNA. Alternatively, a special centrifuge and rotor can be used that allow a picture to be taken of the liquid sample while it is spinning. Because DNA absorbs ultraviolet (UV) radiation, it is possible to use UV illumination to obtain a picture that is a representation of the amount of DNA present as a function of position within the gradient. By scanning the photographic negative with a densitometer, it is possible to obtain a curve that is essentially a plot of the UV absorbance of the DNA (i.e., the concentration) as a function of position along the gradient. Rownd and co-workers used the latter technique. A sample set of densitometer scans can be seen in Figure 10-4. The larger peak with the lower density is the bacterial DNA, and the material of higher density is the R plasmid.

When scans such as these were originally examined, it was noted that there tended to be two peaks of plasmid DNA. By comparing the profiles obtained with various R plasmids, it was determined that tetracycline resistance and self-transfer functions were usually associated with the less dense DNA, and that the other antibiotic resistance cistrons tended to be associated with the more dense DNA. These observations, made in several

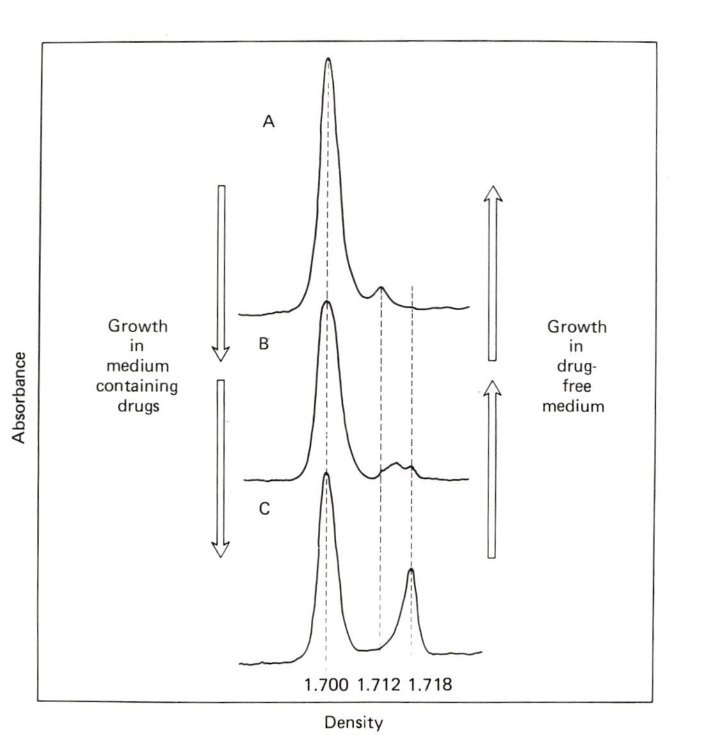

Figure 10-4. Systematic changes in the DNA density profile of an R plasmid. R plasmid R100 DNA was extracted from *Proteus mirabilis* cells grown in either drug-free or drug-containing medium. Profiles A, B, and C were obtained by sampling at different times after inoculation of the medium. The large peak of DNA on the left at a density of 1.700 is bacterial DNA, and the smaller peaks to the right are various forms of R plasmid DNA. From Rownd, R.H., Perlman, D., Goto, N. (1974). Structure and replication of R-factor DNA in *Proteus mirabilis*, pp. 76–94. In: Schlessinger, D. (ed.) Microbiology 1974. Washington, D.C.: American Society for Microbiology.

laboratories, led to the suggestion that R plasmids consisted of two parts: a **resistance transfer factor** (RTF), which coded for all the transfer functions and tetracycline resistance, and an **r-determinant** (r-det), which coded only for antibiotic resistance. However, one unexplained fact remained. In the presence of an appropriate antibiotic such as chloramphenicol, the density of the plasmid DNA increased significantly beyond that normally observed. After removal of the antibiotic, the density gradually returned to its original value (Fig. 10-4). This phenomenon was observed only in *Proteus*.

In order to explain the changes observed in R100, Rownd and his co-workers postulated the existence of a process they called **transitioning** and which is essentially a model for cistron amplification. The basic assumptions of the model are that a large plasmid is more of a retarding force on cell growth than is a small plasmid, and that the amount of protein that can be produced by a cell is proportional to the number of copies of the appropriate DNA sequence present in the cell. These two assumptions lead to a contradiction for the cell. Maximum antibiotic resistance is incompatible with rapid cell growth. Therefore cells that have grown for a long period in the absence of antibiotic will have been selected for the presence of a small-sized plasmid. When antibiotic is added, r-determinant DNA is duplicated to provide extra copies of the DNA, increased transcription, and more mRNA for translation. The observations of Figure 10-4 can be explained if the r-determinant DNA is denser than the RTF. Extra r-determinant copies cause the density of the plasmid to shift to a heavier value. Genetic studies have provided an explanation for how transitioning occurs.

Genetic Analysis of Plasmid R100

Genetic and physical maps of R100, prepared in a manner similar to that used for the F plasmid, provide evidence to account for the transition phenomenon. The genetic map is shown in Figure 10-5 and should be compared with the F plasmid map from Figure 9-11. The *tra* cistrons are found in the same order and relative orientation on both maps. Note also that each region of R100 that codes for antibiotic resistance is flanked by paired insertion elements. This configuration is characteristic of transposons, and in fact the tetracycline resistance domain was quickly shown to be transposon Tn*10*, which is flanked by IS*10* elements in an inverted orientation. More recently Arber and his co-workers have demonstrated that the r-determinant region is also a transposon, Tn*2671*, which is bounded by IS*1* elements in a direct repeat configuration.

One possible interpretation of transitioning is that the addition of antibiotic selects for spontaneous variants in which the Tn*2671* has transposed itself to give extra copies of the r-determinant. Removal of the antibiotic then selects for the opposite phenomenon, transposon-catalyzed deletion

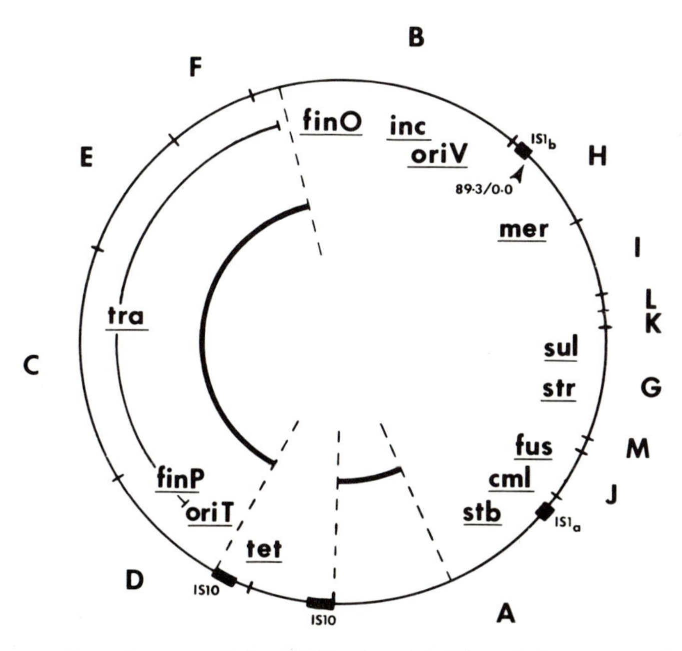

Figure 10-5. Genetic map of the R100 plasmid. The circle represents the R100 DNA molecule and the positions of known insertion elements (black rectangles). The total length of the DNA is 89.3 kilobase pairs, with the coordinate system beginning to the right of $IS1_b$. The letters surrounding the map designate *Eco*R1 restriction fragments (see Chapter 14). The abbreviations within the circle indicate various cistrons that have been identified. Resistances to mercuric ion, sulfonamides, streptomycin/spectinomycin, fusidic acid, chloramphenicol, and tetracycline are indicated by *mer, sul, str, fus, cml,* and *tet,* respectively. The other cistron symbols are as follows: *oriT,* origin of transfer; *oriV,* origin of vegetative (normal) replication; *finO* and *finP,* fertility inhibition; *stb,* stability of the plasmid within a host cell; *inc,* incompatability with other plasmids of the same type; and *tra,* conjugal transfer. Because of the increasingly large number of *tra* cistrons that have been mapped, no attempt has been made to indicate their individual positions. Instead, the *tra* region is indicated by an arc. The innermost heavy line represents the region that is approximately 90% homologous with plasmid F. The interruption in the region of F homology is caused by the presence of the Tn*10* transposon. Redrawn from Dempsey, W.B., Willetts, N.S. (1976). Plasmid cointegrates of prophage lambda and R factor R100. Journal of Bacteriology 126:166–176; and from Dempsey, W.B., McIntire, S.A. (1979). Lambda transducing phages derived from a *finO*$^-$ R100:: λ cointegrate plasmid: proteins encoded by the R100 replication/incompatability region and the antibiotic resistance determinant. Molecular and General Genetics 176:219–334. Figure furnished by Drs. S. McIntire and W. Dempsey.

of r-determinants. However, Peterson and Rownd demonstrated that if the IS*1* flanking elements of Tn*2671* were cut off and replaced by artificially prepared direct repeats, transitioning was still observed. The conclusion seems to be that the r-determinant is indeed a transposon, but the transitioning function may be catalyzed by a special enzyme unique to the *Proteus* system. This enzyme requires only a direct repeat for its activity and can use either the normal IS*1* repeats or any suitably sized cloned DNA.

This conclusion is reinforced by observations made with different host cells. As mentioned above, *E. coli* cells yield R100 DNA with only the composite density. On the other hand, although the entire R100 can be transferred into *Salmonella typhimurium*, in the absence of continued antibiotic selection the complete r-determinant is rapidly lost, leaving only a tetracycline-resistant cell. The process is dependent on normal functioning of the *recA* cistron (major recombination pathway) and may again be related to the direct repeats provided by the IS*1* elements.

Arber and his co-workers have shown that it is possible to have the Tn*2671* transposon hop to the phage P7 plasmid DNA. Amplification of the r-determinant can be seen to occur along with inversion of the P7 DNA. They have also created an artificial r-determinant from the antibiotic resistance markers found in plasmid pBR325 using IS*1* elements in the inverted repeat configuration and cloned it into phage P1 plasmid. This construct also amplifies itself and in the process duplicates various regions of the plasmid DNA.

The transposon nature of the antibiotic resistances provides a possible explanation for the origin of R plasmids. Hughes and Datta examined a large number of bacteria that were collected over the years 1917 to 1924, before the advent of antibiotics, and maintained in a lyophilized state until the present time. The strains showed little antibiotic or mercury resistance, but 14% were colicinogenic. Some 24% of the strains carried plasmids capable of mobilizing a normally nonconjugative plasmid that could not be mobilized by F. The authors concluded that many plasmids were available to act as hosts to transposons and that the present R plasmids may have arisen by transposition of antibiotic markers. The remaining question, of course, concerns the origin of the antibiotic resistance genes. Davies and Benveniste suggested that these genes must be present in the original producing strains, and there is some evidence to support the notion that they have been transposed onto preexisting plasmids to create new R plasmids. For example, DNA sequence analysis has revealed significant homologies between the aminoglycoside phosphotransferase cistrons of Tn*5* and Tn*903* and the corresponding cistrons in *Streptomyces fradiae* and *Bacillus circulans*.

The evidence for the evolutionary origin of R plasmids is suggestive. Evidence for the ability of transposons to move rapidly through a population is direct. In the specific case of R100, transposon Tn*10* has been followed by Kleckner and co-workers through a series of genetic exchanges beginning in *S. typhimurium*. Tn*10* moved from R100 onto the DNA of phage P22 using at least 20 integration sites. From P22, Tn*10* then hopped to the *Salmonella* genome itself, integrating into 100 or more sites. Tn*10* was next acquired by phage λ (appropriately deleted to make room for it) yielding five λ*tet* phages. The λ phages then transported Tn*10* to *E. coli*, where an additional 20 integration sites were identified. It is not possible for Tn*10* to have extensive sequence homology with so many sites, and therefore Tn*10* is behaving similarly to phage Mu. Some models for transposition are discussed in Chapter 13.

Conjugal Plasmid Interactions

As comparative studies of F and R plasmids continued, it soon became apparent that F transfer was, in general, more efficient. The ability of most R^+ cells to transfer their plasmids was found to be less than 1% that of F^+ cells. There were exceptions among the R plasmids, however, whose transfer was as efficient as that of F. Moreover, even those plasmids that normally transferred themselves inefficiently were found to be efficient at transfer, provided they had only recently transferred into their current host cell. The simplest explanation for these observations was that most R plasmids produced a substance that gradually repressed (turned off) the *tra* cistrons. Thus a cell carrying a newly arrived plasmid would have only low levels of repressor and therefore good expression of the *tra* cistrons. However, a cell line that had had the same plasmid for many generations would have maximal amounts of repressor and therefore little *tra* cistron expression.

For the experiments used to establish the various incompatibility groups for the R and F plasmids, many combinations of F and R plasmids in the same cell were prepared. Frequently the R plasmid seemed to have no effect on the F plasmid, but certain R plasmids were observed to reduce the efficiency of transfer of the F plasmid to levels approximating their own inefficient transfer. Such plasmids were said to have the property of **fertility inhibition** (fi^+), which in terms of the repression model presented above means that the repressor produced by the R plasmid also affects the F plasmid. The fi^- R plasmids, those that had no effect on the transfer ability of F, were found to be of two types. One type was observed to be derepressed for its own transfer and therefore presumably did not make a functional repressor. The other was observed to code for I-type pili. As noted above, the I pili indicate the presence of an entirely different transfer system, and therefore it is not surprising that there should be no interaction between the two.

Willetts and co-workers (Finley et al.) analyzed the fi$^+$ phenotype genetically and found that it is determined by two loci, *finO* and *finP*. The FinO protein combines with the FinP RNA to act in a concerted fashion to bind to *fisO* (*traO*, whose sequence is complementary to FinP) and block transcription of *traJ*. Without the TraJ protein, most of the other *tra* cistrons cannot be expressed because they are coordinately regulated as one huge operon such as those discussed in Chapter 12. Without the *tra* functions, self-transfer of the plasmid is impossible. The F plasmid has been shown to be a spontaneous *finO* mutant due to insertion of an IS*3* element, which accounts for its high level of fertility.

The general model for fertility inhibition is shown in Figure 10-6. Initial transfer of the R plasmid is presumed to occur from the rare cell that has expressed the *tra* functions. Once the first transfer has occurred, further transfer is rapid in an epidemic type of spreading initiated by the newly created R$^+$ cells (which are derepressed for several generations). Although plasmids tend to spread through a population of cells of their own accord, the process is considerably enhanced if antibiotic selection is applied. This finding constitutes a strong argument against the indiscriminate use of antibiotics.

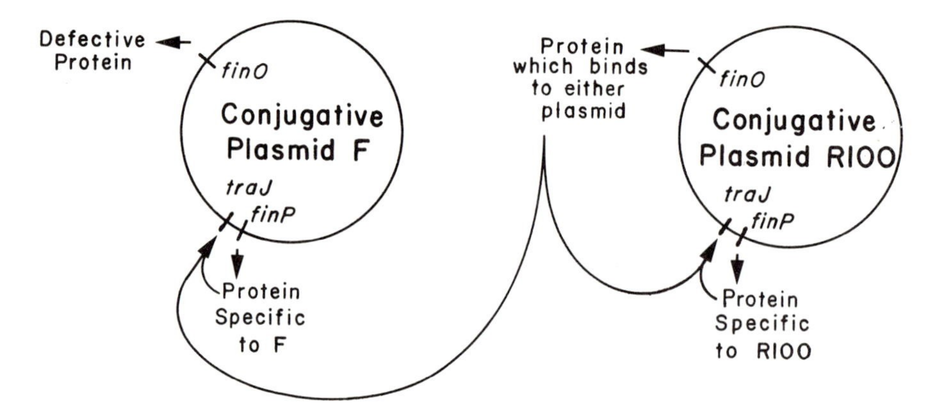

Figure 10-6. Model for the inhibition of F plasmid fertility by R100 as conceived by Willetts. The *fisO* cistron is a regulatory element for expression of the fertility functions and is common to both plasmids. It allows expression of these functions unless two products are simultaneously bound to it. These products (one protein molecule and one RNA molecule) are coded by the *finO* and *finP* cistrons, respectively. In the F plasmid the *finO* cistron product is defective so no fertility inhibition is observed. However, R100 produces a functional protein from *finO* that works with either plasmid. Therefore the F plasmid, in the presence of R100, represses its fertility functions.

Summary

The bacterial kingdom seems to be full of plasmids. They can be conjugative or nonconjugative, but all fulfill the usual plasmid definition. Among the conjugative plasmids are several that frequently integrate into the host cell chromosome to give high frequency donors. This integration can often be enhanced by inserting transposons or bacterial DNA into the plasmid. The conjugation systems that have been well studied show many variations on the central theme. Like *E. coli*, both *Pseudomonas* and *Streptococcus* have conjugative systems that require the formation of mating aggregates, but they use different mechanisms. No plasmid-specific pili have been observed in the case of *Pseudomonas*, and *Streptococcus* uses a system of pheromones to cause release of an aggregation factor that acts to clump the cells. In *Streptococcus* there is also an unusual system of conjugative transposition whose mechanism is not understood but that does not seem to involve plasmids at all. Conjugal DNA transfer in *Streptomyces* is plasmid-dependent but often occurs bidirectionally from a single point of origin. The difference between that type of conjugation and normal Hfr function remains to be clarified. Less well studied conjugative systems have been identified among the colicin and R plasmids.

The *Agrobacterium* conjugative system based on the Ti plasmid serves to transfer DNA not to another bacterium but into a plant cell. It serves as the basis for an unusual form of parasitism and has excited much interest because of its potential for genetic engineering. A portion of the Ti plasmid, the T-DNA, is inserted into a plant chromosome, and other DNA cloned into the T-DNA sequence can be carried along.

Evolutionary change in plasmids is apparently often caused by the insertion and deletion of transposons. Similarities and differences among plasmids can be identified by the usual hybridization and mapping techniques as well as by the fact that some plasmids can coexist within a cell and others cannot. The incompatibility groups thereby defined serve to subdivide most plasmids into manageable families. Even plasmids that are not incompatible may nevertheless interact. A good example is the F plasmid whose transfer can be inhibited by closely related plasmids. The inhibition is mediated through the *fin* system.

References

General

Chater, K.F., Hopwood, D.A. (1983). Streptomyces genetics, pp. 230–286. In: Goodfellow, M., Mordarski, M., Williams, S.T. (eds.) The Biology of the Actinomycetes. Orlando, Florida: Academic Press.

Clewell, D.B., Gawron-Burke, C. (1986). Conjugative transposons and the dissemination of antibiotic resistance in streptococci. Annual Review of Microbiology 40:635–659.

Helinski, D.R., Cohen, S.N., Clewell, D.B., Jackson, D.A., Hollaender, A., eds. (1985). Plasmids in Bacteria. New York: Plenum Press.

Holloway, B.W., Morgan, A.F. (1986). Genome organization in *Pseudomonas*. Annual Review of Microbiology 40:79–105.

Levy, S.B., Novick, R.P. (1986). Antibiotic Resistance Genes. Ecology, Transfer, and Expression. Cold Spring Harbor, N.Y.: Cold Spring Harbor Laboratory.

Lyon, B.R., Skurray, R. (1987). Antimicrobial resistance of *Staphylococcus aureus:* genetic basis. Microbiological Reviews 51:88–134.

O'Brien, S.J. (1986). Genetic Maps, vol. 4. Cold Spring Harbor, N.Y.: Cold Spring Harbor Laboratory.

Riley, M., Anilionis, A. (1978). Evolution of the bacterial genome. Annual Review of Microbiology 32:519–560.

Schaberg, D.R., Zervos, M.J. (1986). Intergeneric and interspecies gene exchange in gram-positive cocci. Antimicrobial Agents and Chemotherapy 30:817–822.

Stachel, S.E., Zambryski, P.C. (1986). *Agrobacterium tumefaciens* and the susceptible plant cell: a novel adaptation of extracellular recognition and DNA conjugation. Cell 47:155–157.

Stephens, K. (1986). Pheromones among the prokaryotes. CRC Critical Reviews in Microbiology 13:309–334.

Specialized

Chardon-Loriaux, I., Charpentier, M., Percheron, F. (1986). Isolation and characterization of a linear plasmid from *Streptomyces rimosus*. FEMS Microbiology Letters 35:151–155.

Christie, P.J., Korman, R.Z., Zahler, S.A., Adsit, J.C., Dunny, G.M. (1987). Two conjugation systems associated with *Streptococcus faecalis* plasmid pCF10: identification of a conjugative transposon that transfers between *S. faecalis* and *Bacillus subtilis*. Journal of Bacteriology 169:2529–2536.

Courvalin, P., Carlier, C. (1987). Tn*1545:* a conjugative shuttle transposon. Molecular and General Genetics 206:259–264.

Finlay, B.B., Frost, L.S., Paranchych, W., Willetts, N.S. (1986). Nucleotide sequences of five IncF plasmid *finP* alleles. Journal of Bacteriology 167:754–757.

Hughes, V.M., Datta, N. (1983). Conjugative plasmids in bacteria of the "preantibiotic" era. Nature 302:725–726.

Iida, S., Kulka, I., Meyer, J., Arber, W. (1987). Amplification of drug resistance genes flanked by inversely repeated IS*1* elements: involvement of IS*1*-promoted DNA rearrangements before amplification. Journal of Bacteriology 169:1447–1453.

Leroux, B., Yanofsky, M.F., Winans, S.C., Ward, J.E., Ziegler, S.F., Nester, E.W. (1987). Characterization of the *virA* locus of *Agrobacterium tumefaciens:* a transcriptional regulator and host range determinant. EMBO Journal 6:849–856.

Mori, M., Tanaka, H., Sakagami, Y., Isogai, A., Fujino, M., Kitada, C., Clewell, D.B., Suzuki, A. (1987). Isolation and structure of the sex pheromone inhibitor, iPD1, excreted by *Streptococcus faecalis* donor strains harboring plasmid pPD1. Journal of Bacteriology 169:1747–1749.

Peralta, E.G., Hellmiss, R., Ream, W. (1986). *Overdrive*, a T-DNA transmission enhancer on the *A. tumefaciens* tumour-inducing plasmid. EMBO Journal 5:1137–1142.

Sinclair, M.I., Maxwell, P.C., Lyon, B.R., Holloway, B.W. (1986). Chromosomal location of TOL plasmid DNA in *Pseudomonas putida*. Journal of Bacteriology 168:1302–1308.

Veluthambi, K., Jayaswal, R.K., Gelvin, S.B. (1987). Virulence genes *A*, *G*, and *D* mediate the double-stranded border cleavage of T-DNA from the *Agrobacterium* Ti plasmid. Proceedings of the National Academy of Sciences of the United States of America 84:1881–1885.

Yoshioka, Y., Ohtsubo, H., Ohtsubo, E. (1987). Repressor gene *finO* in plasmids R100 and F: constitutive transfer of plasmid F is caused by insertion of IS*3* into F *finO*. Journal of Bacteriology 169:619–623.

Chapter 11
Plasmid Molecular Biology

Plasmids face many of the same problems as bacterial chromosomes. They must replicate themselves, synthesize appropriate gene products, and segregate at least one copy of the plasmid into each daughter cell at cell division. However, plasmids also maintain incompatibility functions that prevent other, similar plasmids from taking up residence in the same cell and in many cases can transfer their DNA to other cells. They exist in a characteristic low or high copy number relative to the bacterial chromosome. This chapter considers the various methods by which plasmids accomplish these tasks.

Plasmid DNA Replication

General Considerations

Replication for the most part follows the same pattern as discussed for the bacterial chromosome or for some of the phages, but there are a few differences. Certain plasmids seem to display variations on the general theme. For example, the ColE1 and R100 plasmids replicate unidirectionally rather than bidirectionally. In *Bacillus subtilis* and *Staphylococcus aureus* there are several small R plasmids, typified by pC194, that can be isolated in either single-strand or double-strand form. The general presumption is that the single-strand form is some sort of replication intermediate such as a rolling circle in which the leading strand synthesis is temporally separated from the lagging strand synthesis.

Replication functions can be subdivided into two parts, those that are concerned with initiating a new round of replication and those that are concerned with the actual synthesis of the new DNA strands. Temperature-sensitive (ts) mutations affecting these processes likewise have two general phenotypes. If at high temperature the mutation blocks the initiation process, those replication forks that have already begun continue until the replication is completed, but no new forks are added. With respect to the bacterial chromosome, this type of mutation is called a slow stop mutation. The other type of mutation blocks the elongation process at high temperature and is therefore designated a rapid stop mutation. Plasmids interact with the host DNA replication system in varying ways, but in general the smaller plasmids use more host functions, and the larger plasmids provide more of their own replication functions. Regardless of plasmid type, all fulfill the criteria set down in Table 11-1.

It is fairly easy to demonstrate that at least some plasmids supply portions of their replication machinery. Consider the case of integrative suppression by the F plasmid as discussed in Chapter 9. A *dnaB* ts mutation has blocked initiation of new rounds of chromosome replication. The formation of an Hfr strain restores the ability of the chromosome to replicate, but the origin of replication is now within the F plasmid rather than at *oriC*. This result indicates that initiation of replication by F is independent of *dnaB* function. In a related example, the intercalating dye acridine orange prevents replication of the F plasmid but leaves chromosomal DNA replication intact. If the culture is treated with the dye for a sufficient period of time, **curing,** or loss of the plasmid, results. A similar curing effect is seen if the culture is incubated above 42°C but below about 46°C (the maximum temperature for sustained growth of *Escherichia coli*). Even the large plasmids require some host cell functions for initiation of DNA replication. Both phage P1 (in the plasmid form) and the F plasmid require normal *dnaA* function in order to replicate. By way of contrast, ColE1 plasmids do not code for any proteins required for their replication.

Most plasmids as well as the bacterial chromosome require protein synthesis as a part of the initiation process. The presumption is that a

Table 11-1. General characteristics of plasmid replication

1. There is autonomous replication independent of chromosomal replication and cell division cycles.
2. There is a wide spread in time intervals between replication events.
3. Replication occurs throughout the cell cycle.
4. Characteristic copy numbers are defined by plasmid, host, and growth conditions.
5. Accidental deviations from normal copy number are adjusted.
6. Copy mutants exist. Some are dominant and some are recessive to the wild type.
7. Cloned DNA may cause inhibition of plasmid replication or incompatibility.

Adapted from Nordström (1985).

particular, unstable protein must be present in sufficient quantity to allow replication to begin. If an antibiotic such as chloramphenicol that blocks protein synthesis is added to the culture, DNA synthesis comes to a gradual stop. Such a requirement is not, however, universal. In the case of the ColE3 plasmid and its derivatives, the addition of chloramphenicol has exactly the opposite effect. The plasmids do not require protein synthesis for initiation and can use the existing replication enzymes. Relieved of the competition for resources, the plasmids continue to replicate for many hours, increasing the copy number to extraordinarily high numbers, e.g., 1000 per cell. This phenomenon of **amplification** has been a great boon to scientists working with cloned DNA, and many standard cloning vectors have been based on the ColE3 plasmids.

For low copy number plasmids and the bacterial chromosome itself there is a characteristic DNA/cell mass ratio that seems to be invariant when averaged over an entire culture, which implies that DNA replication is always triggered at approximately the same point in the growth cycle. The control is apparently much more precise for bacterial chromosomes than for plasmids. In a strain in which the R1 plasmid has been inserted into *oriC*, the origin of chromosome replication, the timing of the replication event becomes nearly random with respect to the rest of the cell cycle.

Tresguerres and co-workers examined an *E. coli* F^+ culture and compared it to a similar Hfr culture derived by integrative suppression of a *dnaA* ts mutation. At high temperature they found that the DNA/cell mass ratio was lower, indicating that the F plasmid replication system recognizes a signal different than that of the bacterial replicon. In a related experiment, Lycett and Pritchard showed that the F plasmid origin of replication may be functional in Hfr strains, depending on the actual site of insertion. An Hfr strain with transfer origin near *trp* (close to *terC*, the replication terminus) had a slight excess of copies of the *trp* DNA over what was observed for an Hfr strain with an insertion near *oriC*. They interpreted this observation to mean that the relatively low number of F plasmid copies near *terC* occasionally triggers the plasmid replication origin, whereas an F plasmid inserted near *oriC* never achieves the correct DNA/cell mass ratio to activate its replication origin.

The study of plasmid replication has been greatly facilitated by DNA splicing technology. Many experiments are now carried out on **miniplasmids**, constructs that have been prepared by shortening naturally occurring plasmids so as to eliminate nonessential functions. Generally speaking, the miniplasmids contain only enough DNA to self-replicate and to provide markers (usually antibiotic resistance) for detection of the plasmid. They also allow the experimenter to rearrange components from different systems in order to study their interaction. The essential components appear to be an origin of replication and the necessary sequences for the synthesis of the RNA primer.

Control of Copy Number

The copy number of plasmids is obviously a function of the rate at which new rounds of replication are initiated. Plasmids with relatively high copy numbers demonstrate little if any regulation of initiation of DNA replication. Low copy number plasmids, on the other hand, are highly regulated and have the additional requirement that their initiation step must be coordinated with the cell division cycle. If it were not, cytokinesis might occur at a time when there were insufficient copies of the plasmid to supply at least one complete plasmid DNA molecule to each daughter cell.

Measurement of copy number was not originally a trivial procedure. Simple extraction of plasmid DNA from cells in order to measure its relative amount is complicated by entanglement of plasmid and bacterial DNA. If any plasmid DNA breaks off during extraction, the copy number for the plasmid is underestimated. The most satisfactory methods for estimating copy number involve either using radioactive probes and Southern blots of gently extracted DNA or looking at the relative rates of reversion of identical mutant alleles located on the plasmid or the bacterial chromosome. In the latter case it is assumed that the physical location of the gene does not affect its probability of reverting, and therefore the number of plasmid revertants divided by the number of chromosomal revertants should equal the copy number.

Typical low copy number plasmids are the IncF plasmids including the F plasmid (IncFI subgroup) and plasmids R100 (NR1) and R1 (IncFII subgroup). Plasmids ColE1 and the *Staphylococcus aureus* plasmid pT181 behave similarly to the IncFII group. There are two fundamentally different strategies used by these plasmids for controlling their copy number.

Plasmid R1 (R100) codes for a RepA protein that is absolutely required for its unidirectional DNA replication. The number of initiations of replication is proportional to the RepA protein concentration. It is mainly a *cis*-acting protein, meaning that it preferentially acts on the DNA molecule from which it was produced. The *repA* cistron can be transcribed as part of either of two large mRNA molecules (*copB* RNA and *copT* RNA) that are initiated at different promoters but have the same 3′-ends. It is downstream from a cistron designated *copB* (Fig. 11-1). In this region of the plasmid DNA there are promoters on both strands, so transcription is possible in either direction. The smaller, leftward transcript is designated *copA* RNA and is of course complementary to a portion of either larger transcript. The CopB protein acts as a repressor to inhibit synthesis of *copT* RNA (regulation at the level of transcription), and *copA* RNA binds to some hairpin loops in the *copT* molecule just upstream from *repA*, preventing ribosomes from synthesizing the RepA protein (regulation at the level of translation). It is assumed that pairing begins at this loop and continues to form a more extensive, duplex structure that is energetically more favorable than the structure formed by either individual molecule.

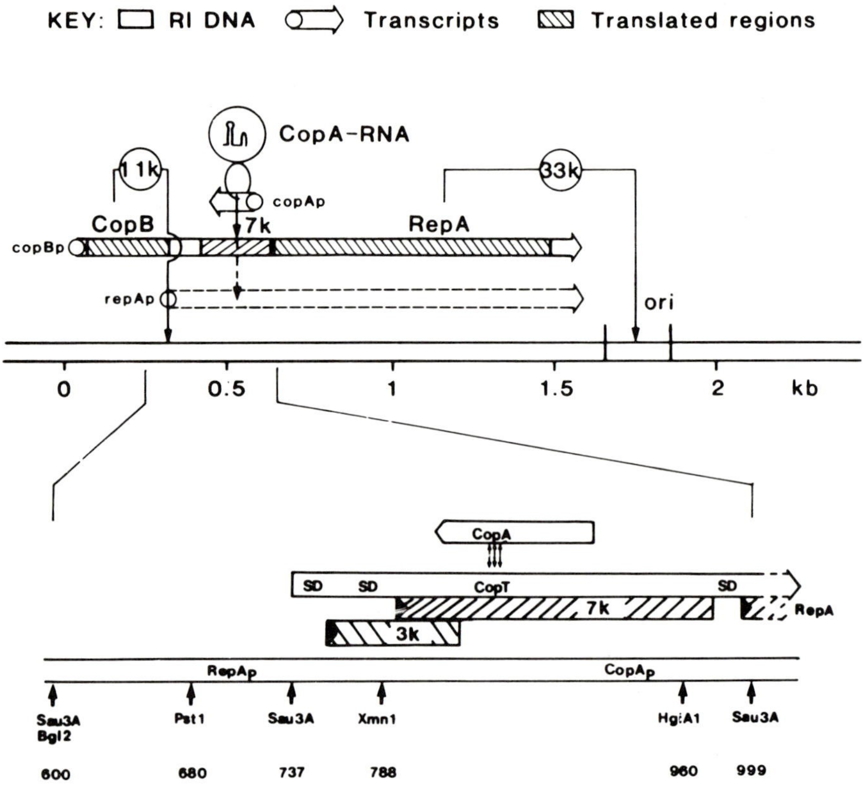

KEY: RI DNA Transcripts Translated regions

CopA-RNA
11k
33k
copAp
CopB
7k
RepA
copBp
repAp
ori
0 0.5 1 1.5 2 kb

CopA
SD SD CopT SD
7k
RepA
3k
RepAp
CopAp
Sau3A Bgl2 Pst1 Sau3A Xmn1 HgiA1 Sau3A
600 680 737 788 960 999

Figure 11-1. Copy number control region of plasmid R1. Note that the main transcript including the *repA* cistron is transcribed from one strand and the *copA* molecule from the other. The upper drawing shows the transcription/translation patterns and the regulatory loops involved in replication control. The lower section shows the *repA* expression control region in more detail. The locations of some restriction enzyme cleavage sites are given below. Potential ribosome binding sites (SD) are depicted on the RepA mRNA. The approximate position for primary recognition between CopA and CopT RNA sequences is shown by arrows. Hatched areas indicate polypeptide translation regions. Note that so far translation of the putative 3k protein has not been demonstrated. The position of the two promoters can be seen between the lines representing the DNA. From Wagner, E.G.H., von Heijne, J., Nordström, K. (1987). Control of replication of plasmid R1: translation of the 7k reading frame in the RepA mRNA leader region counteracts the interaction between CopA RNA and CopT RNA. EMBO Journal 6:515–522.

Translation of the 7k open reading frame eliminates the hairpin loop and prevents binding of the *copA* RNA, so *repA* is more readily transcribed.

When the R1 plasmid transfers to a new cell, neither *copA* RNA nor CopB protein is available in quantity, and the plasmid initially replicates rapidly. This overreplication facilitates establishment of the plasmid in the cell line. The larger number of plasmid molecules soon yield significant quantities of the two regulators, and the copy number reaches its normal value. That the two regulators do not carry equal regulatory weight is shown by mutagenesis experiments. If *copB* is inactivated, the copy number increases somewhat. If *copA* synthesis is prevented, **runaway replication** occurs in which the plasmid essentially replicates out of control and rapidly fills the cytoplasm with plasmid DNA molecules.

Just as the R1 plasmid has its RepA protein, so the F plasmid has an essential protein for its bidirectional replication, the E protein, the product of the *repE* cistron. Like the tryptophan repressor (see Chapter 12), this protein is **autoregulated** in that high concentrations of the protein seem to inhibit further transcription of its cistron. Unlike R1, the IncFI F plasmid (and the P1 prophage replicating as a plasmid) display a copy number control system that does not seem to involve the synthesis of proteins. Copy number mutants of mini-F plasmids have been isolated and map to two positions. The *copA* mutations map to the E protein cistron, and the *copB* mutations map to the region near the origin of replication and the region coding for the carboxy-terminus of protein E (Fig. 11-2). DNA sequence analysis of this region has shown that there are five 19- to 22-bp direct repeats.

By deleting two or more of these repeated sequences, it has been possible to demonstrate that the copy number of the F or P1 plasmid is in-

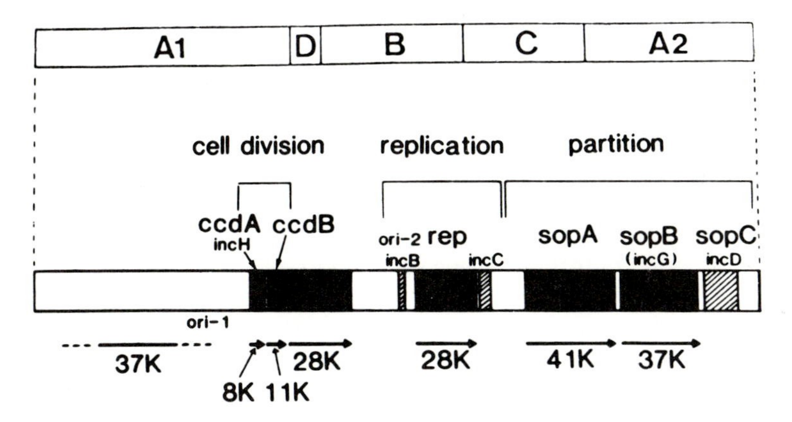

Figure 11-2. Incompatibility and copy number control regions on the F plasmid. Shown is a physical and functional map of an *Eco*RI-generated DNA fragment of the F plasmid. The *sop* region is essential for plasmid partitioning and therefore is also denoted as *par*. Note the extensive overlap of the incompatability functions with the other regulatory regions. The filled areas of the map show the locations of protein coding regions. Hatched areas are direct repeats, 19 or 22 bp in *incB* and *incC* and 43 bp in *incD*. The arrows show the direction of transcription of the proteins. From Hiraga, S., Ogura, T., Mori, H., Tanaka, M. (1985). Mechanisms essential for stable inheritance of mini-F plasmid, pp. 469–487. In: Helinski, D.R., Cohen, S.N., Clewell, D.B., Jackson, D.A., Hollaender, A. (eds.) Plasmids in Bacteria. New York: Plenum Press.

versely proportional to the number of repeats. A comparison of the repeated sequence with that of the start region for protein E transcription reveals that there are some striking similarities. As an autoregulated protein, E must bind near its own promoter, and therefore it seems likely that E also binds to the direct repeats. Because E is essential for F replication, it must be present at a reasonable concentration. To the extent that the direct repeats are present, they bind up E and prevent it from starting a new replication fork. If fewer repeats are present, more E molecules are available to initiate replication and the copy number increases. Thus replication is controlled without the use of an inhibitory protein or RNA molecule. The ability to delete the repeated sequences without loss of replication ability indicates that they are involved only in control, not in initiation of replication.

Partitioning

Partitioning is the act of segregating DNA molecules of each type so that there is at least one DNA molecule for each daughter cell. It is the prokaryotic equivalent to the movement of chromosomes during anaphase of mitosis. The possible use of the membrane attachment site for segregation of the bacterial chromosome was discussed in Chapter 1. By analogy to eukaryotic systems, the DNA site that is presumed to attach to the membrane has been referred to as a centromere. High copy number plasmids do not rquire any special mechanism for partitioning. Unless the plane of cell division is grossly off center, it is almost impossible that at least one plasmid DNA molecule would not be correctly situated.

The situation is different when the low copy number plasmids in the IncFI and IncFII groups are considered. If the copy number is only one or two, precise control must be exercised to prevent all the plasmid DNA molecules from being inherited by only one daughter cell. For example, a potential problem for a low copy number plasmid can arise if two plasmid DNA molecules undergo a single recombination event to generate a larger circle of DNA. In such an instance it is conceivable that all of the plasmid DNA might end up in only one daughter cell. The R46 plasmid circumvents this problem by coding for a *per* (plasmid-encoded recombinase) function. The recombinase is an enzyme that catalyzes site-specific recombination of plasmid DNA somewhat like the *loxP* function of phage P1. The effect of the enzyme is to take a large DNA molecule containing two identical sites and convert it to two smaller ones (Fig. 11-3). There is a suggestion that a similar phenomenon may occur in the ColV2-K94 plasmid. Linear multimers of ColE1 DNA have been reported in cells that carry the *recB* and *recC* mutations. Nothing is yet known about the partitioning function in this situation.

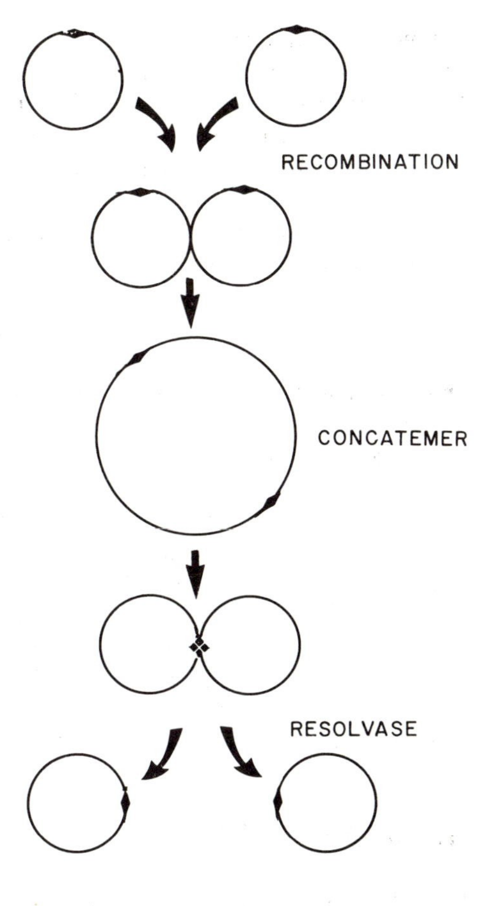

Figure 11-3. Effect of a resolvase enzyme on a concatemeric DNA molecule. The concatemer is assumed to form by a generalized recombination mechanism. Resolvases recognize a particular site (diamond) on a given DNA molecule but catalyze an exchange of strands only when two such sites are present. The reactions may be intermolecular or intramolecular. The latter reaction generates two circular DNA molecules where there had been only one catcatemer, as in this example.

One straightforward approach for resolving partitioning problems has been observed with the R1 plasmid. As reported by Gerdes and co-workers, there is a region on the R1 plasmid designated *parB* that can be subdivided in two parts. One portion codes for the *hok* (host killing) function and the other for the *sok* (suppressor of *hok*) function. The model is that the Hok product is routinely present in the cytoplasm. However, the effects of Hok are blocked by Sok in a manner analogous to the way in which a colicin-producing cell is protected against toxic effects. The functionality of the Sok product is assumed to be short-lived, so if a cell is segregated that does not contain the R1 plasmid the residual Hok product kills it. Therefore clones carrying both the *hok* and *sok* regions do not harm the cell, but a clone of just the *hok* region is lethal.

A related approach is employed by the F plasmid via its *ccd* (coupled cell division) function. Two proteins are encoded with this region. It appears that *ccdB* inhibits cell division and that *ccdA* serves as a suppressor of *ccdB* function. When the number of copies of F DNA in a cell drops below a critical threshold, the CcdB function prevents furthercell division until the plasmid number increases and CcdA is once again predominant. By this means production of a plasmid-free cell is prevented.

The other partitioning function used by F is the *sop* (*s*tability *o*f *p*lasmid) function. Three discrete sites have been identified: *sopA*, *sopB*, and *sopC*. Similar functions designated *par* have been identified in plasmid R1. The *sopC* site is a *cis*-acting region that can help to properly segregate an otherwise unstable miniplasmid. It is apparently the location on the DNA that is partitioned and as such is the functional equivalent of a centromere. The *sopA* and *sopB* cistrons code for specific proteins that further improve the stability of an unstable plasmid and can act in *trans*. DNA sequence analysis indicates that the *sopC* (*parC*) region is a series of eleven or twelve 43-bp direct repeats with no intervening spaces. Each repeated sequence contains internal 7-bp inverted repeats. It can be shown that the SopA and SopB proteins (also known as ParA and ParB) bind to this region and presumably link it to the membrane. Only two of the repeated sequences are required to stabilize the F miniplasmid. Interestingly, there is a small plasmid found in yeast, the 2μ plasmid (named for its size), that displays a similar structure of multiple 62- to 63-bp direct repeats. It is therefore possible that the F partitioning system is widely applicable. Skurray and his co-workers demonstrated that there is a second *cis*-acting partition element in F that they designated ParL. Little is known about the mode of action of this locus.

Incompatibility

One of the intriguing problems of plasmid biology has been how a resident plasmid can prevent one of its close relatives from establishing itself in the cell. As might be expected, the control of another plasmid seems to be exercised via the replication system. The same systems that control copy number can also be used for incompatibility functions.

The case of the incompatibility functions for the F plasmid has been particularly well studied. Genetically, two F plasmid-carrying strains can be mated provided one is phenocopied. If an F′ culture is mated to an Hfr, the progeny are generally like the original Hfr. The occasional exception is a cell carrying two plasmids integrated at different sites on the genome. Such a cell is called a double male, and, oddly enough, some of them are stable. If an F′ culture is mated to a culture carrying a different F′, the resultant cells may carry one or the other of the original F′ plasmids or else may have acquired a single giant plasmid that seems to have resulted from a fusion event. Studies of cloned F plasmid DNA have shown that there are two regions capable of serving as origins of replication. The bidirectional origin is *ori-1* or *oriV*, and the unidirectional origin is *ori-2* or *oriS*.

Experiments with mini-F have identified four incompatibility loci, *incB*, *incC*, *incD*, and *incE*, located near *ori-2*. The *incB* function is to exclude other F plasmids, for if a second plasmid carries a different *incB* locus there is no incompatibility. The DNA sequence for this region includes the four 14-bp direct repeats that constitute the *copB* locus. Acridine orange inhibits replication via the *incC* locus, which is apparently a binding site that controls replication and consists of a series of direct repeats. There are 8 bp within the repeated sequences that could bind protein E, based on a comparison with the regulatory region for protein E. The *incD* locus is the same as *sopC* and therefore affects the partitioning function. The *incE* cistron seems to prevent plasmid replication via a second replication origin. The various control loci are summarized in Figure 11-2. The model for IncFI incompatibility then becomes twofold. The E protein is required for initiation of replication; and under normal circumstances all of the E protein is tied up by the resident plasmid. A newly arrived plasmid therefore is not able to initiate its replication and establish itself. Similarly, the extra plasmid is not able to bind the proteins necessary for its partitioning and cannot establish itself properly. On the other hand, integrated plasmids are replicated and segregated by the host system and therefore are not incompatible.

Conjugal Functions

Although the F plasmid is certainly not the only conjugative plasmid, it is one of the best studied and can mobilize a wide variety of other, normally nonconjugative plasmids. Therefore this section focuses exclusively on F and its known functions.

Pilus Production

Cells carrying the F plasmid have two types of filamentous structure on their surfaces. The **fimbriae**, or common pili, are the products of the *pil* cistrons found on the bacterial chromosome. They are firmly attached to the cell and difficult to break apart, even in the presence of a strong detergent such as sodium dodecyl sulfate (SDS). By contrast, the F, or sex, pili are longer, thinner, and more loosely attached to the cell. They can be removed by blending the culture and are easily disrupted by SDS. Because they are required for mating pair formation, any plasmid that is pilus-defective is also transfer-defective (*tra*). Table 11-2 summarizes many of the known *tra* functions. Most of the *tra* cistrons are transcribed as a single large mRNA that is under the control of the *traJ* protein and the cellular integration host factor normally involved in phage λ activity (see Chapter 12). Mutants in *traJ* fail to express the other *tra* functions.

The structural cistron for F-pilin, the protein component of pili, is *traA*

Table 11-2. Genes of the transfer region

- A. Translated genes
  - 1. Transfer genes
    - a. Pilin synthesis
      - *traA* (prepropillin)
      - *traQ* (prepropillin hydrolase)
    - b. Pilus assembly
      - *traB, traC, traE, traF, traG, traH, traK, traL, traP, traU, traV, traW*
    - c. Mating pair stabilization
      - *traG, traN*
    - d. DNA transactions
      - *traD, traM*
      - *traI* (helicase I)
      - *traY* (oriT-nuclease)
      - *traZ* (oriT-nuclease)
  - 2. Transfer-related genes
    - a. Surface exclusion
      - *traS, traT*
    - b. Regulation
      - *finP, traJ*
- B. Nontranslated genes
  - 1. Transfer gene (DNA transaction)
    - *oriT*
  - 2. Transfer-related genes
    - a. Transcriptional promoters
      - *traIp, traJp, traMp, traTp?, traYp*
    - b. Repressor binding site
      - *traJo*

Adapted from Clark, A.J. (1985). Conjugation and its aftereffects in *E. coli*, pp. 47–68. In: Halvorson, H.O., Monroy, A. (eds.) The Origin and Evolution of Sex. New York: Liss.

and is located in a cluster with the other *tra* functions (Fig. 11-4). The DNA sequence codes for the 121 amino acids of the pilin plus 51 extra amino acids at the amino-terminus. Processing the protein to normal size requires *traQ* activity. Physically the unassembled pilin is associated with the inner membrane of the cell, and it is thought that this pool of material makes possible the rapid assembly of pili. The location of the pili does not appear to be random but, rather, is associated with zones of adhesion where inner and outer cell membranes touch. It has also been proposed that the pili "retract" by disassembly, thereby facilitating the conjugal cell contacts and incidentally providing a route of infection for the male-specific bacteriophages f1, f2, Qβ, MS2, and M13. Although the pili must be present for male-specific phage infection, a female-specific phage such as ϕII cannot infect male cells even in the absence of pili.

The early stage of mating pair formation is inhibited if the other cell

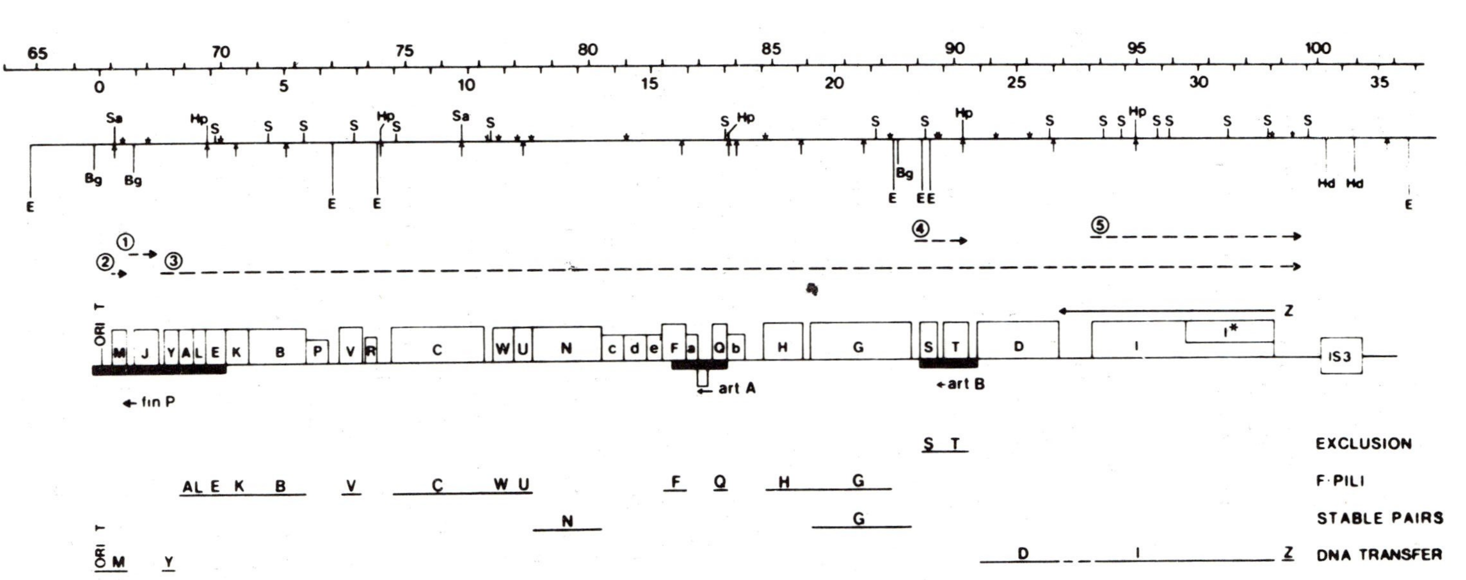

Figure 11-4. Physical, genetic, and functional map of the F transfer region. **(Top)** Kilobase coordinates for F and for the *tra* region. The *oriT* nick site is 0 (F 66.7) placed 142 bp from a *Bgl*II site; the IS*3* junction is 33.3 (F 100/0). Restriction sites shown are *Bgl*II (Bg), *Eco*RI (E), *Hin*dIII (Hd), *Hpa*I (Hp), *Sal*I (Sa), *Sma*I (S), *Eco*RV (small downward arrows), and *Hin*dII (small upward arrows). **(Center)** Boxes indicate the placement and size of *tra* (capital letters) and *trb* (small letters) genes (the number of gene assignments has exceeded 26, and a new mnemonic has been created); a small box height implies an uncharacterized function. Positions based on product size may be inexact. An arrow indicates uncertainty in *traZ* placement. Dotted arrows represent *tra* transcripts; small arrows below indicate the position and direction of upstream promoters (additional promoters may also be present). **(Bottom)** Functional assignments of nonregulatory *tra* region products. From Ippen-Ihler, K.A., Minkley, E.G., Jr. (1986). The conjugation system of F, the fertility factor of *Escherichia coli*. Annual Review of Genetics 20:593–624.

has both the TraT outer membrane protein and the TraS inner membrane protein; this phenomenon has been called **surface exclusion.** When cells are phenocopied, these proteins disappear. The mating aggregates that form are stabilized by the products of the *traN* and *traG* cistrons. Generally speaking, those conjugative systems that have long, flexible sex pili are capable of mating pair formation in liquid culture. Those that do not are limited to matings on solid surfaces such as membrane filters.

Transfer DNA Replication

Because only a single strand of donor DNA is transferred, only one strand needs to be nicked and replicated. The site of the nick is a separate origin of replication, *oriT*. It is a 373-bp fragment that when cloned converts a nonmobilizable plasmid into a high frequency donor. Four *oriT* sequences are known among the IncF plasmids. Willetts and co-workers examined a number of other *oriT* regions. That for IncQ is basically the same as ColE1, whereas that from IncN has 13 direct repeats of 11 bp, three pairs of 10-bp inverted repeats, and an (A+T)-rich region.

In the case of the F plasmid, normal *traY* and *traZ* sequences are required for the nicking activity. The *traY* cistron codes for a small membrane protein, but DNA sequence data do not show an open reading frame in *traZ*. Instead, the *traZ* region may reflect the existence of a bifunctional protein, as missense mutations in *traI* (which maps in approximately the same spot) have variable effects. If they are located near the carboxy-terminus, they affect only the helicase activity; but if they are located in the amino-terminus, they do not. It has been suggested that the amino-terminus of the TraI protein is the nicking activity. The TraY activity is more plasmid-specific than the TraZ activity with respect to the ability to nick foreign *oriT* sequences.

The experiments discussed in Chapter 9 have revealed that it is the preexisting donor DNA that is transferred to the recipient, and the DNA synthesis that occurs during transfer replaces it. The transfer process is independent of strand replacement, although lack of replacement is lethal to the donor cell. It has been suggested but not proved that the TraM protein may be bound to the origin region DNA. The only case in which it is known that a protein is attached to the transferred conjugative DNA is that of the *sog* protein of ColIb. In that case radioactive labeling experiments have shown that as much as 0.9% of protein extracted from recipient cells originates in the donor. The function of this protein is discussed below.

The physical transfer of the donor DNA requires normal helicase I (*traI*) activity. Mutant cells form mating aggregates, but no transfer can occur. A model for this activity that was proposed by Willetts and Wilkins is presented in Figure 11-5. The driving force for the transfer is assumed to be the action of the helicase enzyme as it unwinds the donor duplex.

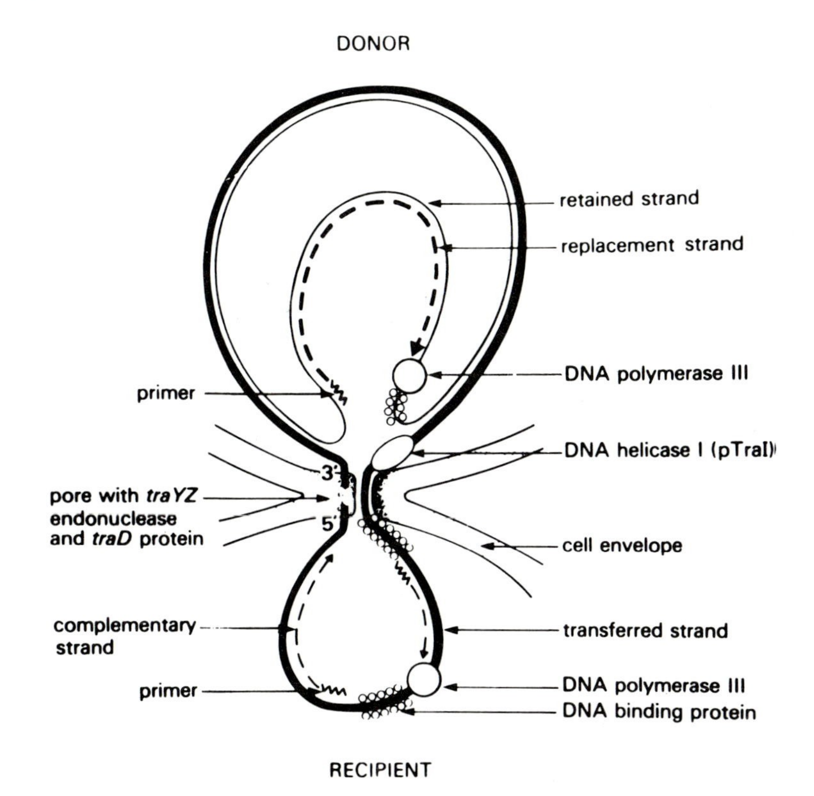

Figure 11-5. Model for the conjugative transfer of F. A specific strand of the plasmid (thick line) is nicked at *oriT* by the *traYZ* endonuclease and transferred in the 5′ to 3′ direction through a pore in the cell membrane, perhaps involving the *traD* protein, formed between the juxtaposed donor and recipient cell envelopes. The plasmid strand retained in the donor cell is shown by a thin line. The termini of the transferred strand are attached to the cell membrane by a complex that includes the endonuclease. DNA helicase I (*traI* product) migrates on the strand undergoing transfer to unwind the plasmid duplex DNA; if the helicase is in turn bound to the membrane complex during conjugation, the concomitant ATP hydrolysis might provide the motive force to displace the transferred strand into the recipient cell. DNA transfer is associated with synthesis of a replacement strand in the donor and of a complementary strand in the recipient cell (broken lines); both processes require de novo primer synthesis and the activity of DNA polymerase III holoenzyme. The model assumes that a single-strand binding protein coats DNA to aid conjugative DNA synthesis; depending on the nature of the pore, this protein might even be transferred from donor to recipient cell, bound to the DNA. From Willetts and Wilkins (1984).

Under normal conditions there is concomitant DNA synthesis in both donor and recipient cells.

The rolling circle model for DNA replication is generally taken as appropriate for strand replacement, and it predicts that the nicked strand will serve as its own primer for synthesis. At least in *dnaB* mutant cells that is not the case. There the replication process is sensitive to rifampin, implying the existence of RNA primers. Single-strand DNA is normally stabilized by a single-strand binding protein (Ssb), and F plasmid DNA is no exception. The F plasmid itself codes for its own Ssb activity in addition to the standard *E. coli* protein.

Mobilization is the process of producing a single-strand DNA molecule for transfer. In the recipient cell there is a corresponding process of **repliconation** in which the plasmid DNA is prepared for replication by synthesis of a complementary strand and circularization. The nature of the priming reaction used by the F plasmid is open to dispute, but the *sog* protein from ColIb-P9 mentioned above serves as the primer for complementary strand synthesis in that system. It is possible that a similar phenomenon may yet be identified in the F plasmid. Circularization of F cannot occur by recombination of concatemers if each round of rolling circle replication must be separately primed. The model of Figure 11-5 suggests that the 5′-end of the transferred DNA is held bound to a membrane protein that also recognizes the 3′-end when it arrives. As the two ends are held in close proximity, they are joined to make a covalently closed circle. Such a reaction is known to be possible in the case of phage ϕX174 using the cistron A protein.

The newly circularized plasmid is now ready to follow the usual pattern of replication and partition described above. The various events necessary for conjugal plasmid transfer are summarized in Figure 11-6.

Summary

Plasmids that are not integrated into the bacterial chromosome must carry out a number of tasks to ensure their survival. They must replicate themselves at least as often as the chromosome itself. They must arrange for their equal distribution throughout the cytoplasm so that when cell division occurs each daughter cell receives one or more copies of the plasmid. Some plasmids basically overreplicate their DNA and in essence saturate the cell with plasmid DNA molecules, thereby ensuring their inheritance by daughter cells. Such high copy number plasmids represent a significant drain on the cell resources, and the larger plasmids must regulate themselves or leave their host cells at a significant disadvantage in the competition for nutrients. Plasmids also display incompatibility functions designed to prevent closely related plasmids from displacing a resident plasmid.

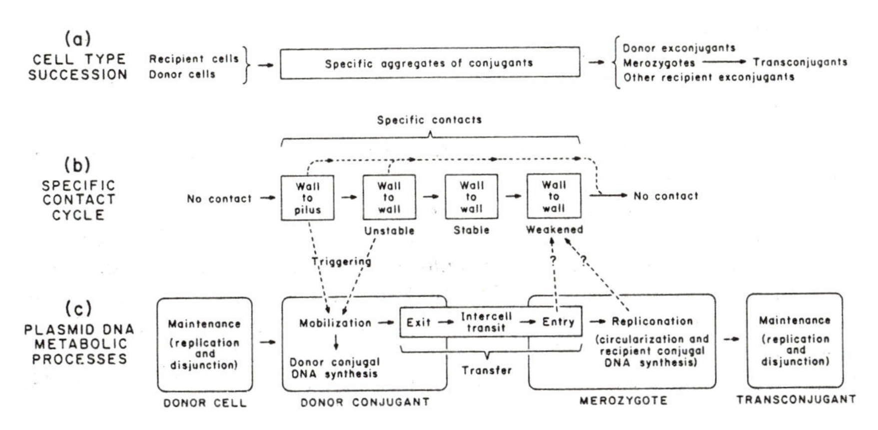

Figure 11-6. Summary of conjugal events that lead to plasmid transmission at the cellular and two subcellular levels. **(a)** Succession of cell types. An exconjugant is any cell that has been part of a mating aggregate but that has not undergone any genetic change, whereas a merozygote is a recipient cell that has received DNA from a donor cell. **(b,c)** Surface and DNA metabolic events underlying the succession. Any of the indicated four types of specific contacts may be present while donor and recipient cells are in mating aggregates. The main pathways are shown by solid arrows and alternative pathways by dashed lines. DNA metabolic events are shown for only four of the cell types shown in **a**. Surface events are related to DNA metabolic events by a process called triggering, which may occur following formations of either of the first two types of specific contact. From Clark, A.J., Warren, G.J. (1979). Conjugal transmission of plasmids. Annual Review of Genetics 13:99–125.

Replication control, copy number, and incompatibility are interrelated phenomena. Most plasmids produce one or more proteins that trigger the start of replication. Synthesis of this protein can be controlled at two levels: transcription and translation. Transcriptional control involves the binding of a protein repressor (which may be the molecule itself) so as to block RNA polymerase function. Translational control is exercised by small complementary RNA molecules that are transcribed from the noncoding strand of DNA. When bound to the mRNA, they block the ribosome binding site and prevent translation. Incompatibility is basically determined by the same genetic loci as the copy number. The regulatory elements prevent the newly arrived plasmid from significant replication and therefore from establishing itself.

Conjugative plasmids must synthesize their surface pili in order to form mating aggregates. Once the cell surfaces touch, a nick is introduced into the plasmid DNA at *oriT*, and a single strand of DNA is displaced by the action of a DNA helicase enzyme and transferred into the recipient cell. Donor and recipient cells then synthesize complementary strands to restore the plasmid DNA to its normal state.

References

General

Hiraga, S. (1986). Mechanisms of stable plasmid inheritance. Advances in Biophysics 21:91–103.

Ippen-Ihler, K.A., Minkley, E.G., Jr. (1986). The conjugation system of F, the fertility factor of *Escherichia coli*. Annual Review of Genetics 20:593–624.

Nordström, K. (1985). Control of plasmid replication: theoretical considerations and practical solutions, pp. 189–214. In: Helinski, D.R., Cohen, S.N., Clewell, D.B., Jackson, D.A., Hollaender, A. (eds.) Plasmids in Bacteria. New York: Plenum Press.

Novick, R.P. (1987). Plasmid incompatibility. Microbiological Reviews 51:381–395.

Scott, J.R. (1984). Regulation of plasmid replication. Microbiological Reviews 48:1–23.

Willetts, N., Wilkins, B. (1984). Processing of plasmid DNA during bacterial conjugation. Microbiological Reviews 48:24–41.

Specialized

Biek, D.P., Cohen, S.N. (1986). Identification and characterization of *recD*, a gene affecting plasmid maintenance and recombination in *Escherichia coli*. Journal of Bacteriology 167:594–603.

Derbyshire, K.M., Willetts, N.S. (1987). Mobilization of the non-conjugative plasmid RSF1010: a genetic analysis of its origin of transfer. Molecular and General Genetics 206:154–160.

Dodd, H.M., Bennett, P.M. (1986). Location of the site-specific recombination system of R46: a function necessary for plasmid maintenance. The Journal of General Microbiology 132:1009–1020.

Gerdes, K., Rasmussen, P.B., Molin, S. (1986). Unique type of plasmid maintenance function: postsegregational killing of plasmid free cells. Proceedings of the National Academy of Sciences of the United States of America 83:3116–3120.

Golub, E.I., Low, K.B. (1986). Unrelated conjugative plasmids have sequences which are homologous to the leading region of the F factor. Journal of Bacteriology 166:670–672.

Hiraga, S., Jaffe, A., Ogura, T., Mori, H., Takahashi, H. (1986). F plasmid *ccd* mechanism in *Escherichia coli*. Journal of Bacteriology 166:100–104.

Imber, R. (1987). Regulation of expression of the cloned *repE* gene from the F plasmid of *Escherichia coli*. Gene 52:1–9.

Lane, D., de Feyter, R., Kennedy, M., Phua, S-H., Semon, D. (1986). D protein of miniF plasmid acts as a repressor of transcription and as a site-specific resolvase. Nucleic Acids Research 14:9713–9728.

Lane, D., Rothenbuehler, R., Merrillat, A-M., Aiken, C. (1987). Analysis of the F plasmid centromere. Molecular and General Genetics 207:406–412.

Loh, S.M., Ray, A., Cram, D.S., O'Gorman, L.E., Skurray, R.A. (1986). Location of a second partitioning region (ParL) on the F plasmid. FEMS Microbiology Letters 37:179–182.

Lycett, G.W., Pritchard, R.H. (1986). Functioning of the F-plasmid origin of replication in an *Escherichia coli* K12 Hfr strain during exponential growth. Plasmid 16:168–174.

Merryweather, A., Rees, C.E.D., Smith, N.M., Wilkins, B.M. (1986). Role of *sog* polypeptides specified by plasmid ColIb-P9 and their transfer between conjugating bacteria. EMBO Journal 5:3007–3012.

Mori, H., Kondo, A., Ohshima, A., Ogura, T., Hiraga, S. (1986). Structure and function of the F plasmid genes essential for partitioning. Journal of Molecular Biology 192:1–15.

Te Riele, H., Michel, B., Ehrlich, S.D. (1986). Single-stranded plasmid DNA in *Bacillus subtilis* and *Staphylococcus aureus*. Proceedings of the National Academy of Sciences of the United States of America 83:2541–2545.

Tresguerres, E.F., Nieto, C., Casquero, I., Canovas, J.L. (1986). Host cell variations resulting from F plasmid-controlled replication of the *Escherichia coli* chromosome. Journal of Bacteriology 165:424–427.

Wagner, E.G.H., von Heijne, J., Nordström, K. (1987). Control of replication of plasmid R1: translation of the 7k reading frame in the RepA mRNA leader region counteracts the interaction between CopA RNA and CopT RNA. EMBO Journal 6:515–522.

Womble, D.D., Rownd, R.H. (1986). Regulation of IncFII plasmid DNA replication: a quantitative model for control of plasmid NR1 replication in the bacterial cell division cycle. Journal of Molecular Biology 192:529–548.

Chapter 12
Regulation

In the preceding chapters there has often been some discussion about the use of a repressor to regulate transcription. Repressors are but one type of regulatory element commonly found in bacteria. This chapter expands on the role of the repressor and introduces many new regulatory elements. Implicit in the discussion that follows is the fact that protein and RNA molecules do not last indefinitely in the cell. Proteases and nucleases are produced that slowly degrade those molecules. It is this slow turnover of macromolecules that makes possible the gradual alteration of phenotypes.

Protein degradation normally occurs at a low rate but is considerably more rapid at lower growth rates (when nutrients are in short supply and recycling is important). The exception to this statement is the case of the abnormal protein. Proteins that are truncated by a nonsense mutation or unfolded as the result of a missense mutation are often unstable in a cell. Apparently some proteases are capable of specifically recognizing aberrant structures and attacking them. For example, missense mutations in the Cro protein (discussed below) that unfold the structure have been shown to make the protein susceptible to proteolytic attack.

Degradation of mRNA is less well understood but seems to involve two steps. First, the structure of the mRNA is altered so that it can no longer be translated, even though blotting experiments can show that the coding sequences are still present. Evidence from studies with the *lac* operon RNA (see below) indicates that cuts may occur in regions that are not being actively translated (unprotected by ribosomes). Later the large fragments of RNA are digested and the bases reused. Newbury and coworkers have reported that about 25% of all *Escherichia coli* RNA transcripts contain a highly conserved sequence designated REP (repetitive extragenic palindromic) at their 3′-end. The Rep moiety is capable of forming a stem-and-loop structure that prevents single-strand-specific exonucleases from attaching to the RNA and thereby helps to stabilize the upstream RNA.

In a sense, regulatory mechanisms can be considered energy-conserving processes because the synthesis of every kind of macromolecule requires the input of energy from ATP or an equivalent molecule. To the extent that a cell synthesizes unneeded protein, DNA, or other molecules, it handicaps itself in the competition with other cells for nutrients and space. A cell has available to it several mechanisms or levels of regulation that vary in sensitivity. The most basic but least sensitive control is at the level of transcription. A less dramatic effect is seen with regulation at the level of translation, and the most sensitive control is posttranslational, the determination of whether the gene product is allowed to act. This last type of control is mediated by biochemical rather than genetic processes, and therefore only the first two types of regulation are considered in detail in this chapter.

It can be difficult to study the expression of a particular cistron. Often the protein is not enzymatically active, is unstable, or is inconvenient to assay. Much of the recent experimentation in regulatory genetics has involved the use of **gene fusions,** where a known, indicator cistron is hooked up to the regulatory region of another. A gene fusion may be prepared by DNA splicing techniques, or it may involve the use of phage Mu derivatives to deliver the indicator DNA more or less randomly to the fusion site. Commonly used systems are Mud*lac* or λp*lac*Mu. In both cases Mu ends are provided to insert the *lacZ* cistron coding for β-galactosidase near an existing promoter. The β-galactosidase protein is a large, stable protein and can have extra amino acids added to its amino-terminus without major loss of enzymatic activity.

Regulation of Simple Functional Units

Operon, the Basic Regulatory Unit

The original concept of the operon came from Jacob and Monod in Paris for which they won a Nobel Prize. The concept grew out of some observations Monod had made while still a graduate student. If a culture of *E. coli* is grown in a medium containing both glucose and lactose as potential carbon and energy sources, a biphasic growth pattern is observed. During the first burst of growth, all of the glucose and none of the lactose is used. After a somewhat variable lag period that begins when the glucose in the medium is exhausted, the cells begin to grow again, this time using the lactose. Monod called this phenomenon **diauxie** and reasoned that the lag period represented a time of change in the regulatory state of the cell. Enzymatic analysis of cell extracts indicated that only low levels of the enzymes necessary for lactose utilization were produced while glucose was present, but the enzyme levels increased in a coordinated fashion upon exhaustion of the glucose. In a similar vein, later experiments showed that the enzymes for tryptophan biosynthesis were produced in the absence of tryptophan but not in its presence. Phrased in contemporary terminology, the conclusion to be drawn from the lactose and tryptophan experiments was that genes affecting the same process (e.g., lactose utilization) were coordinately regulated.

Genetic mapping data indicate that, in *E. coli* but not in *Pseudomonas aeruginosa,* the coordinately regulated cistrons are contiguous. Moreover, the group of coordinately regulated cistrons has a definite orientation, or polarity, that can be demonstrated by introducing a polar mutation. These nonsense mutations have the effect of causing loss of function in cistrons downstream from the site at which they map. For example, if a group of cistrons has the map order ABCDE, a typical observation might be that a polar mutation in A may reduce or eliminate functions in B, C, D, and E, whereas a polar mutation in C would have no effect on A or B. The implication of these results is that the mutation is interrupting some process that begins at cistron A and proceeds in a linear fashion through B, C, and D toward E, thereby establishing a functional gradient.

Taking all of the above observations into account, Jacob and Monod in 1961 proposed a new genetic unit, the **operon**. In molecular biologic terms, the operon consists of a group of cistrons, usually coding for related functions, that are transcribed as a unit (beginning at A in the example above) to produce a polycistronic mRNA molecule. Coordinate regulation of the enzymes encoded within an operon is ensured because translation of the mRNA yields all of the enzymes in sequential fashion, beginning with A and ending with E. If translation or transcription stops prematurely owing to the presence of an abnormal terminator signal (a nonsense mutation), production of enzymes whose genetic information is located distal to (transcribed after) the point of mutation depends on the ability of translation/transcription to restart. Failure to restart leads to the observed polar effects.

In functional terms, the operon is conceived as containing certain definite genetic elements. It must have one or more structural cistrons that are transcribed into RNA (which may be rRNA, mRNA, or tRNA). The transcription must begin at a definite site(s) and end at a definite site(s). There must be the opportunity for some sort of regulator molecule(s) to interact with the operon and affect transcription. There are several ways to fulfill these criteria, and the rest of the discussion in this section deals with specific examples.

Lactose Operon

Most of the early work in Paris concerned the utilization of lactose, and for many years this operon was the one most intensively studied. It is a fairly simple operon consisting of three structural cistrons designated *Z*, *Y*, and *A*. The *lacZ* cistron codes for the enzyme β-galactosidase, which catalyzes the hydrolysis of lactose to glucose and galactose. The *lacY* cistron codes for a galactoside permease that provides transport functions for a variety of sugars, including lactose, melibiose, and raffinose. The *lacA* cistron codes for thiogalactoside transacetylase, an enzyme of uncertain function that may play a role in detoxifying certain thiogalactosides. All three proteins are normally present in trace amounts in the cell, but when the cell is growing on lactose, the enzyme levels increase up to 1000-fold. The process of stimulating the increase is called **induction**, and the enzymes of the lactose operon are therefore considered inducible. Any compound, e.g., lactose, whose presence in the medium results in induction is said to be an **inducer**. After the inducer in the medium is exhausted owing to the action of the lactose enzymes, synthesis of the lactose enzymes is once again repressed, and the cell returns to the original state.

The opposite of an inducible-repressible enzyme is a **constitutive** enzyme, an enzyme that is produced at a constant rate under all conditions. Constitutive production of an enzyme implies a lack of control mechanisms, and certain types of mutation can render the production of lactose enzymes constitutive instead of inducible. Among them is a class of mutations that map in a cistron called *lacI*, which is located adjacent to the lactose *Z, Y,* and *A* cistrons. Nonsense mutations have been observed to occur within the *lacI* cistron, and because these mutations can be suppressed by standard tRNA suppressors the RNA transcribed from the *lacI* cistron must be translated into protein. Although a *lacI* cell is constitutive for *lacZ,Y,A* expression, a merodiploid cell that is F′ $lacI^+$/F^- *lacI* is inducible (the $lacI^+$ cistron is *trans*-dominant). These observations confirm that the *lacI* cistron codes for a protein repressor that exerts negative control over the lactose operon (i.e., prevents transcription), even if the operon is not located on the same piece of DNA as the $lacI^+$ cistron.

The repressor must interact with the operon in some fashion to prevent transcription. The site at which this interaction occurs is called the **operator** and is defined genetically by another class of constitutive mutations. These mutations (called o^c) that map between the *lacI* and *lacZ* cistrons are *cis*-dominant; i.e., the phenotype of a *lac* o^c cell cannot be affected by the presence of a functional *lacI* cistron in the cell. The interpretation given to this observation is that the operator mutation prevents the repressor from binding to the operator, and hence transcription continues unabated. Note that because the operator is only a binding site and produces no diffusible product a *cis-trans* test cannot be applied to it, and therefore it cannot be considered a cistron.

If the presence of the repressor protein on the operator prevents transcription and the absence of the repressor permits it, induction must consist in removal of the repressor from the operator, presumably owing to a change in the shape of the repressor resulting from interaction with the inducer. Various molecules having a β-galactoside linkage such as that found in lactose have been shown to function as inducers in vivo and in vitro. Some of these molecules, unlike lactose, are not degraded by β-galactosidase and therefore are termed **gratuitous inducers**. Among these compounds are thiomethyl-β-D-galactopyranoside (TMG) and isopropyl-β-D-thiogalactopyranoside (IPTG), both of which have frequently been used for studies on induction. Because they are not degraded, their concentration does not change even if the cells grow for many generations. One class of *lacI* mutations, i^s, increases the relative ability of the repressor to bind to the operator and inducer such that induction of the *lac* operon is no longer possible, although basal enzyme levels are still produced.

The *lac* repressor was the first example of a class of proteins that Monod and Changeaux called allosteric proteins. These proteins have two stable minimum energy configurations, and each configuration has a characteristic activity (or lack of activity) associated with it. In the case of the *lac* repressor, one allosteric configuration of the protein binds to the operator as a tetrameric complex and the other does not. The shift in configuration of an allosteric protein is triggered by an allosteric effector molecule, a small molecule that binds to a special site on the allosteric protein. For *lac* repressor, the effector is any of the inducer molecules. The inducer does not actually compete with the operator for the binding of the protein but, rather, can shift the allosteric equilibrium regardless of the amount of operator DNA available. Therefore the inducer need be present only in small amounts.

The independence from inducer concentration is important, for the actual chemical inducer of the lactose operon is not the sugar lactose but a derivative of it called allolactose. This molecule is produced in trace amounts by the action of β-galactosidase on lactose and results from a shift in the glycosidic bond from carbon 4 to carbon 6 on the galactopyranoside ring. In other words, the disaccharide moiety is not broken down by the enzyme but merely altered. One implication of this observation is that a minimal amount of β-galactosidase must be present at all times, or induction of the lactose operon cannot occur. Another is that a sugar does not have to be broken down by β-galactosidase to be an inducer. In addition to the gratuitous inducers mentioned above, melibiose is an inducer that requires only the permease function, as it is broken down via a different enzymatic pathway. Raffinose, on the other hand, is not an inducer but is transported by the permease. Only cells constitutively expressing the *lac* operon are able to grow on raffinose.

In the original operon model, it was assumed that the RNA polymerase bound to the operator in order to begin transcription and that repression was a simple competition between polymerase and repressor for the same DNA site. However, Ullman and Monod isolated *lac* mutations that affected the level to which the lactose enzymes could be induced (the maximum amount of enzyme that could be produced) but not the actual inducibility of the operon. These properties are expected if the mutation affects the ability of the polymerase to bind to the DNA in order to initiate transcription. A convention developed that "up" promoter mutations are those that bind the polymerase more efficiently (more enzyme produced), and "down" promoter mutations are those that bind the polymerase less efficiently and therefore produce less enzyme. The promoter mutations map between the *lacI* and *lacZo* sites and are designated *lacZp*. The "Z" designation indicates that both operator and promoter are "upstream" from the *Z* cistron.

The last regulatory element to be discovered was also the one that finally explained the first observations. This phenomenon was shown by Magasanik to be part of a larger group of regulatory events called **catabolite** (or **glucose**) **repression.** The basic concept was that some component released during the breakdown (catabolism) of glucose resulted in the inhibition of ancillary enzyme systems such as the lactose operon. However, the actual mediator of catabolite repression had not been discovered until Sutherland and co-workers identified a small molecule, 3′,5′-cyclic adenosine monophosphate (cAMP), as a regulatory element in animal cells and bacteria. Perlman, Pastan, and co-workers further showed that addition of this compound to growing *E. coli* cultures relieved catabolite repression and allowed induction of a variety of operons including lactose, although the cells grew poorly under these conditions. Later experiments demonstrated an inverse relation between the amount of glucose in the cell and the amount of cAMP. Apparently glucose or a metabolite inhibits the adenyl cyclase enzyme.

Once again mutations were used to define the role of the new regulatory element. Mutations lying in the promoter region could relieve the requirement for cAMP (i.e., the cells became insensitive to catabolite repression). Another type of mutation that mapped well away from the lactose operon resulted in the inability of cAMP to activate the lactose operon (as well as some others). This new genetic locus (*crp*) was shown to code for a protein variously called the catabolite activation protein (CAP), cAMP receptor protein (CRP), or catabolite gene activation protein (CGA). The requirement for this protein could also be alleviated by mutations mapping in the promoter region. The CRP and cAMP moieties act as **positive regulatory control elements** because in their absence transcription cannot be increased above basal levels.

The regulatory region of the lactose operon has been completely sequenced, as has the *lacZ* cistron. Comparisons with other known transcription regions have been made, and a comprehensive model (Fig. 12-1) has been developed. Normal promoters have a consensus sequence at −10 bases from the RNA start and another consensus sequence at −35 bases from the start. These regions must be separated by 17 base pairs (bp) for normal function. A careful examination of the region between the end of the *lacI* cistron and the beginning of *lacZ* reveals that there are in

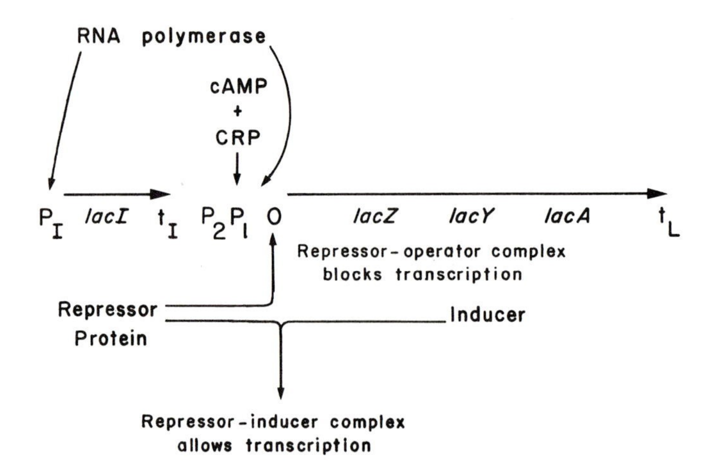

Figure 12-1. Regulation of the lactose operon. Horizontal arrows represent mRNA transcripts of the cistrons listed below them. Promoter sites are indicated by P; transcription terminator sites are indicated by t; and the repressor binding site is indicated by O. Note that cAMP and CRP are necessary for the binding of the RNA polymerase to P_1 but not to P_I (the repressor promoter) or P_2.

fact two candidates for promoters in this region. Promoter 1 has a good match to the consensus sequence in the −10 region (Table 12-1) but a relatively poor match in the −35 region. Promoter 2 has mismatches in both regions and would be expected to operate at low efficiency. It overlaps promoter 1 in the sense that the −10 site of promoter 2 is located in the −35 site of promoter 1 (Fig. 12-2).

In the absence of CRP and cAMP, promoter 1 is nonfunctional, and transcription initiates at promoter 2. This situation provides the essential low levels of β-galactosidase to process the inducer molecule and of galactoside permease to bring the inducer molecule into the cell. When lactose is present in the medium, a small amount is transported into the cell, and some of that is converted to allolactose. The allolactose causes a shift in the structure of the *lac* repressor so it no longer can bind to the operator. The operator sequence is located so it overlaps the −10 site for promoter 1. When the repressor is bound to the operator, RNA polymerase cannot attach. Even if the repressor is not bound, RNA polymerase still may not attach because of the defect in the −35 site of promoter 1. Just upstream (toward the 5′-end) of the promoter 1 sequence is a binding site for CRP that overlaps the −35 site of promoter 2. CRP attaches to this site when in the presence of cAMP. The bound CRP blocks promoter 2 and enables the RNA polymerase to initiate effective contacts with the DNA and form the open transcription complex using promoter 1. Transcription continues until either the supply of the inducer is exhausted as a result of the β-galactosidase activity or the cAMP concentration drops below the effective

Table 12-1. Some regulatory sequences found in *E. coli* operons

	−10 sequence	−35 sequence
Consensus	TATAAT	TTGACA
lac Operon		
	−10	−35
Promoter 1	TATGTTG	GCTTTACACT
	−30	−55
Promoter 2	TTACACT	TCACTCATT
gal Operon		
	−10	−35
Promoter 1	TATGGTT	TGTCACACTTT
	−15	−40
Promoter 2	TATGCTA	ATGTCACACTT
CRP binding		
Consensus	AANTGTGANNTNNNTCANATW	
	−75 −55	
lac	CAATTAATGTGAGTTAGCTCACT	
	−50 −40	
gal	AATTTATTCCATGTCACACTTTTCG	

The usual base abbreviations are used, plus N for any base and W for adenine or thymine. Only the antisense strands are shown. The numbers above the sequence are the number of bases prior to the start of the mRNA when promoter 1 is used.

level. Site-directed mutagenesis of promoter 2 has shown that it has little or no role to play in the activation of promoter 1 by CRP.

Note that although both positive and negative regulatory elements are present, the *lac* operon (defined as the promoter, operator, *Z,Y,A* cistrons, and terminator sequence) is considered to be under negative control because a protein repressor is produced. The *lacI* cistron is not part of the *lac* operon, as it has its own promoter and terminator sequences. It is normally constitutively expressed at a low level. Several up-promoter mutations have been isolated that cause a considerable increase in the amount of repressor produced. Strains carrying these mutations induce

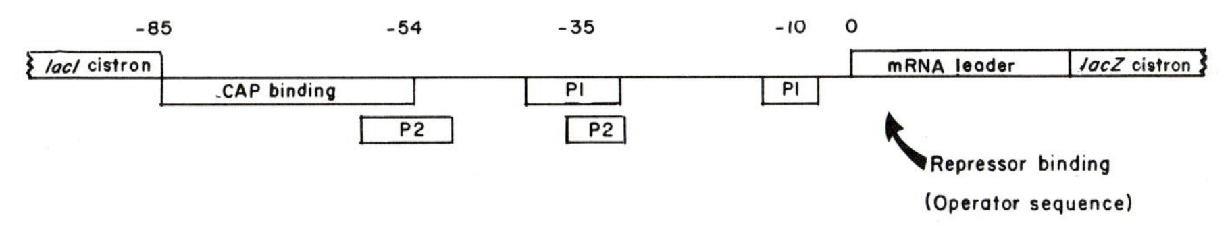

Figure 12-2. Organization of the regulator region for the *lac* operon. Coding segments are shown by boxes above the horizontal line, and DNA binding sites are shown by boxes below the horizontal line. The scale across the top indicates the number of bases before the start of the mRNA molecule when promoter 1 is used.

the lactose operon poorly if at all but are excellent sources of repressor protein. Additional sequence details of both operons can be found in the book entitled *The Operon,* which is listed in the references.

Galactose Operon

Utilization of the sugar galactose as a sole carbon and energy source involves three proteins coded within the *gal* operon. First, the *galK* cistron codes for a kinase enzyme that converts galactose to galactose-1-phosphate. Next, the *galT* cistron codes for a transferase enzyme that attaches the phosphorylated galactose to uridine diphosphoglucose (UDPG) to give uridine diphosphogalactose (UDPgal) plus glucose-1-phosphate. Finally, the *galE* epimerase enzyme converts UDPgal to UDPG (uridine diphosphoglucose) and the cycle repeats. The *galU* cistron, which is not located in the *gal* operon, codes for a pyrophosphorylase enzyme that forms UDPG from UTP and glucose-1-phosphate to initiate the cycle.

Genetic analysis of the *gal* operon reveals that, like the *lac* operon, it too is a negative control system possessing most of the same properties as the *lac* operon, but with several different features. The repressor protein is encoded by the *galR* cistron, which maps at a considerable distance from the *gal* operon itself (near *lysA*, see Figure 9-3 or inside the front cover). One prediction from this observation is that the binding of the *gal* repressor–operator interaction is probably tighter than in the case of the *lac* repressor. Any protein is synthesized near its coding site, as translation follows quickly upon transcription in prokaryotes. If the binding site for the protein is nearby, it is not difficult for the protein to find it. However, if the binding site is located at some substantial distance from the coding site, the protein must be able to readily identify the site from among a large mass of heterogeneous sequences. Therefore more efficient binding to the site is required when it is finally located.

In actuality, footprinting experiments have shown that there are two *gal* repressor binding sites, one external operator located 5′ to promoter 2 at base -60 (O_E) and the other internal operator located within the *galE* cistron at base $+55$ (O_I). Both operator sites are fully functional, and appropriate mutations in those sites can prevent repressor binding and thereby raise the background level of GalE product eight- or tenfold. The *gal* operon remains somewhat inducible even after introduction of a single operator mutation because of the effect of the second operator site.

Once again there are two promoter sequences observed, one providing for a background level of synthesis (Fig. 12-3). The two promoters overlap by about 5 bp, with promoter 2 preceding promoter 1. Promoter 1 has all the properties of *lac* promoter 1, except that the presence of the repressor has only a 10- or 15-fold effect instead of the 1000-fold effect seen for *lac*. The higher level of residual enzyme synthesis indicates that promoter 2 is more efficient than the comparable *lac* promoter. The relatively large

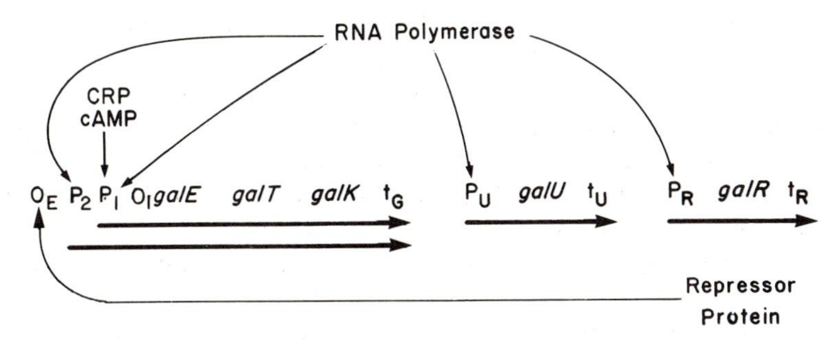

Figure 12-3. Regulation of the galactose operon. The basic operon is the cluster of three cistrons shown at the left, although the two unlinked cistrons are also important. The horizontal arrows indicate the sizes of the mRNA transcripts. The *galR* cistron codes for a protein repressor that binds at either operator (O) site and prevents transcription from promoter P_1. Promoter P_2 is a low-efficiency promoter that maintains a background level of enzymes suitable for providing the UDPgal required for cell membrane biosynthesis. In the presence of galactose, CRP, and cAMP, transcription is initiated from promoter P_1, providing the higher level of enzyme activity necessary for the use of galactose as a sole carbon source. The *galU* cistron is not involved in the regulatory process but is necessary for the transport of galactose across the cell membrane.

amount of UDPgal epimerase production is necessary to permit the continued conversion of UDPG to the UDPgal required for cell wall biosynthesis.

The base sequence for the binding site for CRP (Table 12-1) lies upstream of promoter 1 and within promoter 2. When the CRP binds to the DNA, it physically blocks promoter 2 but activates promoter 1. A second CRP molecule then binds upstream of the first, presumably to further stabilize the RNA polymerase complex and aid in the formation of the open complex. The work of Shanblatt and Revzin has shown that the contacts between protein and bases are the same in the CRP-stimulated promoters as they are in conventional promoters.

As genetic analysis of the *gal* operon proceeded, the first evidence was obtained that indicated the existence of a new mode of regulation involving various insertion elements that was applicable to any operon. The insertion elements IS*1*, IS*2*, and IS*3* had been shown to move from place to place on the *E. coli* genome. The sites of their insertion appeared random, but frequently their arrival in an operon had drastic effects on the regulation of that operon. For example, IS*1* insertion always produced highly polar mutations in the *gal* operon, regardless of whether the insertion occurred near the beginning or the end of a cistron. The behavior of IS*2* and IS*3* seemed even more inexplicable. When inserted in one orientation (I) they acted like a highly polar mutation in the same manner as IS*1*. However, when inserted in an inverted orientation (II) at certain sites they produced constitutive expression of all *gal* cistrons more distant from the beginning of the operon than the point of the insertion.

In order to account for the IS*2* and IS*3* observations, it is necessary to assume that they contain both start and stop signals. These signals must be of opposite orientation so insertion in one direction yields a terminator signal and a polar mutation whereas insertion in the other direction yields a promoter. Sequence analysis has shown that IS*2* (and presumably IS*3* as well) does not carry a complete promoter but, rather, a good −35 site and a 17-bp spacer. The insertion site itself must provide the −10 site. For this reason, orientation II effects can only be seen as the result of insertions at specific points within the *gal* operon. These sites are located away from the normal promoters, and therefore transcription is independent of the normal regulatory system.

Transcription termination within the IS elements can be shown to be of the Rho-dependent type. Therefore the termination (and its concomitant polarity effect) can be prevented or reduced by mutations in the *rho* cistron. Such mutations are true polarity suppressors, as they do not affect the function of the cistron into which the IS element has inserted but, rather, permit expression of all cistrons distal to the point of insertion. In the absence of Rho, mRNA synthesis continues and the ribosomes can reattach at the beginning of the next cistron and continue protein synthesis.

Tryptophan Operon

Biosynthesis of the amino acid tryptophan is a complex process that begins with the compound chorismic acid (a product of the *aro* cistrons) and proceeds via anthranilic acid and indole. The process is catalyzed by three enzymes whose subunits are encoded within five cistrons (A through E). The products of the *trpA* and *trpB* cistrons form the enzyme tryptophan synthetase, and the products of the *trpE* and *trpD* cistrons form the enzyme anthranilate synthetase. The *trpC* cistron codes for the enzyme indoleglycerophosphate synthetase.

The genetic map of the *trp* operon seems conventional, although the structural cistrons are arranged in reverse alphabetical order (Fig. 12-4). As in the case of the *gal* operon, the *trpR* cistron, which codes for a protein repressor, is located at some distance from the operon it regulates. In this case *trpR* is located at minute 100 and the *trp* operon at minute 27 on the *E. coli* genetic map (see Figure 9-3 or inside the front cover). Once again discrete promoter and operator sites can be defined by mutations, and the operator lies between the promoter and the structural cistrons. Promoter strength is affected by two A•T-rich blocks of DNA located upstream from the promoter (at −50 and −90). Deletion of these regions greatly reduces promoter efficiency. When strains carrying polar mutations in cistrons *E* or the first part of *D* are examined, a second low efficiency promoter can be identified that can provide constitutive transcription of the *trpC, B, A* cistrons.

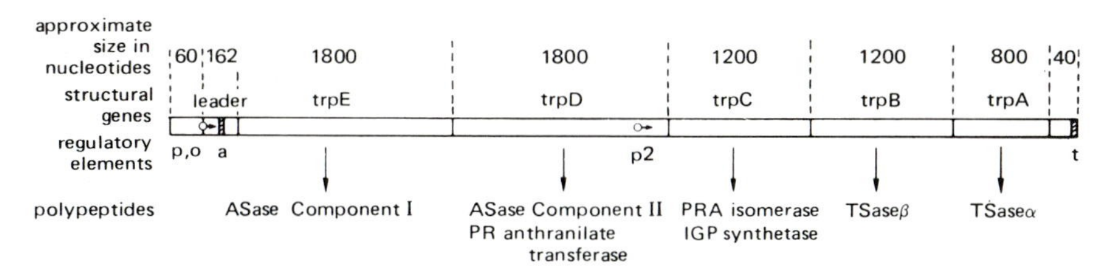

Figure 12-4. Regulation of the tryptophan operon. The operon is drawn approximately to scale with the nucleotide size indicated above each cistron. Abbreviations are as follows: p, promoter; o, operator; a, attenuator; t, terminator; ASase, anthranilate synthetase; PRA, phosphoribosyl anthranilate; IGP, indole glycerophosphate; TSase, tryptophan synthetase. Note the existence of a secondary promoter located near the right-hand end of the *D* cistron. From Platt, T. (1978). Regulation of gene expression in the tryptophan operon of *E. coli*, pp. 263–302. In: Miller and Reznikof (1978).

Despite the similarity to the *lac* and *gal* operons, the *trp* operon has several novel genetic features that should be considered. Unlike the sugar operons, there is no involvement of cAMP or CRP in the regulation of the *trp* operon. Instead, there is regulation of the production of the biosynthetic enzymes according to the amount of tryptophan available to the cell. This regulation requires a means of detecting the presence of tryptophan in the cytoplasm and stopping mRNA transcription whenever the concentration of tryptophan is sufficiently high.

One detection system is by means of the repressor molecule. Binding studies show that the repressor protein, as coded by the *trpR* cistron, is an **aporepressor** and does not bind to the *trp* operator unless it is complexed with tryptophan itself or a structural analog of it. The complex is the actual inhibitory element, and its formation is dependent on the presence of the endproduct of the biochemical pathway, tryptophan. The repressor complex also acts on *aroH* and *trpR* (autoregulation). Thus in cells supplied with excess tryptophan, radioimmunoassay reveals the presence of 120 repressor dimers, but in the absence of tryptophan there are 375 repressor dimers, most of which are not functional owing to the lack of tryptophan. It should be noted in passing that tryptophan also acts as an allosteric inhibitor of the enzyme anthranilate synthetase (see Chapter 14), thereby providing two levels of regulation. When tryptophan is present in excess, the first enzymatic step in the biochemical pathway is blocked, and the newly formed repressor complex prevents further synthesis of the *trp* operon mRNA. Repression of the operon reduces the amount of enzyme present by about 70-fold.

It seems that the *trp* operon has sufficient regulatory mechanisms for its needs, but Yanofsky and co-workers, who were engaged in sequence analysis of the operator and promoter DNA, discovered yet another mechanism. In every *trp* mRNA there is a region that does not code for any enzymes and that consists of the portion of the DNA lying between the end of the promoter–operator region (site at which the RNA polymerase binds) and the start codon for the *trpE* cistron (site where the ribosome binds). This region is called the **leader sequence** and consists of 162 base pairs (Fig. 12-4). The genetic elements within the leader sequence can be identified by appropriate footprinting experiments.

Yanofsky's group showed that if purified *trp* operon DNA obtained from an appropriate plasmid was transcribed by an in vitro system, two RNA products were observed. One product was the expected long RNA molecule that carried the information from the *trp* cistrons. The second, unexpected product consisted only of the first 140 bases of the leader sequence. In other words, premature termination of the RNA transcript had occurred. A comparison of the relative amounts of the two transcripts indicated that 85 to 90% of all transcripts initiated in the presence of excess tryptophan terminated at the early site. This site was called an **attenuator** after a similar site identified in the histidine operon by Kasai, and the termination process was denoted attenuation. By analogy to operators and promoters, the attenuator is designated *trpEa*. When tryptophan is absent, attenuation is rare, and 75 to 90% of all transcripts are completed. The combination of repression and attenuation allows for regulation of tryptophan enzymes over a 600-fold range.

Several lines of evidence indicated that attenuation was not just an experimental artifact. Deletion mutations that removed the region of DNA near the attenuation site increased basal enzyme levels eight- to tenfold without affecting the inducibility of the operon. Small RNA molecules that seemed to correspond to the attenuated transcripts could be isolated from normal cells. Furthermore, some mutations in the *rho* cistron that affected the normal RNA transcription termination process also seemed to prevent attenuation. It was observed as well that cells carrying nonsense-suppressing tRNA molecules produce more tryptophan enzymes than do cells that lack the suppressors.

Sequence analysis of the leader RNA has shown that it can fold in several ways immediately after transcription to yield various stem-and-loop structures (Fig. 12-5). Just at the beginning of the first potential loop (bases 27 to 68) there is an open reading frame coding for a small polypeptide of 14 amino acids. It is not well translated in vitro owing to its secondary structure, which blocks access of the ribosome in a manner similar to that seen with the phage MS2 coat protein, but the polypeptide has been detected in vivo. Among the amino acids contained within this polypeptide are two adjacent tryptophan residues. Tryptophan is a relatively little used amino acid, and two tryptophan residues in succession are rare. Yanofsky and his colleagues suggested that this polypeptide is the key to regulation by attentuation.

Their basic model is presented in Figure 12-5. After transcription is initiated, the RNA polymerase can be shown to pause at base 92 of the

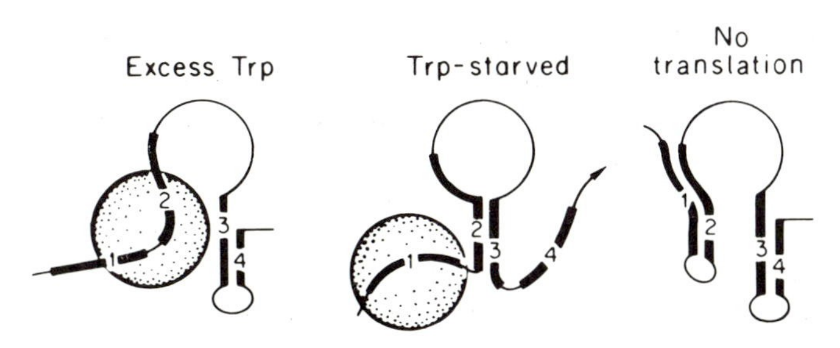

Figure 12-5. Model for attenuation in the *E. coli* tryptophan operon. The coiled structure is the initial portion (leader sequence) of the tryptophan mRNA that lies between the promoter site and the translation start site. Potential regions of hydrogen bonding are numbered. The large shaded circle represents a ribosome. Under conditions of excess tryptophan, the ribosome translating the newly transcribed leader RNA synthesizes the complete leader peptide. During this synthesis, the ribosome masks regions 1 and 2 of the RNA and prevents the formation of stem-and-loop 1-2 or 2-3. Stem-and-loop 3-4 is free to form and signals the RNA polymerase molecule (not shown) transcribing the leader region to terminate transcription (attenuation). Under conditions of tryptophan starvation, charged $tRNA^{Trp}$ is limiting and the ribosome slows down or stalls at the adjacent Trp codons in the leader peptide coding region. Because only region 1 is masked, stem-and-loop 2-3 is free to form as regions 2 and 3 are synthesized by the RNA polymerase. Formation of stem-and-loop 2-3 precludes the formation of stem-and-loop 3-4, which is required as the signal for transcription termination. Therefore RNA polymerase continues transcription into the structural cistrons. Under conditions in which the leader peptide is not translated because of genetic alterations or starvation for amino acids occurring before tryptophan in the leader peptide, stem-and-loop 1-2 is free to form as regions 1 and 2 are synthesized. Formation of stem-and-loop 1-2 prevents the formation of stem-and-loop 2-3, thereby permitting the formation of stem-and-loop 3-4. This step signals transcription termination. From Oxender, D.L., Zurawski, G.,Yanofsky, C. (1979). Attenuation in the *E. coli* tryptophan operon: role of RNA secondary structure involving the tryptophan codon region. Proceedings of the National Academy of Sciences of the United States of America 76:5524–5528.

leader sequence. During this pause the RNA already synthesized may form one of several loops. The loops are created from the sequences numbered 1 through 4 in the Figure. It is assumed that the 1-2 loop (the protector loop) is the most stable and always forms if possible. It can occur if protein synthesis is totally blocked, and there is no ribosome translating the leader polypeptide. Regions 3 and 4 then form a terminator loop of the usual type, and the paused RNA polymerase terminates transcription. If tryptophan is present in excess and translation is allowed, the ribosome has no difficulty translating two successive tyrptophan codons and physically covers both regions 1 and 2. Under these conditions the terminator loop forms as before. However, if tryptophan is in short supply, the ribosome translating the leader polypeptide pauses at the two consecutive tryptophan codons and lags behind the polymerase. In that event regions 2 and 3 can pair to form an antiterminator loop, and the paused RNA polymerase continues on to make a full-length mRNA molecule. In accord with this model, mutagenesis of the 1-2 loop to reduce its stability by reducing its hydrogen-bonding capability results in shorter RNA polymerase pauses and therefore less termination.

Although most of the experiments with attenuation have been done in vitro, Yanofsky and his collaborators demonstrated that appropriately sized transcripts are also produced in vivo. They suggested that the pausing may serve the general function of preventing the ribosome from lagging too far behind the RNA polymerase complex. If a large gap should develop, there is a chance that random folding of the RNA might stimulate rho to terminate the transcript. Normal translation of the mRNA by a series of ribosomes would forestall access of rho to the RNA.

Polar mutations in the tryptophan operon demonstrate some effects related to the concept of keeping the ribosome–RNA polymerase gap small. Specifically, polar mutations in *trpE* have a tenfold greater effect on *trpD* than on *trpCBA*. The difference results from a separate and efficient ribosome binding site located just before *trpC*, whereas *trpD* has a relatively inefficient one that has a 1-base overlap with the *trpE* termination codon. Apparently the *trpD* ribosome binding site is difficult to locate unless ribosomes release their nascent protein chain right beside it (i.e., translate to the end of the *trpE* cistron). Oppenheim and Yanofsky proposed that this phenomenon be called translational coupling. A similar effect is observed for the *trpA* and *trpB* cistrons.

The tryptophan operon in *Bacillus subtilis* has a similar pattern of regulation by attenuation, but the control of the attenuation process is different. Once again several loops are possible in the leader sequence (Fig. 12-6). The C:D loop is a terminator loop; and the A:B loop, which overlaps part of segment C, prevents its formation. In this case the regulatory protein again is an apoprotein requiring tryptophan for functionality. When the two are complexed, they bind to region A of the leader, preventing formation of the A:B loop and allowing the terminator loop to form. The binding of the regulator can be graphically demonstrated by cloning the leader sequence in a high-copy-number plasmid. Cells carrying the clone are derepressed for tryptophan synthesis because nearly all of the available repressor is bound to the cloned DNA, leaving the normal tryptophan operon fully expressed.

Histidine Operon

The histidine operon may be the most complex single operon studied. Most of the data come from experiments using *Salmonella typhimurium* in the laboratories of Ames, Hartman, and Roth, but similar results have

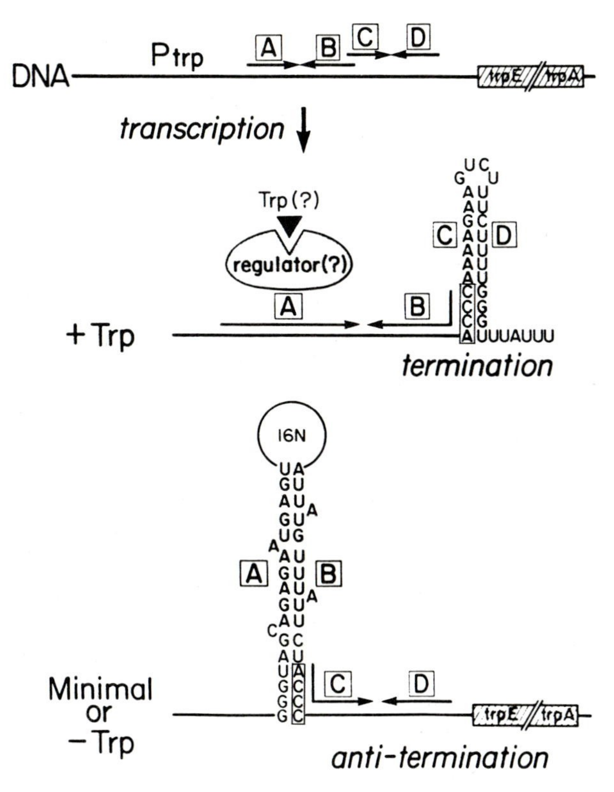

Figure 12-6. Regulated transcription termination model of the *B. subtilis trp* operon. Boxed letters indicate segments, and boxed nucleotides indicate the overlap of segments B and C. From Shimotsu et al. (1986).

been reported for *E. coli*. The operon consists of nine cistrons arranged so that the first cistron codes for the first enzyme in the pathway, but the other cistrons have no particular correspondence to the biochemical order (Fig. 12-7). Mutations have been mapped that have the properties associated with an operator and promoter. As is the case for the tryptophan operon, a second promoter has been identified, but not localized, in the middle of the operon.

The regulation of histidine biosynthesis is different from that of tryptophan in that the addition of histidine to a culture inhibits synthesis by only 20%. Instead, inhibition of specific enzymes within the pathway can result in stimulation of synthesis beyond that normally observed. Consistent with these observations is the fact that no cistron has been identified that codes for a protein repressor. Rather, there seem to be five cistrons influencing regulation, but all have other primary functions. They have been identified by the use of histidine analogs, compounds such as 1,2,4-

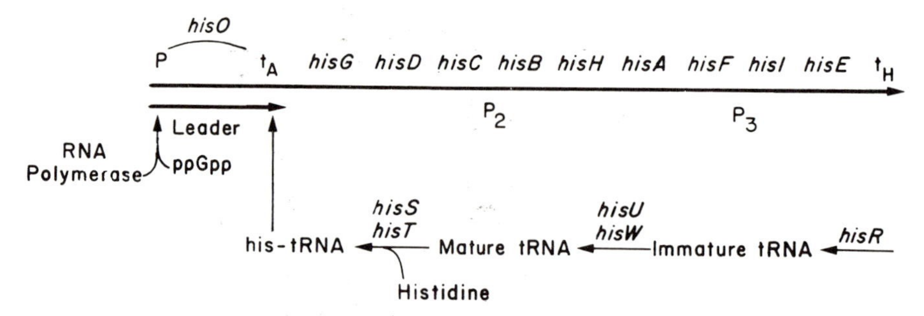

Figure 12-7. Regulation of the histidine operon in *S. typhimurium*. The basic operon is the large transcript shown at the top of the figure. The horizontal arrows indicate that two mRNA transcripts are possible from this segment of DNA. The shorter transcript includes only the leader portion of the operon and none of the structural cistrons. The longer transcript covers the entire operon. The control point that determines the size of the transcript is t_A, the attenuator site. In the presence of charged histidine tRNA, mRNA synthesis is terminated at t_A. In the absence of charged histidine tRNA, a small hypothetical peptide encoded within the leader sequence is assumed to be poorly translated, resulting in lack of attenuation. The five other histidine cistrons (which are individually transcribed and regulated) are involved in production of the histidine tRNA molecules. Initiation of RNA polymerase binding to the promoter site is facilitated by the presence of ppGpp. Many of the mutations that map within the leader region have the characteristics of operator mutations, and this region is sometimes referred to as *hisO* or *hisGo*.

triazole-3-alanine, which resemble histidine sufficiently well to cause repression of histidine biosynthesis but not well enough to be charged onto tRNA. Cells resistant to the repressing effects of analogs have been found to carry mutations in one of the *hisR, S, T, U,* or *W* cistrons.

All of these cistrons are basically involved in the production of histidinyl-tRNA molecules. The *hisR* cistron codes for the actual tRNA molecule. The *hisS* cistron codes for the synthetase enzyme that charges the tRNA. The *hisU* and *W* cistrons code for enzymes to mature the tRNA. The *hisT* cistron codes for an enzyme that modifies two uridine residues in the anticodon loop of the tRNA molecule so that they become pseudouridines. The effect of mutations in the *R, S, U,* and *W* cistrons is to decrease the amount of charged histidinyl tRNA in the cell and indirectly to reduce the amount of pseudouracil-modified tRNA in the cytoplasm.

The level of charged histidine tRNA in the cell is important because an attenuator region similar to that described for the *trp* operon has also been observed in the *his* operon. In the case of the *his* operon the leader region is approximately 250-bp in length, and contains in its middle a sequence coding for a hypothetical peptide that would contain 16 amino acid residues, including seven histidine residues in a row. As in the case of the *trp* attenuator, transcription termination seems to occur at a point beyond the peptide coding region. Attenuation is once again assumed to be prevented when ribosomes are arrested or slowed in their translation of the leader peptide due to a shortage of histidinyl tRNA containing a pseudouracil modification (i.e., *hisR, U, T,* or *W* mutants are constitutive). Thus the model presented in Figure 12-5 for the *trp* operon seems to apply also to the *his* operon. There are two internal promoters in the *his* operon. Promoter 2 is reported to be regulatable, but not by histidine.

Hartman and co-workers, using F′ plasmids carrying *his* operons deleted for the attenuator sequence, demonstrated the existence of a positive regulator for the *his* operon, ppGpp. It is the product of two reactions in which guanosine 5′-triphosphate is first reacted with ATP to yield guanosine 5′-triphosphate-3′-diphosphate (pppGpp) and then dephosphorylated to yield guanosine 5′-diphosphate-3′-diphosphate (ppGpp). The first reaction is catalyzed by a ribosome-associated protein encoded by the *relA* cistron, and the second reaction appears to be catalyzed by several uncharacterized enzymes. Cells do not produce ppGpp all of the time but, rather, only when ribosomes have tRNA molecules bound to them that are correctly paired to the codon on the mRNA but uncharged. When ppGpp is produced, it causes general inhibition of DNA and RNA synthesis and, indirectly, protein synthesis. Cells with normal *relA* function therefore reduce metabolic activities under amino acid starvation conditions (**stringent control**). Cells mutant for *relA* fail to synthesize ppGpp and therefore continue to make RNA and DNA even when starved for amino acids (**relaxed control**). The product of the *spoT* cistron gradually breaks down ppGpp, so that it must be continually produced in order to maintain its repression of nucleic acid synthesis.

In the histidine operon, however, the role of ppGpp is exactly opposite of what is usually ascribed to it. In a strain lacking an attenuator and carrying a *spoT* mutation to elevate the ppGpp levels, the amount of histidine biosynthetic enzymes is increased as much as twofold over normal. This stimulation apparently occurs at the promoter site, and the presumption is that ppGpp facilitates RNA polymerase binding. The *his* operon thus has several levels of control, which are summarized in Figure 12-7.

Regulation of Complex Operon Systems

Maltose Regulon

The term **regulon** is applied to combinations of two or more operons (complete with structural cistrons, promoters, etc.) that are coordinately regulated. In the case of the cistrons coding for maltose utilization, two widely separated operons are involved and were originally designated *malA* and *malB* (see the *E. coli* map inside the front cover at minutes 74 and 90). However, fine structure mapping by Hofnung, Schwartz, and co-workers has shown that each so-called operon is in fact composed of two smaller operons, so MalA and MalB are now used to indicate the groups of operons.

The MalB region contains the information coding for transport of maltose into the cell and consists of six cistrons. The exact functions of the *malG* and *malK* cistrons are unknown; *malE* and *malM* code for maltose binding proteins located in the periplasmic space (between the outer membrane of the cell wall and the cell membrane), and *malF* codes for a binding protein located in the cell membrane. The last cistron, *lamB*, has a dual role. It acts as the primary maltose receptor on the outer membrane whenever the concentration of maltose is less than 10^{-4} *M*. However, the *lamB* protein's principal claim to fame is as the λ receptor, the site at which the λ phage binds to initiate infection. Lack of this protein makes a cell resistant to the phage. The MalA region codes for regulation and for maltose-hydrolyzing enzymes. The product of the *malQ* cistron is amylomaltase, which acts to hydrolyze maltose to yield glucose and a glucose polymer. The polymer is then hydrolyzed to glucose-1-phosphate by a phosphorylase encoded by the *malP* cistron. The *malT* cistron produces a protein that acts as the positive regulator (i.e., promotes transcription) for the MalA and MalB regions. The properties of cells carrying various combinations of *mal* mutations are given in Table 12-2. Note that *malT* function is absolutely required for expression.

Genetic mapping data for the MalA region indicate that *malT* is a separate operon, followed by a promoter site and the *malP* and *malQ* cistrons. The promoter consists of two elements, one to bind the RNA polymerase and one to bind the positive regulator, a system that is analogous to that used in the *lac* operon. A mutation in the promoter site can relieve the requirement for the activator complex, which apparently consists of the MalT protein in association with maltose (Fig. 12-8).

Table 12-2. Phenotypes of cells carrying various combinations of *mal* mutations

	Phenotype	
Genotype	Maltose	λ
Wild type	+	S
malT	−	R
malP or *malQ*	−	S
malK	−	S or R
lamB	+	R
malE, malF, malG, or *malM*	−	S

+, ability to utilize maltose; −, inability to utilize maltose; S, sensitivity to the phage λ; R, resistance to phage λ.

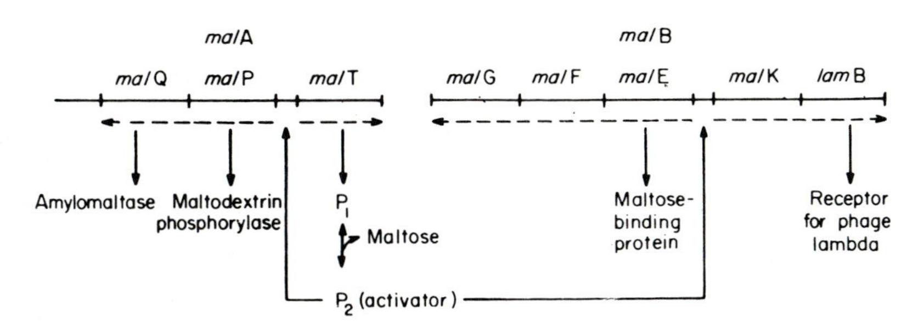

Figure 12-8. Maltose regulon in *E. coli*. The *mal*A and *mal*B regions are located at 74 and 90 minutes, respectively, on the genetic map of *E. coli* (see Figure 9-3). The products of cistrons *mal*F, G, and K have not yet been purified; but like the products of *mal*E and *lam*B, they are involved in the transport of maltose and maltodextrins. Cistron *mal*T is a positive regulator whose product, in the presence of maltose, activates the expression of the three maltose operons. The directions of transcription of the three *mal* operons are indicated by broken arrows. Adapted from Débarbouillé, M., Schwartz, M. (1979). The use of gene fusions to study the expression of *mal*T, the positive regulator gene of the maltose regulon. Journal of Molecular Biology 132:521–534.

In the MalB region the situation is somewhat different. The cistrons map next to one another, but polar mutations in *malK* affect only *lamB* and *malM* expression (making the cell λ-resistant), whereas polar mutations in *malE* affect only *malF* and *malG* expression. The simplest explanation for these observations is to assume that a promoter site is located between *malE* and *malK* and that transcription can proceed in either direction (Fig. 12-8). This type of arrangement is called **divergent transcription** and is not unique to maltose. The *malT* activator complex must be necessary for activation of the promoter, but the way in which it acts in the MalB region may be slightly different from the way it acts in the MalA region because no *malT*-independent mutations have ever been observed in the MalB region.

Heat Shock

All organisms examined, ranging from vertebrates to *E. coli,* show a definite response pattern when shifted to temperatures above their normal growth range. The **heat shock response** is the stimulation of synthesis of a particular subset of cellular proteins, some of which are normally present at low levels and others of which are synthesized de novo. In *E. coli* the heat shock response is an example of a **global regulatory network** in which a series of operons that are not obviously related by function (as are the maltose operons) are coordinately expressed or repressed. The functions that are overexpressed during heat shock include *dnaK, dnaJ, groEL, groES, lysU, grpE,* and *rpoD.* The peak activity is expressed some 5 to 10 minutes after the heat shock.

Expression of the heat shock proteins is controlled by *htpR* (*rpoH*), which is a new σ factor for the RNA polymerase holoenzyme complex. The normal σ protein has a molecular weight of 70,000 daltons and is designated σ^{70}. The new protein is substantially smaller and is designated σ^{32}. Neidhardt and his collaborators theorized that the heat shock proteins each have two promoters, a low efficiency one that responds to σ^{70} and a high efficiency one that responds to σ^{32}. They suggested that the σ^{32} is normally in an inactive configuration but can be triggered into its active configuration by heat shock as well as by some other treatments including ethanol, ultraviolet (UV) radiation, inhibition of topoisomerases, or λ infection. The heat shock proteins automatically return to their normal rate of synthesis after a brief interval because the *rpoD* cistron codes for σ^{70} and is stimulated by heat shock. It therefore gradually displaces σ^{32}, and the situation returns to normal.

Not much is known about what makes heat shock proteins protective for a cell, but there are strong similarities between heat shock proteins found in *Drosophila* and those found in *E. coli*, suggesting a certain generality of function. The DnaK protein may use ATP to disrupt hydrophobic protein aggregates such as might form after partial denaturation of a protein. It plays a similar role during normal λ infection in catalyzing the release of protein P from the λ O protein–host DNA helicase complex (see Figure 6-2). Heat shock in *B. subtilis* is less well studied, but again many of the proteins seem to antigenically match those observed in *E. coli*.

Bacteriophage Lambda

The basic properties of the λ genetic system are described in Chapter 6. Another version of the λ genetic map is presented in Figure 12-9, and a brief review of it might be helpful. The λ DNA as found in the mature virion is a linear double-strand molecule with cohesive ends (*cos*). After entry into the host cell, the λ DNA circularizes to form its replicative structures. If an integrated prophage is formed, the recombination is at *att*, giving different ends to the prophage map. The complete DNA sequence for λ has been determined, and therefore it is possible to relate most genetic loci to specific sequences. It is convenient to consider regulation of vegetative growth as separate from regulation of lysogeny, although in practice the two regulatory systems overlap.

Vegetative growth begins with the arrival of the phage DNA in a cell or with induction of a prophage. The first event observed is synthesis of two short transcripts (immediate early mRNA) that are complementary to different regions on the right arm of the mature phage DNA. These transcripts are represented by the light arrows labeled L2 and 9S in Figure 12-9. They begin at promoters p_L and p_R, respectively, and terminate at t_{L1} and t_{R1}. In vitro experiments have shown that t_{R1} is actually three po-

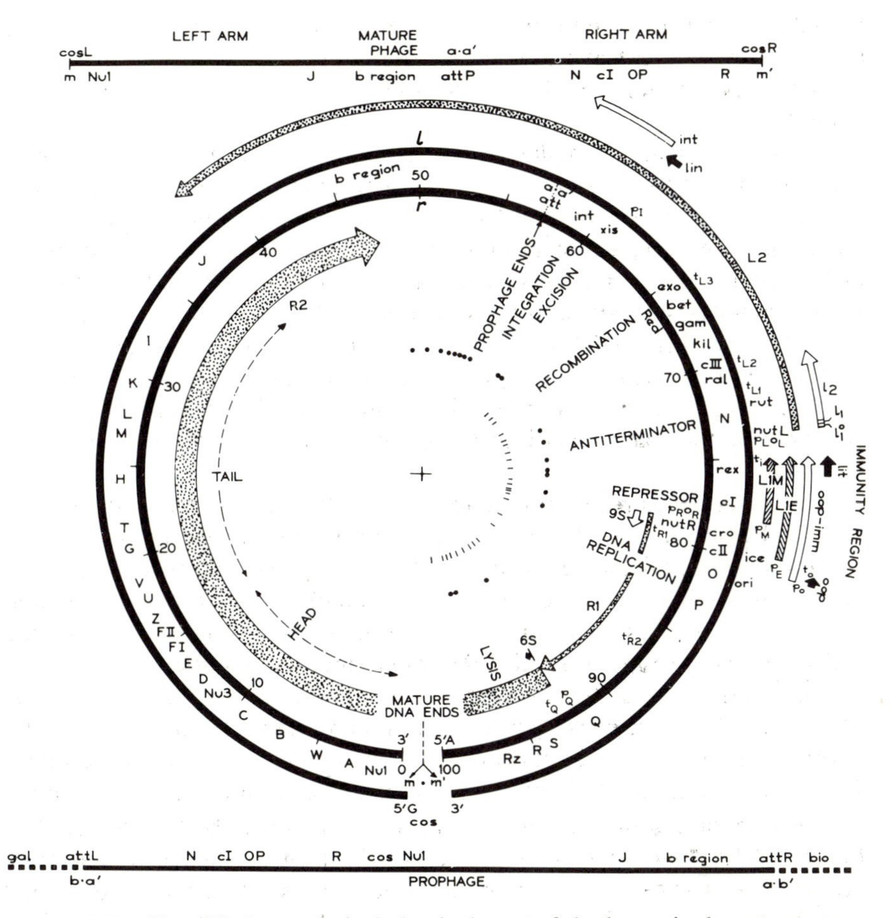

Figure 12-9. Simplified genetic and physical map of the bacteriophage λ genome. The heavy lines represent the complementary *l* and *r* strands. The genome is presented in the form of an open circle obtained by bending downward the left and right termini of the linear λ DNA molecule (drafted above the circle), according to the convention. Strand *l* is transcribed leftward (counterclockwise) and has a 5′-G at its left cohesive end, *m*. Strand *r* is transcribed rightward (clockwise) and has a 5′-A at its right cohesive end, *m*′. Strand *r* displays a higher density in a CsCl-poly(U,G) gradient and a lower density in an alkaline CsCl gradient than does strand *l*. The shaded arrows represent the various leftward and rightward transcripts, always placed next to the complementary coding strand. The prophage map (drawn at the bottom) is a circular permutation of the linear map of mature λ DNA (drawn at the top). The RNA polymerase binding sites are represented by small black circles near the center of the drawing. From Szybalski, E.H., Szybalski, W. (1979). A comprehensive molecular map of bacteriophage lambda. Gene 7:217–270.

tential terminators spread out over 60 bp. All are ρ-dependent, and the third site is the most efficient of the three. The only cistrons whose information is contained within these transcripts are *N* (from L2) and *cro* (from 9S).

As the N protein accumulates, it forms a complex with RNA polymerase carrying one of the host *nus* proteins. The NusA protein is involved in termination of transcripts, and the NusB protein is involved in antitermination. The complex of N and RNA polymerase interacts with the λ DNA at one of two *nut* sites. The *nut* sites are regions that can form a small stem-and-loop structure and are flanked by specific regions designated box A (left) and box C (right). It appears that some of the host proteins interact with these regions. The result of the interaction with a *nut* site is a polymerase complex that reads through the first terminators and extends the transcripts into the coding regions needed for a lytic infection.

The larger transcripts, designated L2 and R1 and shown as shaded arrows in Figure 12-9, code for most of the replication, recombination, integration, and excision functions of the phage. They represent the bulk of the early mRNA production. Several minor transcripts are indicated by the white arrows in the *int* and *OP* regions. In addition to providing DNA replicative functions, the R1 transcript also produces Q product, which functions in antitermination of transcription from the p_Q promoter. This new R2 transcript is synthesized in large amounts and contains the information for head and tail production and cell lysis (the late functions). The antitermination site is at t_Q, and Q is basically a *cis*-acting protein.

The temporal sequence of the events so far described is depicted in Figure 12-10. Note that in several cases the transcription termination is not perfect. For example, R1 transcription tends infrequently to extend into the R2 region. The increased level of R2 transcription provided by p_Q ensures that structural components for a large number of phages are produced. However, various functions in the L2 region tend to be harmful to the host, and the progeny phage yield is not high unless these functions are repressed after they have served their purpose. The repression is accomplished by the small *cro* protein (66 amino acids long) whose symbol is an acronym for *c*ontrol of *r*epressor and *o*ther things. Experiments with β-galactosidase fusions have shown that the Cro protein binds to operators at three sites (o_L, o_R, and near p_E) and prevents or reduces the binding of N protein-modified RNA polymerase. The DNA sequence for o_L and o_R is basically the same and includes three sites at which Cro can bind. Cro binds preferentially to the third site and gradually fills in the other sites if its concentration reaches sufficient levels. Figure 12-11 shows that the Cro binding sites physically overlap the promoter region so that, as Cro is sequentially bound, the leftward and rightward promoters are progressively blocked.

The repressor protein is the regulator of the temperate response of the phage. It must shut down the major lytic functions of the phage while

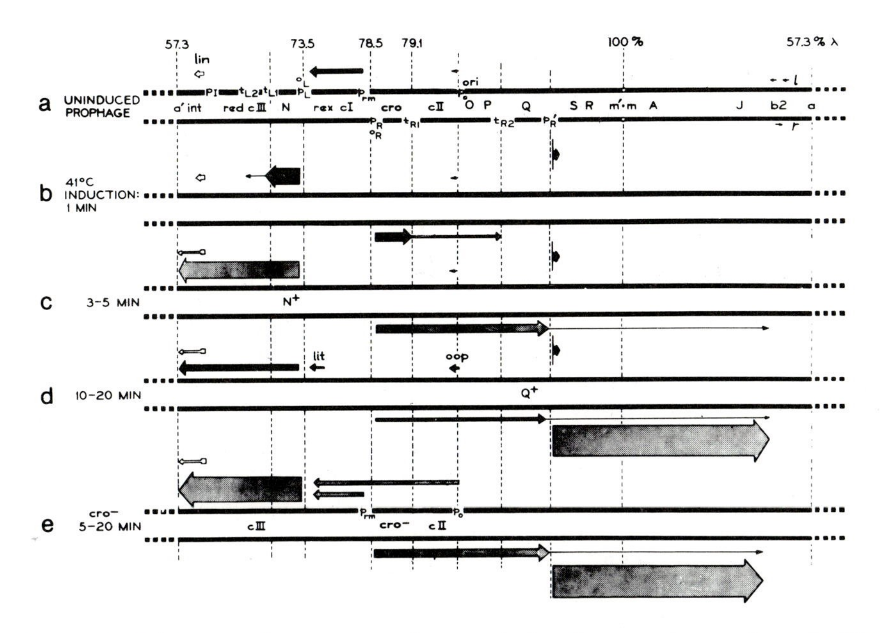

Figure 12-10. Temporal sequence of transcriptional events in prophage λ. The leftward transcripts are coded by the *l* strand and drawn above the λ DNA; the rightward transcripts are coded by the *r* strand and drawn below the λ DNA. **(a)** Transcription in the uninduced prophage. The *c*I-*rex* transcripts correspond to 80 to 90% of the total prophage-specific RNA. **(b)** Immediate early transcription after induction (the same pattern occurs during lytic infection). **(c)** Delayed early transcription. **(d)** Late transcription. **(e)** Decontrolled transcription in an induced *cro*⁻ mutant of λ. The prophage maps are not drawn to scale but, rather, with the immunity region expanded. The numbers in the top line indicate the positions of various sites with respect to the left end (0% λ) and right end (100% λ) of mature λ DNA. The width of the arrows is a measure of the rate of transcription. In the case of the 198 nucleotide 6S RNA transcribed early (arrow under $p_{R'}$), it was found that in vitro synthesis provides 10- to 20-fold more of the 5′-proximal 15 nucleotide sequence (represented by the vertical line in **a**, **b**, and **c**) than of the total 6S RNA. Thus $p_{R'}$ is the strongest λ promoter but is immediately followed by strong termination signals that can apparently be overcome by the *Q* product, with resulting synthesis of late RNA **(d)**. From Szybalski, W. (1977). Initiation and regulation of transcription and DNA replication in coliphage lambda. In: Copeland, J.C., Marzluf, G.A. (eds.) Regulatory Biology (Ohio State University Biosciences Colloquia, no. 2). Copyright © 1977 by the Ohio State University Press. All rights reserved. Used by permission of the author, the editors, and the publisher.

allowing the integration activity necessary for prophage formation. The shutdown is accomplished at some of the same promoters used by Cro. Specifically, the repressor protein binds to o_L and o_R but adds to the multiple binding sites in the reverse order, so that it blocks the leftward and rightward promoters first (Fig. 12-12). Thus if the repressor can be obtained in quantity, a lytic infection can be prevented. The *c*I cistron coding for

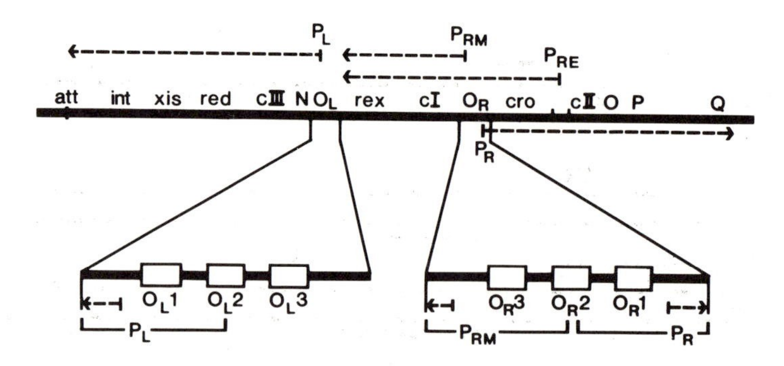

Figure 12-11. Regulation of early λ transcription. Transcription of early λ genes is initiated at p_L and p_R and is subject to repression by repressor acting at o_L and o_R, respectively. Transcription of the *c*I (repressor) and *rex* genes can be initiated either at p_{RM} (the maintenance promoter) or p_{RE} (the establishment promoter). Transcripts are indicated by dashed lines. The expanded diagrams illustrate spatial relations between p_L and three repressor binding sites in o_L, and among p_{RM}, p_R, and three repressor binding sites in o_R. The *cro* protein also binds at o_R and o_L to reduce transcription from p_R, p_{RM}, and p_L. From Gussin, G.N., Johnson, A.D., Pabo, C.O., Sauer, R.T. (1983). Repressor and Cro protein: structure, function and role in lysogenization, pp. 93–121. In: Hendrix, et al. (1983).

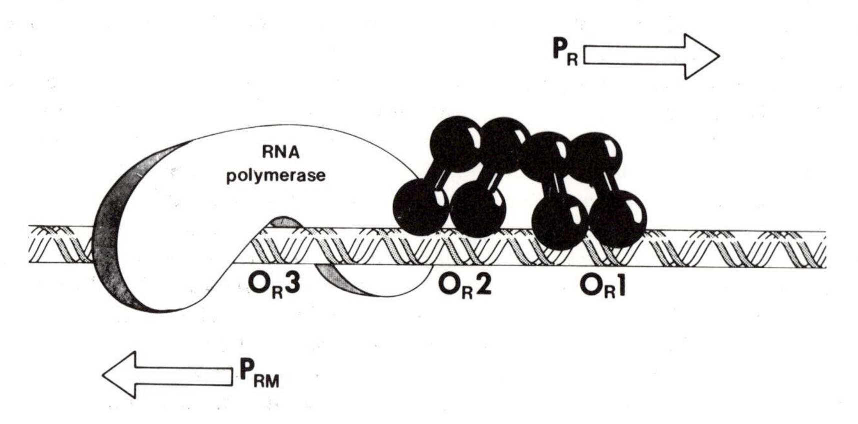

Figure 12-12. Lambdoid phage right operator (o_R) *showing the disposition of bound proteins found in a lysogen.* o_R1, o_R2, and o_R3 are the repressor binding sites; in a lysogen, repressor is bound to o_R1 and o_R2. RNA polymerase binds to the leftward promoter, p_{RM}, but is unable to bind to the rightward promoter, p_R, due to blockage by repressors bound to o_R1 and o_R2. From Bushman, F.D., Ptashne, M. (1986). Activation of transcription by the bacteriophage 434 repressor. Proceedings of the National Academy of Sciences of the United States of America 83:9353–9357.

the repressor can be transcribed from either of two promoters, p_{RE} or p_M. Neither promoter is initially active, so they must require positive regulatory elements.

Among the cistrons contained within the L2 and R1 transcripts are *c*II and *c*III, which are important for the lysogenic cycle. As described in Chapter 6, the products of these cistrons seem to act as regulators of λ repressor expression, but they also stimulate *int* transcription from p_I. Their effect is exerted via the *c*II protein, which binds to the similar −35 regions on both p_I and p_{RE} in the major groove of the DNA helix, i.e., on the opposite side of the helix from the RNA polymerase. Together the *c*II and *c*III proteins form a regulatory circuit with the host *hfl* protein, a protease that degrades the *c*II protein. The *c*II protein, as it accumulates, stimulates the appropriate promoters, but under normal conditions the stimulation is not great owing to the proteolytic action of Hfl. Mutations in that locus confer a *h*igh *f*requency of *l*ysogeny because the resultant high level of Int protein converts nearly all of the phage DNA into integrated prophages. The normal frequency of lysogeny is maintained by *c*III action, which antagonizes the Hfl protein so as to preserve a minimum level of *c*II activity.

The binding of the *c*II protein to p_{RE} initiates the establishment mode of repressor transcription (the L1E transcript). This mode yields relatively large quantities of repressor protein and results in the reverse transcription of the *cro* region. This reverse transcription antagonizes transcription of *cro* and may facilitate a lysogenic response. Note, however, that transcription from p_{RE} cannot be maintained because binding of the repressor to o_L and o_R eventually prevents further production of the *c*II and *c*III proteins. The maintenance mode of repressor synthesis is from p_{RM}, a promoter that overlaps o_R (Fig. 12-11) and yields the L1M transcript. As the repressor binds to turn off rightward transcription, it also activates its own promoter (another example of autoregulation) (Fig. 12-12). Transcription is maintained from this promoter so long as repressor is present in the cell. Should the level of repressor become too high, binding of repressor to the leftmost portion of o_R prevents further transcription until the repressor concentration is reduced. The p_M transcript includes information for *c*I and *rexAB*, the latter being cistrons coding for proteins that prevent phage T4 *r*II mutants from growing. The transcriptional state of the prophage is also shown in Figure 12-10.

An interesting type of regulation was observed when transcription of the *int* region from p_L was examined in a *c*II mutant strain (p_I poorly activated). The observation was that transcription is relatively inefficient unless the *b* region is separated from it (as would be the case for an integrated prophage). The inefficient transcription is an example of **retroregulation**, regulation of a transcript by a downstream sequence. The deleted *b* sequence includes DNA that codes for a stem-and-loop structure of a particular sort (Fig. 12-13). The region that includes this potential

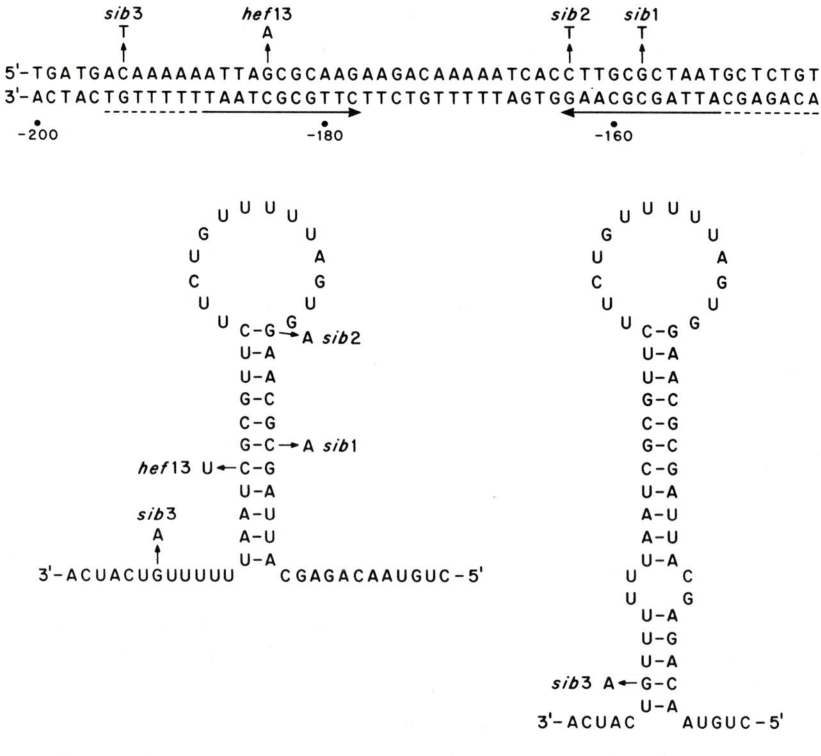

Figure 12-13. The *sib* region. Locations of four *sib* mutations are shown in the DNA sequence (top). Possible secondary structures of the RNA from this region are shown below. The RNA structure on the right can form from p_L RNA but not from p_I RNA because the latter terminates at the last of the six consecutive U bases (position −193). From Echols, H., Guarneros, G. (1983). Control of integration and excision, pp. 75–92. In: Hendrix, R.W., Roberts, J.W., Stahl, F.W., Weisberg, R.A. (eds.) Lambda II. Cold Spring Harbor, N.Y.: Cold Spring Harbor Laboratory.

loop is designated *sib* (*s*itio *i*nhibidor en *b*). The loop formed by the p_L transcript has been shown to be a signal for RNase III processing in vivo and in vitro. The cut RNA may be a substrate for further degradation by the RNase II enzyme, as mutants in this function do not exhibit normal retroregulation. Transcription from p_I, on the other hand, terminates at a different spot and cannot form the same stem and loop. The differing transcription end points are another example of how RNA polymerase can be programmed for certain termination signals while it is initiating transcription.

Both Cro and repressor recognize the same sequence on the DNA, i.e., TATCACCGCCAGAGGTA. Both proteins fold to give a pair of α-helical regions that fit into the DNA major groove, and both bind as dimers. The larger λ repressor protein dimer extends its amino-terminal arms around the helix, thereby protecting more of the DNA structure during chemical alteration experiments. It also forms a loop in the operator DNA that is visible by electron microscopy. Counting from the right in the recognition sequence, Cro keys on base 3, and repressor keys on bases 5 and 8. This observation of similarity in binding site emphasizes the major remaining problem in λ regulation: understanding why certain infected cells form lysogens whereas other cells in the same culture lyse and release phage particles. Ward and Murray focused on the opposing transcripts in the *cro* region. They suggested that minor fluctuations in the efficiency of one transcript may shift the metabolic balance of the cell toward either lysogeny or lysis by emphasizing either the L1 or R1 transcripts.

Nitrogen Fixation

A relatively small group of bacteria, both photosynthetic and nonphotosynthetic, is capable of reducing atmospheric nitrogen and using it for metabolism. Some typical genera that include species known to fix nitrogen are *Rhizobium, Azotobacter, Klebsiella, Clostridium, Rhodopseudomonas,* and *Anabaena*. The nitrogenase enzyme is extremely sensitive to oxygen and has three protein subunits that are encoded by *nifH,D,K,* whose base sequence is highly conserved from organism to organism. Cloning of these cistrons into *E. coli* yields an organism that fixes nitrogen.

The best studied genetic system is that found in *Klebsiella pneumoniae,* which has 17 *nif* genes organized into eight transcription units along a 25-kb fragment of DNA. The regulatory scheme for *nif* is diagrammed in Fig. 12-14 as a part of a global regulatory network. The network regulates not only nitrogen fixation but also histidine and proline utilization and glutamine synthetase, all of which aid the cell in assimilating nitrogen. Overriding control of the network is vested in the *ntr* system. The *ntrB* protein is a kinase/phosphatase that phosphorylates/dephosphorylates NtrC, which is needed for activation of NtrA. NtrA must interact with all of the necessary promoters, including NifA, to turn on transcription; therefore it is a sigma factor. The NifL protein is activated by oxygen to tie up the NifA protein and prevent activation of the *nifHDK* operon.

Nitrogen-fixing systems in other organisms are organized along similar lines, although not necessarily as compactly as in *Klebsiella*. One point of difference is seen in *Anabaena,* a cyanobacterium that carries out oxygenic photosynthesis. The sensitivity of the nitrogenase enzyme to oxygen means that a cell cannot carry out both normal photosynthesis and nitrogen fixation. Instead, certain cells within a filament of cells differentiate to form thick-walled heterocysts where nitrogen fixation can occur. During the differentiation process an 11-kb segment of DNA is recombined from the middle of the *nifD* cistron to yield a normal *nifD*. The presence of the "excison" prevents undifferentiated cells from futile attempts to fix nitrogen.

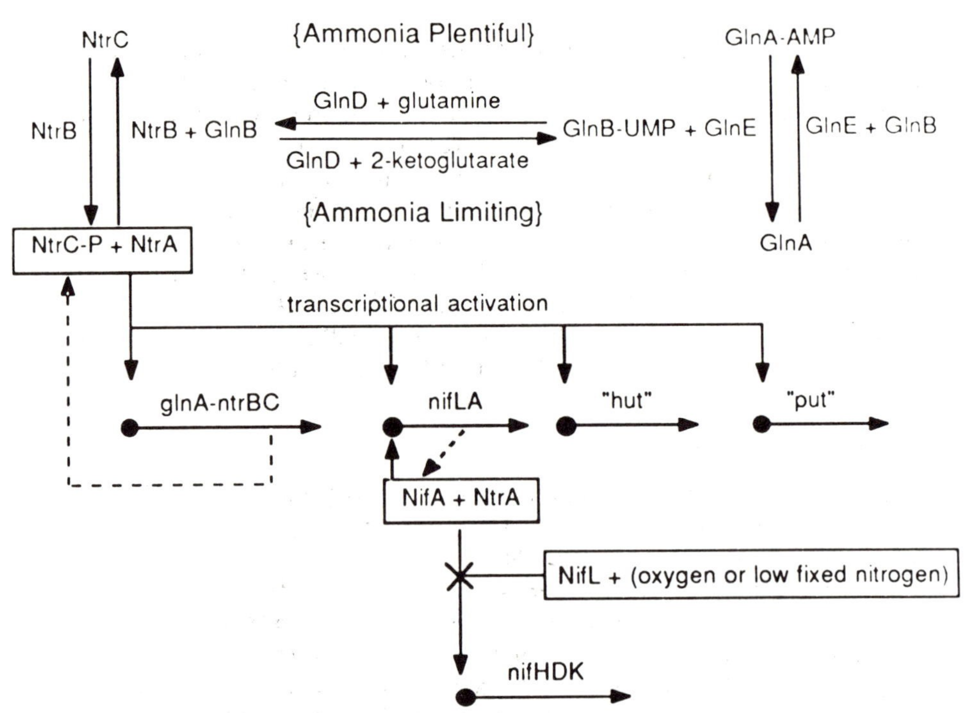

Figure 12-14. Circuitry of *nif* regulation in *Klebsiella pneumoniae*. The *ntr* regulatory cistrons control nitrogen fixation (*nif*) as well as histidine and proline utilization (*hut* and *put*) and glutamine synthesis (*gln*). All of the pathways are involved in nitrogen uptake by the cell. From Gussin, et al. (1986).

Other Regulatory Systems

Regulation by an Invertible Segment

Phages P1 and Mu have a region of DNA that inverts under the control of another locus. In their cases the trait involved is the host range of the virus. This type of regulation is not, however, unique to the viruses. There are several examples known from among the bacteria as well, and one of the best studied is the phenomenon of **phase variation** in *Salmonella*.

Members of the genus *Salmonella* are generally flagellated, and the flagella can be shown to be composed of two antigenically distinct monomers, H1 or H2. At any given time, a cell produces only one kind of flagellum, but there is a low probability (10^{-3} to 10^{-5}) that the cell will switch to the other type. Therefore all cells have the capability to produce both antigenic types of flagellum, but only one type can be expressed at any given moment.

Simon and co-workers investigated this phenomenon by making F′

plasmids carrying the H1 or H2 flagellar antigen determinants and transferring them into *E. coli.* The H1 plasmid expressed its genetic information in the usual fashion. However, the H2 plasmid was variable in its ability to express the *Salmonella* DNA. Heteroduplex analysis of the F′ DNA hybridized to itself indicated that a region of the H2 DNA inverted itself spontaneously. In one orientation the promoter was connected to the H2 locus, and in the other orientation it was not. Transcription of the H2 locus also resulted in expression of the *rh1* cistron, which acted as a repressor of H1 expression. Therefore the two antigens were not able to be expressed simultaneously. The stimulus for the inversion of the H2 locus comes from a cellular factor *vh2*. In the absence of this factor, inversion does not occur. Interestingly, phages P1 and Mu can cause inversion of the H2 region in the absence of a functional *vh2* cistron. The enzyme catalyzing these changes has been designated a DNA invertase.

Similar phenomena are known in *E. coli* with respect to the fimbriae and flagella. Cells are either fully fimbriated or bald. A 300-bp invertible segment of DNA has been identified that carries the promoter for the *fimA* cistron that codes for the fimbrin subunit. Two regulatory loci have been identified, with *fimB* switching synthesis on and *fimE* turning it off. In the case of the flagella, the *vh2* function of *Salmonella* is replaced by the *pin* locus.

Translational Control

Translational control in prokaryotes is much more common than was originally thought. It is a wellknown phenomenon in eukaryotic organisms, but its initial identification was only in viral infections. Phage T4 codes for a modified initiation factor required for proper assembly of the ribosome–mRNA complex. When IF3β is present instead of IF3α, preferential translation of late cistrons is observed. Plasmids too can exhibit translational control. As detailed in Chapter 11, a small RNA molecule transcribed from the noncoding DNA strand is often used to bind to mRNA and inhibit translation.

There are several examples of translational regulation within bacterial operons. One relatively trivial example is *dnaG,* which is located in the middle of a transcript that includes *rpsU* and *rpoD.* Each coding region has its own ribosome-binding site, but less DnaG product is synthesized because "rare" codons are used. A rare codon can be operationally defined as a codon used infrequently by that organism when compared to random probability. They usually, but not always, are associated with a relative paucity of charged tRNA as well.

The best examples for translational control in bacteria have been presented by Nomura, Kjeldgaard, and their co-workers for some of the ribosomal proteins. Most of the ribosomal proteins are encoded in large operons that may contain as many as 11 cistrons. In the case of the largest of these operons, the S10 operon, the product of the third cistron (*rplD*) or L4 protein has at least two regulatory functions. It has been noted that the leader sequence of the S10 mRNA bears some resemblance to the 23S rRNA sequence where L4 normally binds. Thus if L4 is in excess, it can bind to its own mRNA and prevent translation. It can also induce attenuation of the same mRNA, probably by stabilizing a potential terminator loop located 30 bases upstream from the start codon.

Endospore Formation in *Bacillus*

The process of endospore formation is complex and highly regulated. The cells pass through the seven morphologic stages that are listed in Table 12-3. The process is triggered by starvation of the cells for carbon or nitrogen sources and is reversible until the cells reach stage IV.

Endospore formation presents a considerable regulatory problem. The *spo* loci are not grouped but, rather, are scattered about the genome (see Figure 8-6). Therefore some of the models developed for global regulatory networks may be applicable here. *Bacillus* often uses a regulatory strategy involving the use of various σ factors to modify RNA polymerase function. There are eight such proteins that have been identified or inferred from genetic evidence, and three of them are specifically related to sporulation. The transcriptional changes produced in sporulating bacilli must be more gradual that those that occur in a system such as phage T4, because many of the vegetative cistrons must continue to be transcribed in order to provide necessary functions for the developing spore. The use of multiple σ factors may also provide for the necessary timing of the various stages of sporulation.

Table 12-3. Morphologic stages in endospore formation by *Bacillus subtilis*

Stage	Morphology	Affected by mutations in cistron
0	Normal appearance	*spo0*
I	DNA may form a single axial filament	*spoI*
II	Cell membrane invaginates to form a spore septum	*spoII*
III	The spore septum elongates to completely surround the prespore	*spoIII*
IV	Germ cell wall and thick cortex are deposited around the spore	*spoIV*
V	Previously synthesized spore coat proteins are deposited around the spore	*spoV*
VI	Heat resistance develops	*spoVI*
VII	Cell lysis may occur	*spoVII*

Summary

The basic unit of regulation is the operon, a genetic entity consisting of one or more structural cistrons, at least one promoter site and one terminator site, and usually one or more regulatory sites. The immediate product of an operon is a single RNA molecule. This molecule may serve as an mRNA that programs a ribosome to synthesize one or more protein molecules that are the ultimate product. Alternatively, the RNA molecule itself may be the ultimate product, as in the case of tRNA or rRNA molecules. The ability of the promoter to bind the RNA polymerase can be affected positively or negatively by various regulator molecules or by insertion sequences. In the case of the lactose and galactose operons, negative regulatory elements are protein repressors that bind to operator sites, whereas positive regulatory elements are cyclic AMP and the cyclic AMP receptor protein (CRP), which bind near the −35 site of the promoter.

Binding of the RNA polymerase to the promoter DNA does not necessarily guarantee transcription of the operon. In the tryptophan, histidine, and S10 operons the leader sequence that lies between the promoter and the start of the coding sequences contains elements that can form terminator loops. Depending on the binding of an external protein (S10) or the ability of a ribosome to synthesize a leader peptide (tryptophan and histidine), the terminator loop may be able to form. Lack of the external protein or inefficient protein synthesis prevents terminator loop formation.

Termination of transcription of an operon is also a variable process. The RNA polymerase can be programmed while it is initiating transcription at a promoter to recognize only certain termination signals. Two consequences of this capability are antitermination and retroregulation, both of which are observed in phage λ.

Global regulatory networks such as the heat shock proteins, nitrogen fixation, endospore formation, or λ regulation are groups of scattered, coordinately regulated cistrons that are controlled by a plethora of diffusible repressors, activators, and new σ factors. Other unusual kinds of regulation include regulation by inverting segments of DNA, excision of noncoding DNA, and regulation by prevention of translation of the mRNA.

References

General

Apirion, D., Dallmann, G., Miczak, A., Szeberenyi, J., Tomcsanyi, T. (1986). Processing and decay of RNA in *Escherichia coli:* the chicken and egg problem. Biochemical Society Transactions 14:807–810.

Gussin, G.N., Ronson, C.W., Ausubel, F.M. (1986). Regulation of nitrogen fixation genes. Annual Review of Genetics 20:567–591.

Haselkorn, R. (1986). Organization of the genes for nitrogen fixation in photosynthetic bacteria and cyanobacteria. Annual Review of Microbiology 40:525–547.

Hendrix, R.W., Roberts, J.W., Stahl, F.W., Weisberg, R.A., eds. (1983). Lambda II. Cold Spring Harbor, N.Y.: Cold Spring Harbor Laboratory.

Lindahl, L., Zengel, J.M. (1986). Ribosomal genes in *Escherichia coli*. Annual Review of Genetics 20:297–326.

Losick, R., Youngman, P., Piggott, P.J. (1986). Genetics of endospore formation in *Bacillus subtilis*. Annual Review of Genetics 20:625–669.

McFall, E. (1986). *cis*-Acting proteins. Journal of Bacteriology 167:429–432.

Miller, J.H., Reznikoff, W.S., eds. (1978). The Operon. Cold Spring Harbor, N.Y.: Cold Spring Harbor Laboratory.

Pelham, H.R.B. (1986). Speculations on the functions of the major heat shock and glucose-regulated proteins. Cell 46:959–961.

Saunders, J.R. (1986). The genetic basis of phase and antigenic variation in bacteria, pp. 57–76. In: Birkbeck, T.H., Penn, C.W. (eds.) Antigenic Variation in Infectious Diseases. Oxford: IRL Press.

Specialized

Arnosti, D.N., Singer, V.L., Chamberlin, M.J. (1986). Characterization of heat shock in *Bacillus subtilis*. Journal of Bacteriology 168:1243–1249.

Bloom, M., Skelly, S., VanBogelen, R., Neidhardt, F., Brot, N., Weissbach, H. (1986). In vitro effect of the *Escherichia coli* heat shock regulatory protein on expression of heat shock genes. Journal of Bacteriology 166:380–384.

Donnelly, C.E., Reznikoff, W.S. (1987). Mutations in the *lac* P2 promoter. Journal of Bacteriology 169:1812–1817.

Gilson, E., Rousset, J-P., Charbit, A., Perrin, D., Hofnung, M. (1986). *malM*, a new gene of the maltose regulon in *Escherichia coli* K12. I. *malM* is the last gene of the *malK-lamB* operon and encodes a periplasmic protein. Journal of Molecular Biology 191:303–311.

Hochschild, A., Douhan, J., III, Ptashne, M. (1986). How λ repressor and λ Cro distinguish between o_R1 and o_R3. Cell 47:807–816.

Kramer, H., Niemöller, M., Amouyal, M., Revet, B., von Wilcken-Bergmann, B., Müller-Hill, B. (1987). *lac* repressor forms loops with linear DNA carrying two suitable spaced *lac* operators. EMBO Journal 6:1481–1491.

Kuhnke, B., Fritz, H-J., Ehring, R. (1987). Unusual properties of promoter-up mutations in the *Escherichia coli* galactose operon and evidence suggesting RNA polymerase-induced DNA bending. EMBO Journal 6:507–513.

Landick, R., Carey, J., Yanofsky, C. (1987). Detection of transcription-pausing in vivo in the *trp* operon leader region. Proceedings of the National Academy of Sciences of the United States of America 84:1507–1511.

Newbury, S.F., Smith, N.H., Robinson, E.C., Hiles, I.D., Higgins, C.F. (1987). Stabilization of translationally active mRNA by prokaryotic REP sequences. Cell 48: 297–310.

Nishi, T., Itoh, S. (1986). Enhancement of transcriptional activity of the *Escherichia coli trp* promoter by upstream A + T regions. Gene 44:29–36.

Pakula, A.A., Young, V.B., Sauer, R.T. (1986). Bacteriophage λ *cro* mutations:

effects on activity and intracellular degradation. Proceedings of the National Academy of Sciences of the United States of America 83:8829–8833.

Shanblatt, S.H., Revzin, A. (1986). The binding of catabolite activator protein and RNA polymerase to the *Escherichia coli* galactose and lactose promoters probed by alkylation interference studies. The Journal of Biological Chemistry 261:10885–10890.

Shimotsu, H., Kuroda, M.I., Yanofsky, C., Henner, D.J. (1986). Novel form of transcription attenuation regulates expression of the *Bacillus subtilis* tryptophan operon. Journal of Bacteriology 166:461–471.

Tullius, T.D., Dombroski, B.A. (1986). Hydroxyl radical "footprinting": High-resolution information about DNA-protein contacts and application to λ repressor and Cro protein. Proceedings of the National Academy of Sciences of the United States of America 83:5469–5473.

Chapter 13
Repair and Recombination of DNA Molecules

Every organism has mechanisms for maintaining the integrity of its nucleic acid (i.e., for repairing any damage). Nevertheless, most organisms, including even bacteria and viruses, exhibit some sort of genetic exchange. The two processes may seem antithetical, as recombination, the movement of genetic information from one molecule of nucleic acid to another, implies that the nucleic acid undergoes some kind of structural alteration. However, as is discussed in this chapter, many of the steps involved in completion of the recombination process are the same as those involved in repair, and recombination can be looked on as a process in which the potential for damage to the nucleic acid is outweighed by the potential benefit to be derived from the new genetic information.

As a practical matter, only repair and recombination of DNA molecules have been studied, as RNA molecules are sufficiently unstable or imprecisely synthesized to make genetic analysis nearly impossible. The processes that lead to the complete repair of damaged DNA can be subdivided into two groups: those that correct (reverse) the actual chemical alteration and those that first remove a DNA segment that includes the damaged region and then resynthesize it in a corrected fashion. Similarly, there are conceptually two types of recombination process: copy-choice, in which two DNA molecules are used as templates to direct the synthesis of a single recombinant DNA strand, and breakage and reunion, in which physical exchange of segments of DNA occurs.

All four of these processes are discussed in this chapter, and all are found to occur throughout the bacteria and viruses. However, in order to simplify the explanation of these processes, *Escherichia coli* and its bacteriophages provide almost all of the examples to be discussed. Note, however, that cistrons with similar names in other organisms generally carry out the same processes, as considerable effort has been made to maintain a uniformity of genetic nomenclature among the bacteria. As you read this chapter, keep the following comment, by David Stadler, about recombination models in mind.

> The prudent scientist keeps his hypothesis simple. He endows it with only enough complexity to account for the observations. He stands stubbornly by his simple hypothesis resisting the complicating results of other people's experiments, until there can be no further doubt of their validity. Then he retreats a very short way, taking up a new position with only enough added complexity to accommodate the unwanted findings. There he digs in and prepares for the next attack of the anarchists.*

Possible Types of Damage to DNA Structure

There are two general types of damage that can occur to DNA: **extrareplicational** and **intrareplicational**. Intrareplicational damage refers to those mistakes that are made by the various polymerase molecules and result in the insertion of incorrect bases or the omission or addition of a base. The fundamental structure of the DNA duplex is unaltered, although a heteroduplex condition exists; if the damage is not repaired, the next replication fork that passes through the area will yield two different homoduplexes. Extrareplicational damage, on the other hand, refers to damage caused by external agents such as the mutagens described in Chapter 3. The result of a mutagenic treatment may be a chemically modified normal base (e.g., alkylated or deaminated), or there may be damage to the physical structure of the DNA itself. A chemically modified base does not necessarily have a subsequent effect on replication of that molecule. Its *enol* form base-pairing properties may be altered (see Figure 3-6), and a mutation may result from that mispairing, but that is no barrier to the replication fork. Physical damage to the DNA structure, however, is a different matter.

The physical damage may take different forms. For example, x-rays often cause double-strand breaks in a DNA molecule. The problem of reassociating the broken ends so repair can be effected is not a trivial one. The damage produced by ultraviolet (UV) radiation is of two types: a pyrimidine dimer in which two adjacent pyrimidines are joined to form a cyclobutane ring, and a pyrimidine-(6→4)-pyrimidone (Fig. 13-1). Both are examples of intrastrand cross-links. Another type of extrareplicational damage is induced by agents such as mitomycin C that form interstrand cross-links. It is not possible for DNA polymerase III to replicate areas

---

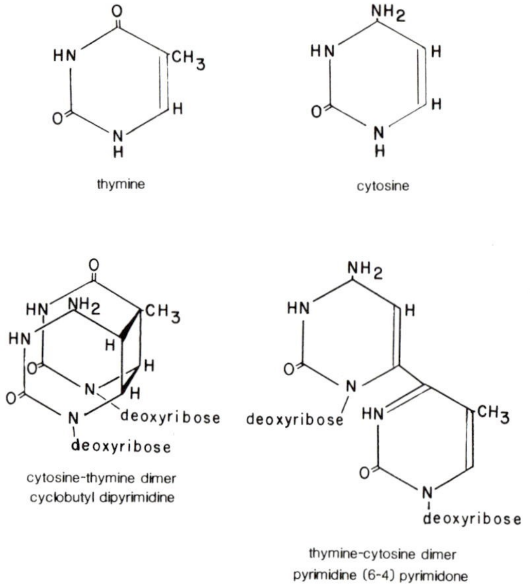

thymine

cytosine

cytosine-thymine dimer
cyclobutyl dipyrimidine

thymine-cytosine dimer
pyrimidine (6-4) pyrimidone

Figure 13-1. Effects of UV radiation on adjacent pyrimidine bases. Two types of structure are formed: cyclobutane dimers (left) and pyrimidine-(6→4)-pyrimidones (right).

of DNA that are cross-linked, although it can restart after the damaged region has been bypassed (Fig. 13-2). Howard-Flanders and co-workers have shown that replication of cross-linked DNA yields daughter molecules containing gaps approximately the size of one or more Okazaki fragments. These gapped molecules cannot be further replicated until repair has occurred, as discussed in the next section.

DNA Repair in *E. coli*

Mismatch Repair

The measured error rate of the DNA polymerase III activities in various bacteria is substantially higher than the observed spontaneous mutation rate by a factor of 100 to 1000. This finding suggests that there is a **mismatch repair system** that corrects polymerase mistakes before they can be converted to permanent changes in the DNA by another round of replication. Mutations affecting this repair system have a mutator phenotype and are localized in the *mutS*, *mutH*, *mutL*, and *mutU* cistrons. All code for proteins, although only the *mutU* product (DNA helicase II) has been characterized. A study of the type and location of the induced mutations caused by inactivating each of these cistrons indicates that they all yield the same

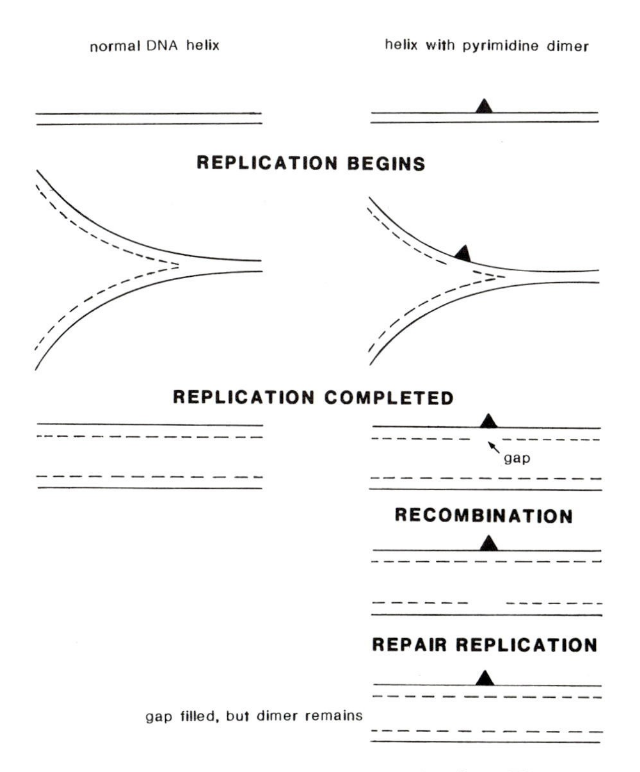

Figure 13-2. DNA damage can prevent DNA replication. The presence of pyrimidine dimers causes the DNA polymerase to stop before the dimer and to resume its synthetic activities some distance away. The result is a gap in the newly replicated DNA strand. If the companion DNA molecule has normal DNA in the corresponding region, recombination can be used to fill in the gap, although the dimer still exists and must be removed at some later time. However, if gapped regions overlap, no repair is possible, and the cell does not have any intact DNA molecules.

mutational hotspots and varieties of mutations and therefore probably are components of the same process. Mismatch repair is often studied by the preparation of artificial heteroduplex DNA molecules that are then introduced into the cell by genetic transformation. The *hex* system discussed in Chapter 8 is an example of a mismatch repair process.

Mismatch repair depends on normal methylation of the DNA. *E. coli dam* (DNA adenine methylase) mutants fail to methylate the sequence GATC and have a pleiotropic phenotype. They have a higher than normal spontaneous mutation rate and are more sensitive to various base analogs. Radman and co-workers have postulated that the role of *dam* is to identify which is the old strand and which is the new. Thus normally methylated duplex DNA just after replication has one strand of the duplex correctly methylated and one not. Correction then normally favors the methylated strand. Artificially prepared heteroduplexes generally confirm this hypothesis. For example, Modrich and co-workers have shown that a heteroduplex that carries no GATC sites undergoes no significant repair. A phage such as λ that has its DNA incompletely methylated (only about 75% under normal circumstances) shows less mismatch repair and a higher mutation rate because the enzymes cannot distinguish the old from the new strand. If the λ DNA is fully methylated in vitro, normal mismatch repair is observed. The process does not work with single-strand loops and has difficulty with transversion mutations, both of which disrupt the normal helical structure.

Basic Observations with UV-Treated Cells

E. coli cultures exposed to UV radiation are the standard for all discussions of DNA repair. The treatment itself is easy to quantitate, the damage that it produces is well characterized, and the potential hazard to the experimenter is easily controlled. The various pyrimidine dimers can be completely reversed by a process called **photoreactivation.** An enzyme designated a photolyase, which consists of a single protein chain and two chromophores and is isolatable as a blue complex, absorbs energy from long-wave (more than 300 nm) UV radiation (maximally at 384 nm). This energy is then used to cleave the pyrimidine dimers to monomers, restoring the DNA to its original state (an example of error-free repair) (Fig. 13-3). Both long and short wave UV can catalyze dimer formation and photolyase activation, but short wave UV gives predominantly dimer formation, and long wave yields primarily photolyase activation.

An alternative repair process, which is also error-free, is termed **short patch repair** or **nucleotide excision repair.** This process involves the removal of part of one DNA strand, including the dimer, resulting in a gap of about 20 bases which is then filled in. Because no radiant energy is required, this type of repair is sometimes called dark repair. The crucial enzymatic activity is that of a **correndonuclease** (correction endonuclease). The enzyme consists of multiple subunits encoded by the *uvrA, B,* and *C* cistrons. The UvrA protein and ATP bind to the damaged DNA. UvrB and UvrC then add to the complex to introduce two nicks into the DNA, one 8 bases upstream and the other 4 or 5 bases downstream of the site of the dimer. The Uvr complex apparently can also act as an ATP-dependent helicase to unwind the damaged region of the DNA. After the nicks have been created, polymerase I can use its 5′→3′ exonuclease activity to degrade the damaged DNA and resynthesize an undamaged replacement using the existing 3′-end as a primer. Complete integrity of the DNA is restored when DNA ligase seals the nicks. Note that this repair process removes entire nucleotides from the damaged area.

A related process is **base excision repair,** which is carried out by DNA

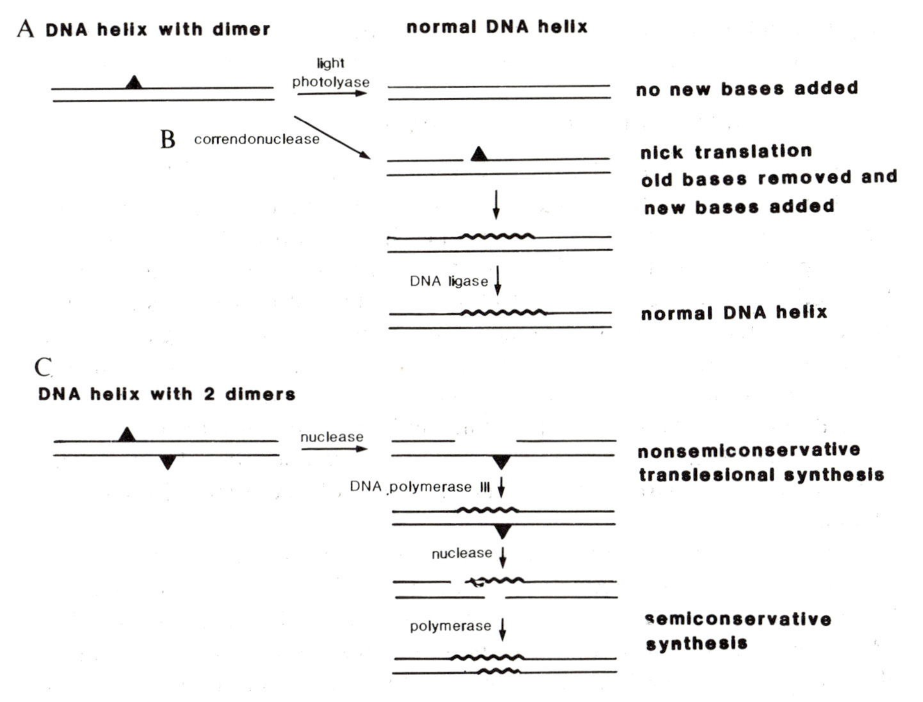

Figure 13-3. DNA repair mechanisms. **(A)** Photoreactivation. The chemical reaction is reversed using energy derived from long wave UV or blue light. **(B)** Short-patch repair. A specific nicking enzyme acts next to a pyrimidine dimer, and the damaged region is replaced by DNA polymerase I using the intact strand as a template. **(C)** Long-patch or SOS repair. The same system as for short-patch repair except that under SOS induction the patch is longer. Again the intact strand is required as a template.

glycosylases. These enzymes recognize erroneous or damaged bases, e.g., uracil in place of thymine, and cleave the sugar–nitrogen bond to remove the base. This action leaves the sugar–phosphate backbone of the DNA intact but with an apurinic or apyrimidinic (AP) site that is easily cleaved by a specific AP endonuclease such as the exonuclease III of *E. coli*.

Photoreactivation, short-patch repair, base excision repair, and mismatch repair enzymes are produced constitutively. They represent the primary cellular defense against DNA damage. However, it is possible for them to be overwhelmed if the amount of damage to the DNA is excessive. In the case of short-patch repair, for example, it is necessary that dimers appearing in opposite DNA strands be located sufficiently far apart that no two regions needing to be patched overlap. If overlaps do occur, the polymerase I lacks a proper template and is unable to complete the excision-repair process. In such a case unrepaired DNA tends to accumulate in the cell, and some other mechanism must be used to eliminate it. The next section presents some pathways by which this action occurs.

The SOS Global Regulatory Network

The process an *E. coli* cell uses to handle large amounts of DNA damage is called inducible or **SOS repair** because it is thought to occur only when the cell cannot complete repairs via the short-patch pathway. The term SOS repair was coined by Radman to signify that the accumulation of damaged DNA in the cell leads to the equivalent of a distress signal that triggers induction of new enzymes. SOS repair is not a single discrete function, as it includes such diverse responses as the ability to repair pyrimidine dimers, induce various prophages, delay septum formation during cell division, shut off respiration, and increase protein degradation (Table 13-1). Because all of these properties are coordinately regulated, they constitute a global regulatory network of the type discussed in Chapter 12.

The easiest way to demonstrate the existence of an inducible repair system is via the **W-reactivation** phenomenon first observed by Weigle. If UV-irradiated λ virions are used to infect normal *E. coli*, the yield of infectious centers is low because the host cells are not able to repair all of the damage in time to permit normal phage functions. However, if the *E. coli* cells are first given a dose of UV radiation and then infected, damage to the irradiated λ DNA is rapidly repaired and the yield of infectious centers greatly increased. It is possible to demonstrate that it is not the UV radiation per se but the DNA damage it causes that leads to repair induction and hence to W-reactivation. For example, if an Hfr or F′ cell is UV-irradiated and then conjugated to an unirradiated F^- cell, the newly transferred damaged DNA causes induction of the repair system in the recipient cell and leads to W-reactivation. A period of about 30 minutes is required for the maximal enzyme levels to be achieved.

A search of various types of *E. coli* mutant cells identified two loci that seemed to be regulators of the SOS response, *lexA* and *recA*. The LexA

Table 13-1. Selected functions of the SOS global regulatory network for which the LexA protein is a repressor

Cistron	Function
lexA	Repressor of SOS cistrons (autoregulated)
recA	DNA synaptase, enhancer of proteolysis
sulA	Inhibitor of septation, normally degraded by Lon protease but overproduced after SOS induction
dnaG	DNA primase, may trigger new rounds of replication
uvrABC	Correction endonuclease, long-patch repair
ssb	Single-strand DNA binding protein

protein functions as a repressor at a variety of operators and is the major regulatory element in the global regulatory network. Several types of mutation have been identified in that protein and, unfortunately, given different names. If the LexA protein cannot be inactivated as a repressor (noninducible), the mutation is *lexA*. A defective repressor protein can be either *tsl* or *spr*, but the mutation maps to the same spot as *lexA*. The latter pair of mutants yield constitutive expression of the SOS response. Unlike the case with the operons discussed in Chapter 12, the LexA protein is not allosterically changed during induction, it is physically cleaved at a specific alanine–glycine bond near the middle of the molecule. This bond is apparently naturally unstable, as cleavage can occur in vitro in the absence of other factors. In the cell, however, cleavage seems to occur as a result of the action of another protein, RecA.

The *recA* locus is the major recombination proficiency locus for the cell, and its protein is normally produced at low levels. The low level synthesis is regulated by LexA repressor acting at the *recA* operator. Mutations in the *zab* locus behave like down-promoter mutations for *recA*. The role of RecA in recombination is discussed below. Its role in the SOS response is to turn on a "protease" activity that rapidly promotes cleavage of the LexA protein. This activity is apparently triggered by the accumulation of large amounts of single-strand DNA such as might occur following attempts at short patch repair. At high temperature, *tif* mutants do not require an inducer to express SOS responses, so their "protease" activity is more easily stimulated. With cleavage of the LexA protein, a wide variety of cistrons is derepressed and the SOS response occurs. As repair progresses, the stimulus of the single-strand DNA is lost, the RecA "protease" activity is lost, and LexA protein regains control of the network.

Members of the SOS network can be identified in one of two ways. They may be found by making fusions of *lacZ* DNA to random promoters using a phage Mu derivative. Such fusions would express β-galactosidase activity upon exposure to UV radiation. They may also be found by examining DNA sequences for LexA binding sites (consensus sequence TACTGTATATA-A-ACAGTA). Some of the sites identified can be found in Table 13-1. Basically, they are associated with inhibition of septation (which prevents segregation of unrepaired DNA and possible death of daughter cells), enhanced recombination (suitable for certain types of repair, see below), and enhanced excision repair (long-patch repair). Long-patch repair seems to involve the same enzymes as short-patch repair, but the length of the patch is now about 100 bases. The inhibitor of septation, *sul*, is unusual in that it is normally synthesized at a reasonable rate but rapidly degraded by an ATP-dependent protease called Lon. The SOS response greatly increases the amount of Sul protein and leads to inhibition of cell division. As LexA regains control, the Lon protein cleaves the Sul inhibitor and allows cell division to resume.

There are some noncellular systems that piggyback on the SOS regulatory network. A good example is the λ prophage. Maintenance of the prophage requires the presence of the λ repressor protein in relatively low concentration. In that protein is another alanine–glycine unstable bond similar to that seen for LexA. When RecA protein is activated to promote cleavage of LexA, it acts on the λ repressor as well. Loss of the repressor induces the prophage and thereby allows the phage to escape from a seriously damaged cell. Production of colicins E1, E2, and Ib as well as cloacin DF13 is similarly activated.

Whereas non-SOS repair is basically error-free, SOS repair characteristically is error-prone. It is this trait that makes UV irradiation mutagenic. The ability to perform mutagenic repair is controlled by two loci, *umuC* and *umuD*, that code for proteins whose function has not yet been identified. If either of these loci is mutated, normal mutagens that require activation of the SOS response do not work. Similar but more mutagenic functions, identified as *muc*, have been identified on certain plasmids such as ColI or R46. Some bacteria such as *Hemophilus influenzae* or *Streptococcus pneumoniae* seem to be natural *umu* mutants and are not mutable by UV radiation. Those mutagens that are independent of SOS (nitrosoguanidine, ethylmethanesulfonate, ICR191) are still operative, as they induce lesions that result in direct mispairing. Echols' group suggested that the mutagenic effect may be due to RecA protein binding directly to single-strand DNA in UV-induced lesions. The bound protein could then attach itself to the polymerase III holoenzyme complex and inhibit the 3′-exonuclease function normally used for correction. A lack of correction would make the synthetic process faster but also leave more mutations in its wake. This explanation correlates well with Tessman and co-workers' suggestion that RecA plus the Umu proteins allows replication past the DNA lesion.

Genetic and Functional Analysis of Recombination

RecA Protein

Mutations affecting the *recA* cistron were first identified by Clark, Howard-Flanders, and co-workers, and it was immediately apparent that the dominant feature of these mutations was their extremely pleiotropic effects. The *recA* mutations take their name from the fact that they reduce generalized recombination to low levels, usually less than 10^{-6}. In addition, they make cells sensitive to UV radiation by preventing the induction of SOS repair; they prevent the induction of prophages such as λ; and they lower the viability of the cell to about 50% of normal.

The *recA* protein is known to have a molecular weight of about 38,000 daltons. The "protease" activity of RecA is independent of its recombinase activity. Large quantities of RecA protein do not necessarily guarantee SOS induction; and, conversely, SOS induction can occur in the absence of overproduction of RecA. It has important direct functions in the repair and recombination of DNA molecules. Radding differentiated two functions of the RecA protein as it interacts with DNA. The first is renaturation, taking two complementary single-strand DNA molecules and promoting their renaturation into a duplex. The second is its **recombinase** activity of promoting **strand invasion.** The RecA protein activities are driven by its enzymatic property of being a DNA-dependent ATPase. In fact, during SOS repair a noticeable drop in cellular ATP is seen in all strains except *recA* or *lexA* mutants.

Strand invasion is subdivided into three phases. During the first phase, **presynapsis,** RecA protein uses the energy from ATP to comparatively slowly polymerize itself along a single-strand DNA molecule (about one RecA protein molecule for every three nucleotides). This step is facilitated by helix-destabilizing proteins such as Ssb, which break up random pairing within the molecule (needed in the proportion of one molecule for every eight nucleotides). At the same time, however, a strand fully coated with Ssb is not available for RecA binding. The nucleoprotein filament that results from presynapsis is polyvalent with multiple weak binding sites for double-strand DNA and resistant to the action of DNase I. It next enters into a **synaptic phase** during which the nucleoprotein filament binds to a DNA duplex first nonspecifically and later specifically with a homologous region. This phase is surprisingly rapid, probably because the multiple filament-binding sites ensure that the two molecules remain closely associated at all times, even though the binding at any single site is relatively weak and transient. Finally, there is a slow **postsynaptic** or strand exchange phase during which the single-strand DNA forms a displacement loop, or D-loop (Fig. 13-4). The helix is unwound, and the single-strand DNA physically replaces a portion of the duplex to create a het-

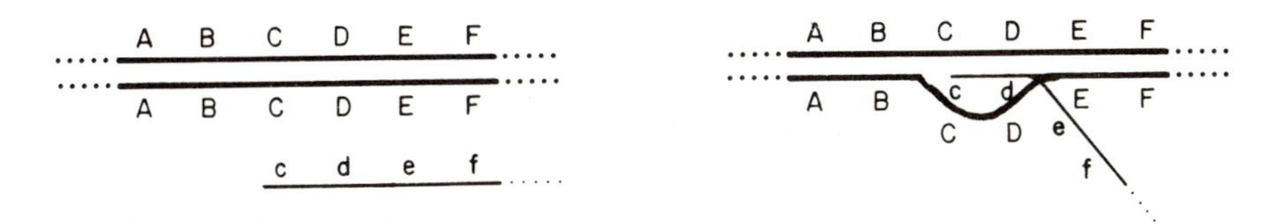

Figure 13-4. D-loop formation. The figure shows a portion of a DNA duplex (thick lines) and one end of a single strand of DNA (thin line). Various genetic markers are indicated by letters. The end of the single strand of DNA may insert into the homologous portion of the DNA duplex and loop out one of the existing strands. Because the resulting loop was caused by insertion of an extra DNA strand, the loop is designated a D-loop. If the loop had been produced by insertion of an RNA strand, it would have been designated an R-loop.

eroduplex region. The displaced strand is of the same polarity as the original single strand and becomes nuclease-resistant owing to protein binding. The complementary strand does not.

The fidelity of the strand displacement reaction is not high. If one of the participating DNA molecules is supercoiled, the average density of mismatched base pairs can be as high as 30% and still give a heteroduplex joint molecule. This capability may allow the RecA protein to drive the strand exchange reaction beyond a damaged region, allowing for recombinational repair. It is, however, error-prone because of the allowable mismatches.

Other Cistrons Affecting Recombination

The original investigations into recombination deficiency produced not only *recA* mutants but *recB* and *recC* mutants as well. The latter two mutant types produced similar phenotypes (moderate UV sensitivity, only 1% of normal recombination proficiency, and cell viability of 30%) and mapped at essentially the same spot on the *E. coli* genome, near *thyA* (see inside the front cover).

The affected enzyme is exonuclease V, an ATP-dependent enzyme with multiple activities. It consists of three nonidentical subunits, α, β, and γ, encoded by *recB*, *recC*, and *recD*, respectively. When presented with linear double-strand DNA, it attacks one strand processively, releasing oligonucleotides. After proceeding into the duplex a distance of several thousand base pairs, the enzyme switches to the opposite (single) strand and reduces it to oligonucleotides. It is the latter function that represents the potential point of interference with rolling circle replication in the case of phage λ. It also has the ability to unwind stretches of DNA as it moves along them and to introduce nicks at about 15 kb intervals. Mutations in *recD* yield cells that lack the nuclease activity but are nonetheless recombination-proficient. These cells had previously been designated *recB*‡.

Exonuclease V participates in DNA excision–repair systems as well as in recombination, although it is not induced as part of SOS repair. In UV-irradiated cells, exonuclease V degrades damaged DNA, but its action is held in check by one of the SOS functions. In the absence of RecA protease activity, extensive DNA degradation is observed—reckless degradation, as one experimenter put it (Fig. 13-5).

Loss of exonuclease V activity can be compensated via two classes of *recBC* intercistronic suppressor mutations, *sbcA* and *sbcB*. The *sbcA* mutations are located within a 27-kb cryptic prophage known as the *rac* (recombination activation) locus. The *rac* prophage cannot produce viable progeny, presumably owing to loss of its late functions, but does have a functional origin for DNA replication and zygotically induces its remaining functions. The result of the induction is that Hfr strains that transfer the *rac* locus into *recBC* recipient strains lacking the *rac* locus produce normal

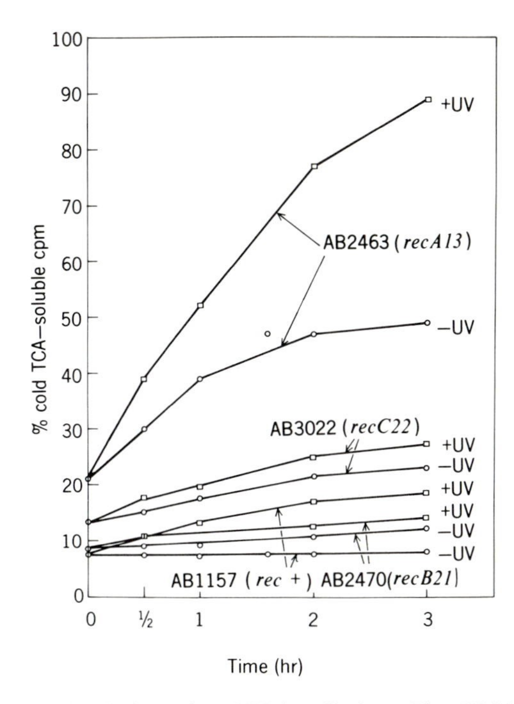

Figure 13-5. DNA degradation after UV irradiation. The DNA in growing cells was labeled with ^{3}H-thymidine, and the cells were then exposed to UV radiation (57 J/m^2). Intact DNA precipitates in the presence of cold 5% trichloroacetic acid (TCA), but individual nucleotides and small oligonucleotides do not. Therefore the increase in the amount of cold TCA-soluble radioactivity after irradiation represents DNA degradation. From Willetts, N.S., Clark, A.J. (1969). Characteristics of some multiply recombination-deficient strains of *E. coli*. Journal of Bacteriology 100:231–239.

numbers of transconjugants. This finding indicates that among the remaining prophage functions is *recE*, which codes for an exonuclease (exonuclease VIII), which replaces exonuclease V. The *sbcA* mutations, then, are alterations that result in the constitutive expression of exonuclease VIII and hence restore recombination proficiency.

The *sbcB* mutations, instead of providing a replacement exonuclease, actually result in the loss of yet another enzyme. In this case the new deficiency is in exonuclease I, but the result is the same as in the case of the *sbcA* mutations, i.e., recombination proficiency. This apparent contradiction seems to arise because loss of the exonuclease I enzyme prevents degradation of some unspecified recombination intermediate, which can then be processed by a new recombination pathway, the *recF* pathway.

The *recF* pathway is of little use in wild type cells except as a mediator of plasmid recombination, as *recF recBC*$^+$ strains are recombination-proficient. However, *recBC sbcB recF* strains are once again recombination-deficient. There are multiple steps necessary for the completion of recombination via the *recF* pathway, and many of these steps have been defined by mutations. The role of the *recF* cistron seems to be similar to that of *lexA*, as the *recF* phenotype (in *recBC sbcB* cells) includes UV sensitivity, deficient gap filling, deficient double-strand break repair, and failure of *recA* protein induction, as well as blockage of recombination. Mutations in *lexA* reduce *recF* expression.

A number of other recombination mutations have been identified, but none is as well characterized as those that have been discussed. Because less is known about these mutations and the subsequent discussion does not depend on knowledge of their properties, further discussion of recombination deficient mutants is omitted.

Postulated Mechanisms for Generalized Recombination

The lack of an organized chromosome as well as meiosis and mitosis means that obligatory pairing of homologous DNA molecules does not occur during cell division in bacteria. Nonetheless, genetic exchange does occur, and presumably the DNA structures at the actual moment of physical exchange are similar in prokaryotes and eukaryotes, even though the methods by which the structures are initiated may not be similar. Therefore all models for generalized recombination in bacteria assume that extensive homology exists between the two DNA molecules involved. The term **homology** does not imply that the base sequences within the two recombining regions are identical, merely that they are essentially similar. Minimum homology for efficient recombination is 40 bp, with greater homology offering increased efficiency.

There is one noticeable difference between models for eukaryotic and prokaryotic recombination that may be more artifactual than real. Eukaryotic models tend to envision reciprocal recombination (exchange of absolutely identical lengths of DNA), whereas prokaryotic models permit reciprocal events but tend to assume that nonreciprocity of exchange is more usual. The reason for the distinction may lie in the relative degrees of map saturation in the two kinds of organism. Because *E. coli* has a much higher density of known genetic markers than any eukaryotic organism, it is much easier to detect nonreciprocal events in *E. coli* than, for example, in *Saccharomyces* or *Neurospora*. Nonreciprocal recombination has indeed been observed in eukaryotes, although it has been given the special name **gene conversion** and may in fact be the normal consequence of recombination or repair. Because gene conversion can be observed only if genetic markers are properly located at the actual exchange site (Fig. 13-6), its comparative rarity is most likely due to the inappropriate location of genetic markers.

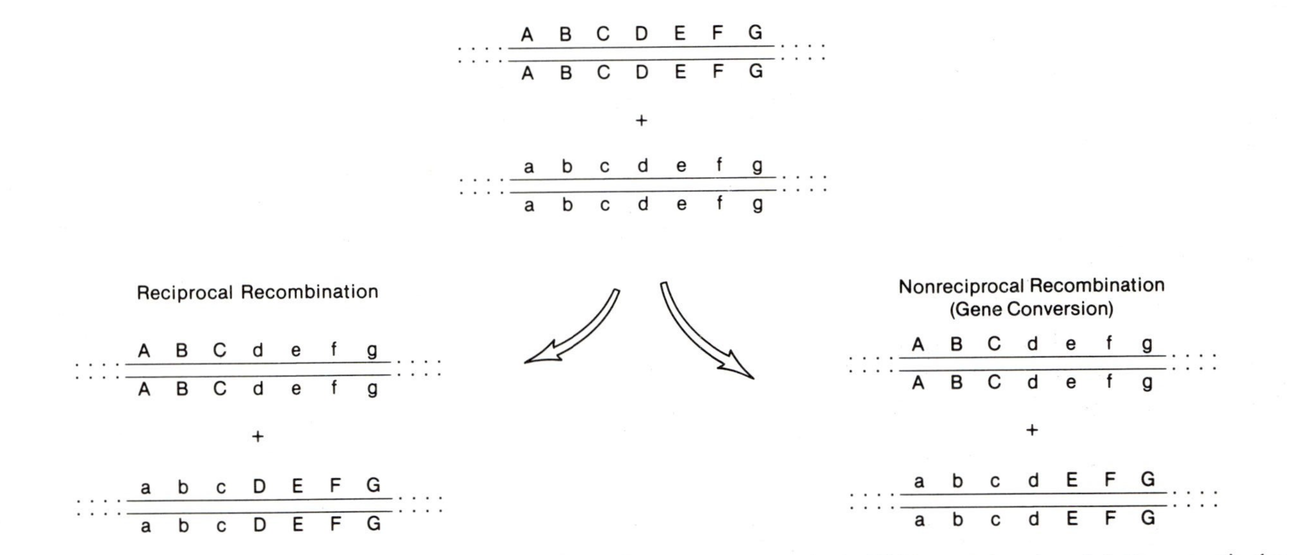

Figure 13-6. Reciprocal and nonreciprocal recombination. Each line represents a single DNA strand and each letter a particular genetic marker. Note that, with respect to markers ABC and EFG, the two types of event are reciprocal. It is only when marker D is checked that nonreciprocity is observed.

The present model for recombination stems from a long line of recombination models presented by many workers and owes much to the work of Radding. The general outline of the model is presented in Figure 13-7. It begins with the known properties of RecA protein and requires generation of a piece of single-strand DNA. One way in which such a piece of DNA might be obtained is via strand displacement, as shown at the upper right. For this reason, nicks in DNA are often considered to be recombinogenic. Gap formation (upper left) is a natural consequence of excision repair and can also lead to D-loop formation. After suitable degradative steps, both pathways result in production of a single-strand exchange. The difference is in whether the DNA synthesis occurs before or after the strand exchange. The joint molecule common to both pathways is two essentially complete DNA helices connected by a single DNA strand that originates in one molecule and terminates in the other. Its formation requires homology at the 3′-end of the invading strand.

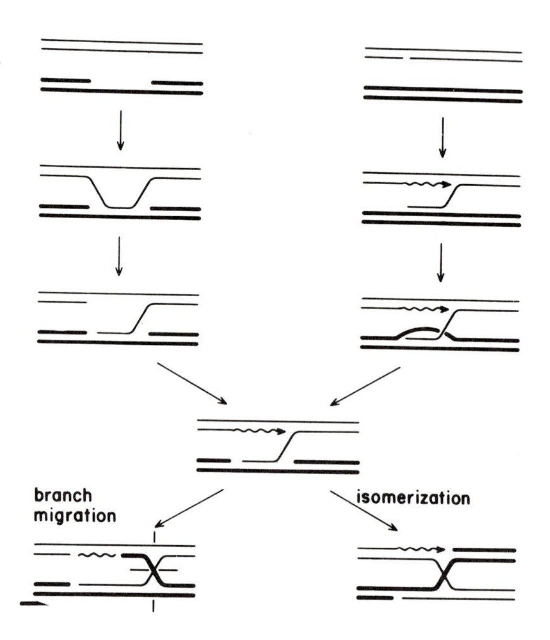

Figure 13-7. Meselson-Radding model (lower three diagrams) and two hypothetical pathways of initiation of strand transfer. Gaps or nicks are assumed to result from repair or other processes. The wavy line in the pathway on the upper right indicates new synthesis displacing an existing strand from a nick, and the arrowheads indicate growing 3′-ends. The horizontal and vertical marks at the lower left show endonucleolytic cuts that would lead, respectively, to the parental or recombinant configurations of the flanking arms of DNA. From Radding, C.M. (1982). Homologous pairing and strand exchange in genetic recombination. Annual Review of Genetics 16:405–437.

The isomerization step indicated at the lower right is a crucial one and is based on some model-building studies by Sigal and Alberts. After isomerization the positions of each member of each DNA duplex are reversed as indicated. At first glance it seems that it is an unlikely event, but in fact the isomerization is merely a rotation of the DNA strands about the longitudinal axis of the paired helices as shown in Figure 13-8. This rotation can be accomplished without displacing any bases from their normal helical configuration and therefore should require little or no energy input. The final structure is important because it now shows an exchange of two DNA strands in the manner corresponding to a model first proposed by Holliday and sometimes called a **Holliday structure.** Note that the structure still may contain nicks and that in the region of the original strand exchange events have been nonreciprocal. A similar Holliday structure (lower left) can be obtained by the action of RecA protein in displacing one strand from the recipient molecule into the donor molecule so as to continue the strand exchange already begun.

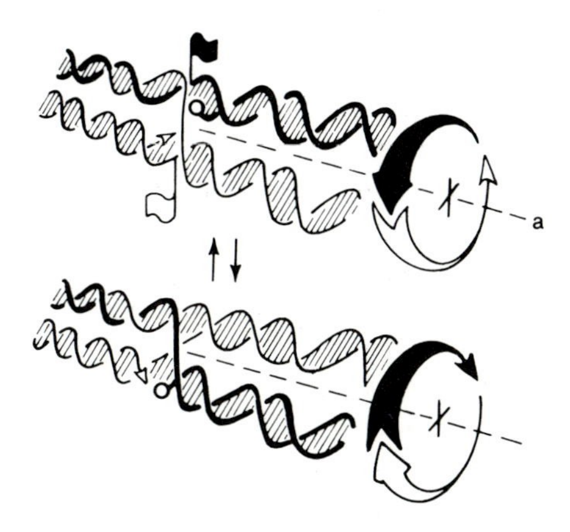

Figure 13-8. Hypothetical isomerization of a one-strand crossover to produce a two-strand crossover. The top diagram represents DNA molecules linked by the crossover of one strand. The structure is identical to the central intermediate in Figure 13-7. The arrowhead represents a 3′-end, and the open circle represents a 5′-end. Isomerization, which can be demonstrated with molecular models, proceeds by rotation of the two arms to the right of the flags about an axis, *a*, that is between the arms and parallel to them. The heavily shaded arm would rotate above the plane of the paper, and the lightly shaded arm would rotate behind the plane of the paper. A phosphodiester bond at the point signaled by the dark flag becomes the cross-connection closer to the viewer, and a bond signaled by the light flag becomes the other cross-connection. The isomerization is reversible unless branch migration supervenes, as in Figure 13-7, or the nick is sealed where the 3′-end (arrowhead) abuts on the 5′-end (open circle). From Radding, C.M. (1978). Genetic recombination: strand transfer and mismatch repair. Annual Review of Biochemistry 47:847–880.

The two bottom structures are susceptible to the process known as **branch migration,** which was originally proposed for phage T4 by Broker and Lehman. It consists in moving the site of the physical crossover along the DNA duplexes in a sliding motion (to the right in Figure 13-7), thereby forcing additional exchange of strands. It of course requires that hydrogen bonds be broken to allow the exchange; but for each base that loses its hydrogen bonds because it is being exchanged from one helix to another, a second base is restored to a condition in which it can again hydrogen bond (albeit with a new partner). Thus there is no net energy change in the state of the molecule.

Three facts are important about the process of branch migration: (1) The exchanges it produces are all reciprocal. (2) If one base in ten is mismatched, branch migration stops. (3) If the molecule is linear (as in the case of individual T4 genomes), the branch can migrate right off the end of the molecule and the structure resolves itself into two intact but recombinant DNA molecules. Of course this situation cannot occur in the case of *E. coli* as the DNA is circular, and it is therefore necessary to postulate the existence of one or more nucleases that resolve the structure into two separate molecules by introducing appropriate nicks as indicated in Figure 13-8. These resolving events could be either reciprocal or nonreciprocal and may involve exonuclease V. The evidence for exonuclease V involvement is that *recBC* strains can carry out recombination to the point where it is possible to transcribe mRNA from the recombinant region of the DNA, but the recombinant DNA structure is incapable of replication unless exonuclease V is supplied. The bottom structures in Figure 13-7 are consistent with these observations. The overall result of the recombination process outlined in Figure 13-4 is two molecules showing a large region of reciprocal exchange with nonreciprocal exchange at one or both ends. Proteins similar in function to RecA have been identified in phage T4 (*uvsX*) and in the fungus *Ustilago maydis (rec-1)*. In the case of RecA, the protein remains on the heteroduplex for at least 30 minutes.

Electron micrographic evidence is also basically in accord with the model. Potter and Dressler have studied the structure of recombining ColE1 molecules. These small circular DNA molecules form visible figure-eight structures during recombination (Fig. 13-9) in which it is possible to see strands exchanging between molecules. Such molecules are consistent with the bottom structures of Figure 13-7 but do not imply anything about the mechanism of formation. In fact, Potter and Dressler reported that the figure-eight structures can form in the absence of DNA replication, which is not consistent with the model in Figure 13-7 and suggests the possibility of alternative modes of recombination.

One interesting alternative, proposed by Wilson, involves the interaction of two DNA molecules without the need for any nicks. In essence he postulated the rewinding of DNA strands in a reverse helix to form a quadruple-strand structure that can undergo a sort of isomerization similar

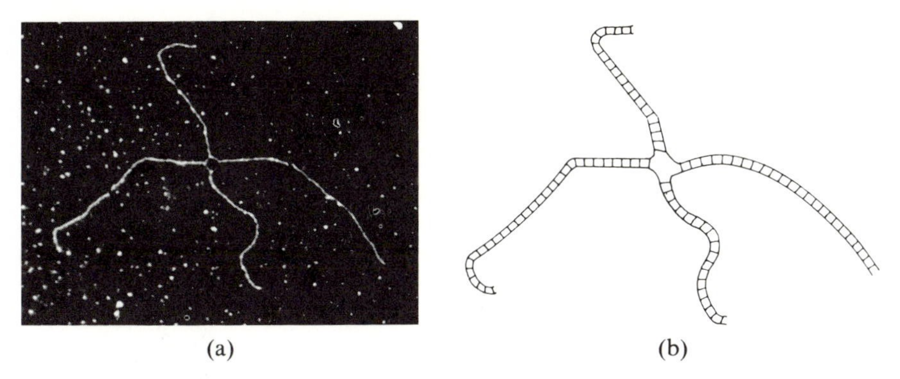

Figure 13-9. Recombining DNA molecules. When plasmids are isolated from recombination-proficient *E. coli,* one finds, at a level of about 1%, fused genomes that appear to be intermediates in genetic recombination. These joint circular molecules have the form of a figure-eight. When cut with a restriction enzyme such as *Eco*RI, the figure-eights assume the shape of the Greek letter chi (χ), thus showing that they consist of two genomes held together at a point of DNA homology. The lengths of the arms of the chi structure are variable, depending on the point of genetic exchange. However, within any one structure, the arms are either of equal length or else both short arms are the same length as are both long arms.

(a) One such molecule composed of two pMB9 DNA molecules undergoing recombination. **(b)** Diagram showing how the DNA strands are thought to have associated. A nicked structure such as this one would resolve itself into two separate DNA molecules by the processes of branch migration with or without isomerization. From Potter, H., Dressler, D., (1978). In vitro system from *E. coli* that catalyzes generalized genetic recombination. Proceedings of the National Academy of Sciences of the United States of America 75:3698–3702.

to that invoked by Radding. The circular DNA molecules compensate for the additional twists by the formation of oppositely oriented supercoils (Fig. 13-10). The enzymology required to produce genetic exchange from this sort of structure is probably similar to that used by Radding but remains a subject for speculation.

It is important to realize that none of the models for generalized recombination explicitly accounts for the possibility of "neighborhood effects" on the incidence of recombination. They tend to assume that initiation of recombination can occur at any point along the DNA duplexes. Consequently, it is assumed that the greater the distance between two markers, the more genetic exchanges are expected to occur between them. It is fairly easy to show that this assumption is not true for increasing distances between very closely linked markers (intracistronic recombination), although it may be true for intercistronic recombination. For example, Norkin measured the yield of transconjugants from matings in which an Hfr strain carrying a *lacZ* mutation that mapped at the operator

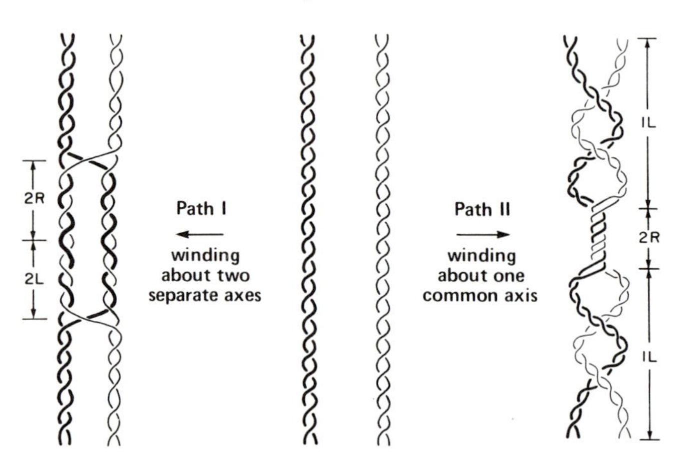

Figure 13-10. Proposed mechanism for the production of reciprocal DNA heteroduplexes. Two helical DNA molecules are shown in the center of the diagram. Path I produces a conventional structure similar to those shown in Figure 13-7 but requires no nicks in the DNA. Instead, coiling in one direction is compensated by coiling in the other direction. Path II produces a unique structure in which a tetrameric region containing two right-handed superhelical turns is produced by introducing two left-handed superhelical turns into the remainder of the molecule, one turn at either end of the tetrameric region. (If a strand on the front surface of a vertically oriented helix points upward and to the right, the helix is right-handed; if it points upward and to the left, the helix is left-handed.) Note that path II does not require any strand breakage in order to form the closely paired tetrameric region. Molecular models indicate that within such intercoiled heteroduplexes various types of hydrogen bonding rearrangements are possible. From Wilson, J.H. (1979). Nick-free formation of reciprocal heteroduplexes: a simple solution to the topological problem. Proceedings of the National Academy of Sciences of the United States of America 76:3641–3645.

proximal end of the cistron was conjugated to an F^- strain carrying various *lacZ* mutations. The results are shown in Figure 13-11. It is easy to see that, although the general tendency is for the number of transconjugants to increase with increasing distance between the markers, there are dramatic exceptions. The exceptional events may reflect the relative ease with which D-loops can be formed.

Interrelation of Repair and Recombination Pathways

In addition to the repair mechanisms discussed earlier in the chapter, there is yet another variety of repair: recombinational repair. This type takes advantage of the gaps corresponding to damaged regions that are left in the newly synthesized DNA by polymerase III (i.e., it is postreplicational repair). Clark and Volkert proposed that these gaps can be filled by either a copy-choice excision repair mechanism or a breakage-and-reunion gap-

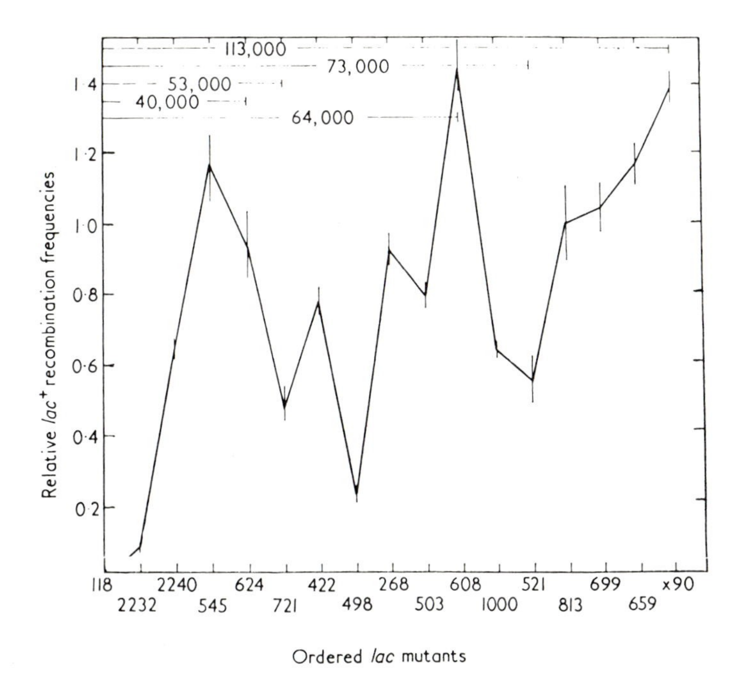

Figure 13-11. Intracistronic recombination frequencies are not necessarily proportional to physical distances between the markers. In this experiment, an Hfr strain carrying the *lacZ118* mutation (which maps close to the *lac* operator region) was crossed to various F^- strains carrying other *lacZ* mutations. The precise order of the *lacZ* mutations, but not their relative spacings, was known from deletion mapping experiments. The relative lac^+ recombination frequency for each mating was determined by normalizing the number of colonies observed on plates selective for Lac^+ to the number of colonies observed on plates selective for Leu^+ (a marker transferred some minutes after *lacZ*). Note that although the general tendency is for the recombination frequency to increase with increasing map distance, there are several dramatic exceptions to the rule. Similar results were also obtained if the Hfr strain carried mutation *X90* instead of *118*. At the top of the figure are presented the molecular weights of the polypeptide fragments produced by some of the terminator mutations used in this experiment. These numbers provide some indication of the physical distance between the various markers. From Norkin, L.C. (1970). Marker specific effects in recombination. Journal of Molecular Biology 51:633–655.

filling repair, resulting in long-patch repair. In both cases the assumed intermediate resembles that of Figure 13-7 with the bridging DNA strand coming from the boundary of the gap in the newly synthesized DNA (Fig. 13-12). Gap-filling repair requires *recF* function; and if the gaps are not filled, the double-strand breaks that develop after replication trigger sister duplex recombination.

As shown in Figure 13-12, the breakage-and-reunion pathway is really

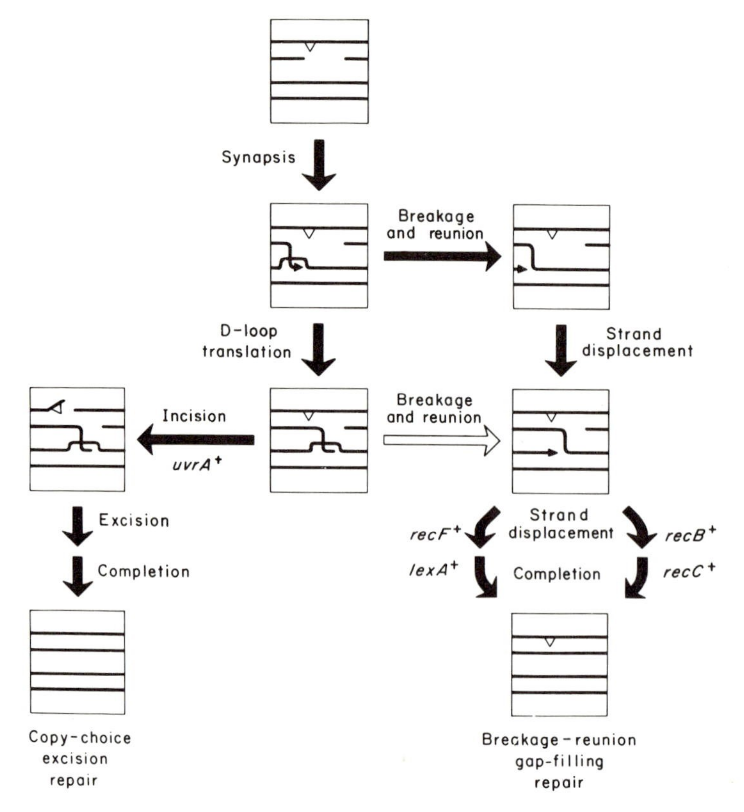

Figure 13-12. Alternative copy-choice excision and breakage-reunion gap-filling pathways of intrareplicational repair. The small triangles represent pyrimidine dimers, and the arrows represent DNA replication. The box at the top contains both portions of an incompletely replicated chromosome segment containing a single pyrimidine dimer. The open arrow indicates a hypothetical pathway. The small arrowhead indicates the point at which DNA synthesis is occurring. Note that the copy-choice pathway fills in the gap with newly synthesized DNA, leaving the lower DNA duplex exactly as it was at the start. The breakage-and-reunion pathway, on the other hand, fills in the gap with old DNA and puts the newly synthesized DNA into the lower DNA duplex. From Clark, A.J., Volkert, M.R. (1978). A new classification of pathways repairing pyrimidine dimer damage in DNA, pp. 57–72. In: Hanawalt, P.C., Friedberg, E.C., Fox, C.F. (eds.) DNA Repair Mechanisms. New York: Academic Press.

not repair in the normal sense of the word, because the thymine dimer remains. All that has been accomplished is to fill in the gap so the DNA molecule can make another attempt to replicate. On the other hand, the copy-choice pathway leads to the formation of two completely normal DNA duplexes and is therefore true repair. Because the long-patch repair depends on *recA* function, it is an example of error-prone repair.

Strong evidence for the existence of a pathway resembling the postulated copy-choice mechanism comes from the experiments of Howard-

Flanders and co-workers. They have shown that if a cell is infected with two phage particles, one carrying nonradioactive damaged DNA and the other carrying undamaged radioactive DNA, the radioactive DNA is cut during the repair process, even though it is undamaged. Moreover, heterologous DNA such as that from an unrelated phage present in the same cell is unaffected. The process has been called **cutting in *trans*** and has been shown in vitro to require the presence of the RecA protein.

Based on the foregoing discussion, it seems evident that the distinction between repair and recombination pathways is somewhat artificial. Some types of repair occur by what is essentially recombination, and all recombination events conclude with repair of the helices to generate intact DNA duplexes. Figure 13-13 is a summary of the various recombination and repair pathways and their interrelations.

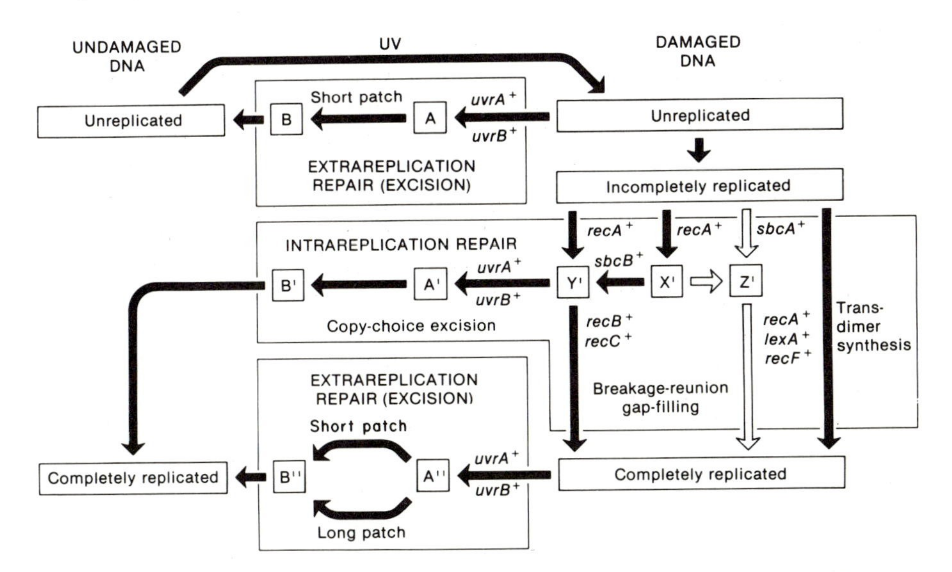

Figure 13-13. Summary of the possible DNA repair pathways in *E. coli*. Undamaged DNA is shown on the left and damaged DNA on the right. For simplicity, the damage to the DNA is assumed to result from UV irradiation. There are two basic ways to repair the damage to the DNA: intrareplicational or extrareplicational repair. However, extrareplicational repair may occur either before the damaged area of the DNA has attempted to replicate (top) or after a replication fork has passed through the damaged area and left a gap in the newly synthesized strand opposite the dimer (incompletely replicated). If the gap can be filled in by one of the recombinational repair processes listed at the right of the figure, extrareplicational repair can remove the dimer (bottom). The pathways indicated by the open arrows are operative only in *sbcA* or *sbcB* cells. The letters A and B refer to classes of excision intermediates, and the letters X, Y, and Z refer to dimer-containing recombination intermediates. Transdimer synthesis results when the proofreading capability of the DNA polymerase is inhibited so that the lack of hydrogen bonding by dimers is not a barrier to replication and no gap is left opposite the dimer. Modified from Clark and Volkert (1978).

Double Site-Specific Recombination

Generalized recombination involves large regions of homologous DNA sequences, but site-specific recombination involves considerably smaller segments of DNA. The essential components for site-specific recombination include an enzyme or enzymes specific for a particular DNA sequence and two DNA duplexes, at least one of which carries the base sequence recognized by the enzyme. Under the proper conditions, reciprocal (or nearly so) recombination occurs within or adjacent to the recognition sequence. If both DNA duplexes must carry the recognition sequence, the process is described as **double site-specific recombination.** If only one DNA duplex must carry the recognition sequence, the term **single site-specific recombination** has been used.

Phage Lambda

Several examples of double site-specific recombination have been discussed earlier. One especially well studied example is that of the integration of λ DNA to form a prophage. The recognition sequence in this case is the *att* site, which was represented in Chapter 6 by the letters PP′. The corresponding region on the *E. coli* genome is *att*λ, which was represented as BB′ (see Figure 6-4). Analysis of these two regions by both genetic and heteroduplex techniques has led to the surprising conclusion that they are not homologous at all. Although the minimum size of the *att* site is about 240 bp, there is a small 15-bp sequence embedded within each *att* site that is the actual point at which the integrase enzyme acts to produce staggered cuts. In recognition of this small, centrally located binding site, the terminology for designating the *att* sites has been changed to POP′ and BOB′, where **O** designates the short homologous sequence. The integration event results in the production of prophage endpoints (BOP′, POB′) that are slightly different from the original sequence. In fact, in vitro experiments have shown that integrase has difficulty binding to the right prophage end (POB′), and the role of the *xis* protein may be to assist the binding of integrase. This situation would account for the genetic observations that only *int* function is necessary for integration but both *int* and *xis* are required during prophage excision unless the level of Int protein is high. In all cases, Integration Host Factor (IHF), the product of the *himA* and *hip* cistrons, is necessary for normal activity. IHF also binds to *c*II DNA, but more weakly than to *att* DNA.

A general diagram for λ recombination at *att* is shown in Fig. 13-14. The *att* region for the phage is 234 bp and can be subdivided into P and P′ arms. The *att* region for the bacterium is only about 25 bp and is subdivided into B and B′ arms. Within each arm, footprinting experiments have identified specific sequences protected by the Int protein: three in the P′ arm, two in the P arm, one each in B and B′ arms. IHF has three binding sites, all within the phage DNA. Two of these sites flank the region of the phage DNA in which exchange occurs. Landy and his collaborators

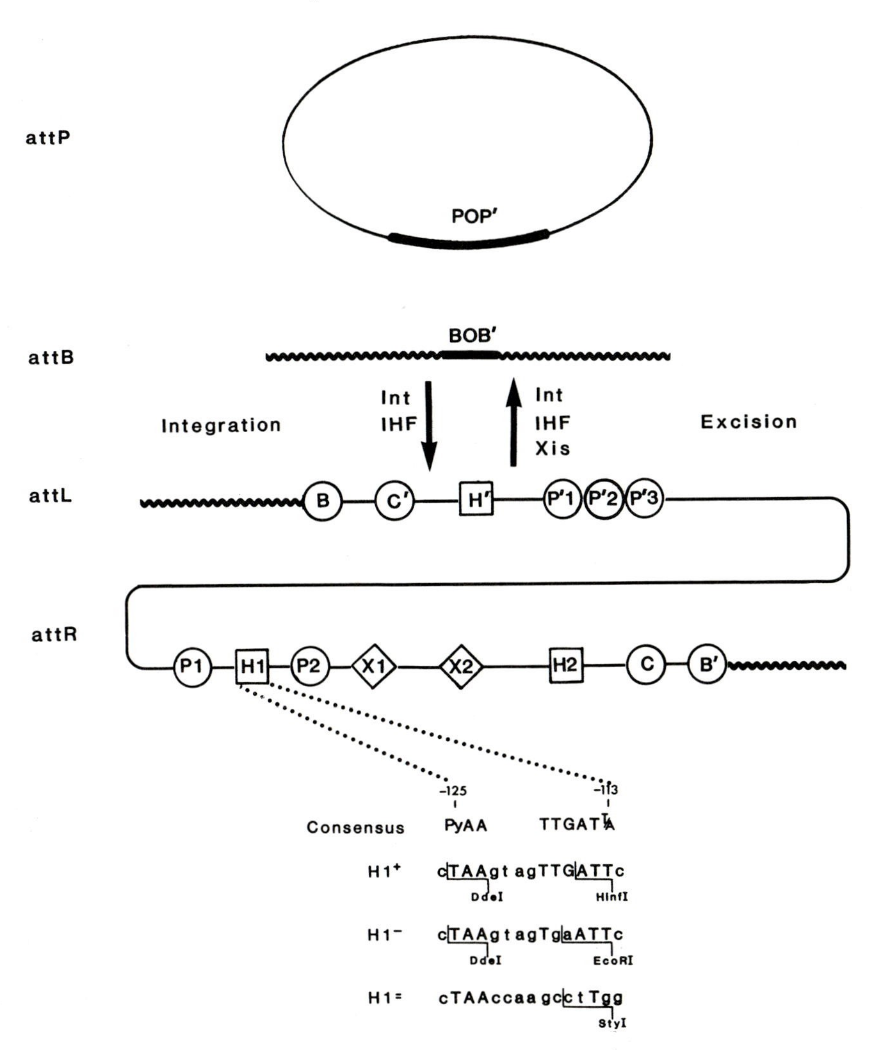

Figure 13-14. Lambda site-specific recombination. The *att* sites involved in the integration and excision reactions as well as the proteins required for each are shown. The locations of the binding sites for Int (○), IHF (□), and Xis (◇) have been deduced from footprinting experiments. The IHF consensus sequence and coordinates relative to the center of the overlap region are shown over the sequences of the various H2 sites. Matches to the consensus sequence are in capital letters. Diagnostic restriction endonuclease sequences for each site are indicated. From Thompson et al. (1986).

have shown that the H1 site appears to be of critical importance in regulating the integration event.

Overall regulation of integration and excision is provided by several mechanisms. The end of the *xis* cistron overlaps the p_I promoter so that when Xis is being produced Int expression is reduced. Furthermore, the weak binding of IHF to the *c*II region can occur only after the *att*P binding sites are all occupied. This requirement ensures that the phage DNA is ready to integrate before the temperate mode of transcription becomes predominant.

If all types of recombination are considered, λ phage has three independent systems available to it: *rec,* the bacterial system for generalized recombination; *red,* the phage system for generalized recombination; and *int,* the phage system for site-specific recombination. The relative contribution of each system can be assessed using the appropriate mutant strains. Data from one such experiment are presented in Table 13-2. The amount of recombination in the *c*I-*R* interval is specific to the *red* or *rec* systems, and the general conclusion is that both systems can catalyze significant amounts of recombination but that *red* is more efficient than *rec* at catalyzing recombination in the interval. All three systems can catalyze exchanges between *J* and *c*I, with *int* causing exchanges between two POP′ sites on free phage DNA. Approximately 26% of the exchanges between *J* and *c*I are observed to be *int*-promoted. Note that when all three recombination systems are mutant, no genetic exchanges are detected.

Table 13-2. Comparison of the relative contributions of the *red, int,* and *rec* systems to phage recombination

Phage		Host	Percent recombination	
red	*int*	*rec*	*J-c*	*c-R*
+	+	+	7.5	3.6
+	−	−	4.1	3.0
−	+	−	2.0	≤ 0.05
−	−	+	1.3	1.3
+	+	−	7.8	3.1
−	−	−	≤ 0.05	≤ 0.05

Average values for recombination were determined in four sets of crosses of the type *susJ* × *c*I *susR*, carried out in two hosts. The J^+ R^+ recombinants were selected and scored visually for *c*I (turbid or clear). The percentage recombination was determined by the frequency of J^+ R^+ turbid (crossover in the interval *J-c*) or J^+ R^+ clear (crossover in the interval *c-R*) × 2 × 100. The interval *J-c* contains *att* and shows generalized as well as site-specific recombination, whereas the interval *c-R* shows generalized recombination only. From Signer, E.R., Weil, J. (1968). Site-specific recombination in bacteriophage λ. Cold Spring Harbor Symposia on Quantitative Biology 33:715–719.

F Plasmid

Another example of double site-specific recombination that was discussed earlier was the integration of the F plasmid to form Hfr cells. The specific sites for integration are the IS elements IS*2*, IS*3*, and γδ. As yet no enzyme analogous to the λ integrase has been identified in connection with these integrations, but the results shown in Figure 9-10 demonstrate the characteristic reproducibility of site-specific recombination. It seems probable that each insertion element codes for its own "integrase."

Resolvases and Invertases

As was discussed in Chapter 11, plasmids often code for a resolvase enzyme that breaks up any concatemeric plasmid DNA molecules so that proper segregation can take place at cell division. The resolvases have also been identified in connection with the transposition of certain transposons, e.g., Tn*3* (see below). They are encoded by the *tnpR* cistron and always act at a specific site (*res*) on a plasmid DNA of unit length. Therefore if two sites occur within the same molecule, it must be a concatemer and should be broken down (see Figure 11-3). Binding of the resolvase to the *res* site induces "kinking" or bending of the DNA that may facilitate the strand exchange.

In a similar vein, the invertases seen in phages Mu and P1 and in the *Salmonella typhimurium* phase variation system are essentially double site-specific recombination enzymes. The sites in this case are the boundaries of the invertible region, which are equivalent, so there is no favored orientation of the invertible segment.

Protein studies have indicated that there are substantial similarities between invertases and resolvases, but there are nevertheless some significant differences in function. The primary difference is that for resolvase to work well the *res* sites must be arranged as directly repeated elements in the concatemer (same orientation), whereas in the case of the invertases the pertinent boundary segments are arranged as inverted repeats (opposite orientation).

Single Site-Specific Recombination

Chi Sites

The double site-specific recombination events discussed above require only limited DNA homology, but the single site-specific recombination events require little if any homology. An interesting example of this situation is phage λ. It is possible to genetically alter λ so that it is *red gam* and therefore can grow only poorly because of an inability to switch to the rolling circle mode for DNA replication, resulting in a lack of concatemeric DNA molecules for packaging. Cells infected with such mutant λ phages produce few progeny and yield only small plaques. However, mutant phages can be found that produce many progeny under these conditions and give rise to large plaques. The phages carry mutations at several sites, all of which are designated **Chi.**

Stahl and co-workers have shown that Chi mutations increase the recombination frequency in their immediate vicinity (10 to 20 kb), but primarily in one direction only. The increase is seen only when the *recA recB* pathway is used, not in the case of *recF* or *red* pathways. Certain *recC* mutations leave recombination proficiency intact but prevent Chi function, indicting that Chi is not the starting point for all recombination events. If the DNA carrying the Chi site is an insertion, so it represents a region of nonhomology, the recombination frequency is still found to be elevated. Therefore Chi is an example of single site-specific recombination. It has been shown to work in conjunction with *cos* or Tn*10*.

Chi sites are found as naturally occurring elements in *E. coli,* the Enterobacteriaceae and Vibrionaceae in general, and Tn*10* but not in λ, Tn*5*, Tn*3*, *Pseudomonas*, *Hemophilus*, *Bacillus*, or *Klebsiella*. It has been suggested that the latter bacteria may have similar elements but with different sequence specificity. All that is necessary to observe Chi sites is some sort of selective procedure to identify them. They may result from as little as a single base change. The necessary DNA sequence for a Chi site is 5′GCTGGTGG, but related sequences may have partial activity. Stahl and co-workers have estimated that Chi sites may be as frequent as one for every 5000 bp in *E. coli* DNA. However, knowledge of the enzymology of the chi effect is nonexistent, and considerably more work is required to explain exactly what is taking place.

Smith proposed that most recombination events begin with binding of exonuclease V to the end of a DNA molecule. The enzyme unwinds the DNA to make large loops until it reaches a Chi site in the correct orientation. At that point it makes a cut and generates a free end, which can be used to trigger recombination.

Transposon Tn*10*

The other principal example of single site-specific recombination is really a group of examples, the transposons. Included here are phenomena such as the highly promiscuous integration of phage Mu DNA, the integration of the R100 plasmid to form an Hfr, and the movement of various pure transposons, e.g., Tn*10*. However, all of these phenomena reduce to single site-specific recombination events catalyzed by the insertion elements bounding various transposons. For example, when R100 integrates, it usually loses its Tn*10* transposon, which means that the entire transfer region of the R plasmid as well as the r-determinant is acting like a large transposon and hopping from one DNA molecule (the R100 plasmid) to another (the *E. coli* chromosome). In order to emphasize the size differences, the process has been called inverse transposition (i.e., the Tn*10* stays where it is, and the rest of the DNA moves).

The molecular mechanisms proposed for transposition basically come down to two varieties. Each involves a transposon located on a donor DNA molecule and a target site that may be located on the same or different DNA molecule. In one model there is DNA replication involved in the process, and an intermediate **cointegrate molecule** is formed if the target site is on a separate molecule. A cointegrate structure is one in which two DNA molecules are fused into one. The other model requires no DNA replication as an intrinsic part of the transposition process. This mode of transposition was first suggested by Berg and is characteristic of several transposons, including Tn*5* and Tn*10*.

Tn*10* is the 9.3-kb transposon encoding tetracycline resistance that is found within the R100 plasmid. It carries inverted IS*10* elements at its boundaries. As detailed in Chapter 10, it has moved into and out of the DNA of a variety of bacteria and their phages and therefore must use an extremely versatile recombination system. A close examination of Tn*10* has revealed several important aspects of its behavior. As shown in Figure 13-15, it can excise itself from a molecule precisely or imprecisely, invert a region of DNA, or delete a region of DNA. Prior to the discovery of transposons, these recombination events would have been classified as examples of illegitimate (nonhomologous) recombination. Lack of replication in this process was demonstrated by Bender and Kleckner, who prepared λ phages carrying slightly different Tn*10* moieties. Strand separation and reannealing were used to make heteroduplexes differing at specific bases. After packaging and infection, when the products of transposition were examined the transposon was still a heteroduplex. This result could happen only if there were no replication as a part of transposition.

Despite this observation, transpositions have been observed in which Tn*10* apparently remained where it was and also appeared in a new site. Bender and Kleckner suggested that this duplication is due to the nature of the repair process, which acts on the donor DNA after transposition is completed. If there is no repair, the double-strand gap destroys the integrity of the donor DNA. Proteins such as RecA are known to have the capability of joining the ends of broken DNA in order to repair damage such as that induced by x-rays. A similar phenomenon might occur in this case, and the result would be precise excision of the transposon. Finally, recombinational repair might be used, in which case the missing transposon DNA would be replaced by transposon DNA replicated from another DNA molecule. Note carefully that in this instance the replication is part of the recombination process and not a part of transposition. The net result would be the appearance of an additional copy of the transposon within the cell.

The transposition frequency for Tn*10* is about 10^{-4} per cell per generation when the cells are growing on minimal medium. At a frequency

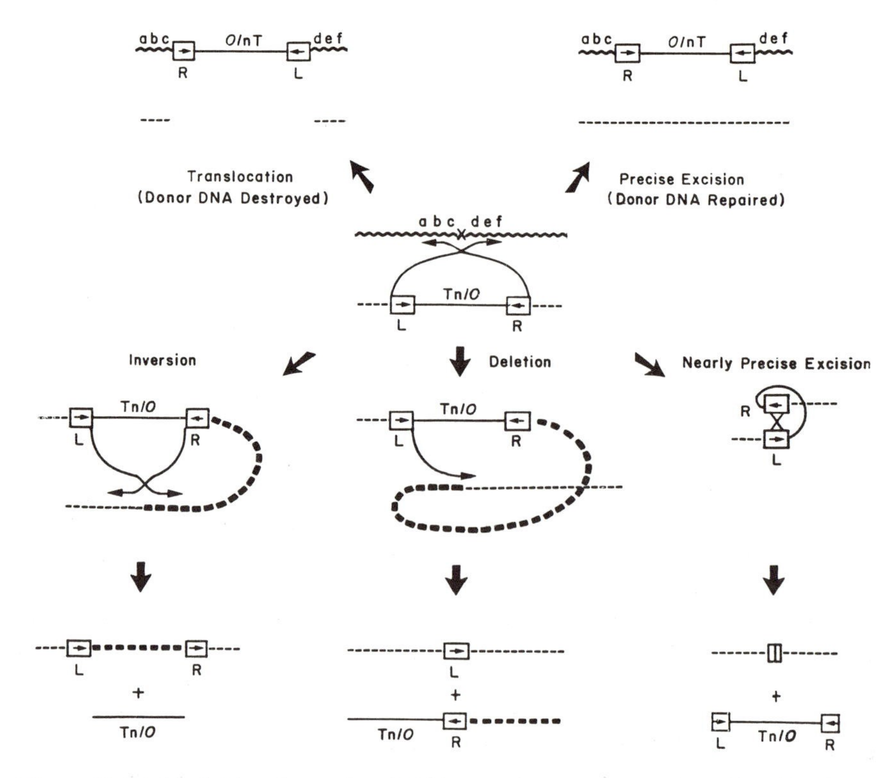

Figure 13-15. Model for the various DNA rearrangements catalyzed by Tn*10*. The upper portion of the diagram shows the interaction between two DNA duplexes, only one of which carries Tn*10*. Because transposition does not involve replication, the donor DNA can be either repaired (precisely or nearly precisely) or destroyed. The lower portion of the diagram shows the interactions that may occur within a DNA duplex that contains Tn*10*. The inverted repeat elements at the ends of Tn*10* are represented by the small boxes, with the arrows indicating the relative orientation. The longer arrows indicate the direction of "attack" of the Tn*10* moiety on the target molecule. The sites of attack are indicated by an "X" or by a change in line thickness. Adapted from Ross, D.G., Swan, J., Kleckner, H. (1979). Physical structures of Tn*10*-promoted deletions and inversions: role of 1400 bp inverted repetitions. Cell 16:733–738.

of 10^{-5} or less, Tn*10* also promotes deletions or inversions and deletions. These rearrangements preferentially occur near the transposon itself, whereas transposition targets are more or less randomly located.

Transposon Tn*10* is, comparatively speaking, a specific transposon. An extensive analysis of the DNA sequence at its insertion sites has revealed that there is some specificity involved. Approximately 85% of all insertions are found at the sequence NGCTNAGCN, where N represents any base. The sequence is symmetric, so there is no preferred orientation. After insertion of the Tn*10* there is duplication of the 9 bp forming the target site. This fact suggests that transposition may involve an enzyme (a **transposase**) that makes offset cuts in the manner of the type II restriction endonucleases. The IS*10* elements are necessary and sufficient for transposition, implying that they code for the transposase. However, the two IS*10* elements are not identical. The left-hand element in the conventional genetic map is vestigial, as all but 13 bp at the tip can be deleted and still give transposition. Similar deletions of the right-hand element cut the frequency of transpositions by 90%.

The normal frequency of Tn*10* transposition is on the order of 10^{-6} to 10^{-7}. Regulation of this rate is obtained by synthesis of complementary RNA in a manner similar to that used to regulate copy number in R plasmids. The start of the transposase coding region in IS*10* is overlapped by a small RNA transcribed from an outwardly directed promoter called p-OUT. Inactivation of that promoter allows extra translation of the mRNA transcribed from p-IN and increases the transposition frequency. Transcription of the transposase cistron from outside the transposon is prevented by a stem-and-loop sequence that would have to be transcribed in order to reach the coding region and that sequesters the AUG codon of the transposase so that translation initiation would be difficult if not impossible. The F128 plasmid can stimulate Tn*10* precise excision after conjugation, possibly by virtue of its single-strandedness acting to stimulate the recombination system.

Phage Mu Transposition

Phage Mu can carry out transposition in either a replicative or a nonreplicative mode. The latter reaction is called the simple insertion mode and occurs during a new infection of a cell. It is a special case because the donor DNA is linear, whereas other types of transposition involve circular, supercoiled DNA molecules. Mu replication is via replicative transposition. Studies on the relations of the strands of Mu to the donor and target DNA have shown that the transposition model proposed by Shapiro is relevant for both reactions (Fig. 13-16).

Consider first the replicative transposition mechanism. In step I, single-strand nicks are made at the ends of the transposon and five base offset nicks are made in the target region. The transposase for Mu is the product of cistron A, and its functionality depends on the presence of the cistron B protein. Transposase by itself is only 1% functional. ATP is required for proper B function, and the strand transfer reaction also requires one host protein, HU. Proper folding of the helices generates the X-shaped structure that is held together by ligating the transposon ends to the offset nicks in the target DNA. When the gaps of structure II are filled in, the typical 5 base repeat of the target DNA is generated. The gaps can serve as primers for DNA replication to yield the two structures shown in IV in Figure 13-16. Note that if the original DNA molecules were circular,

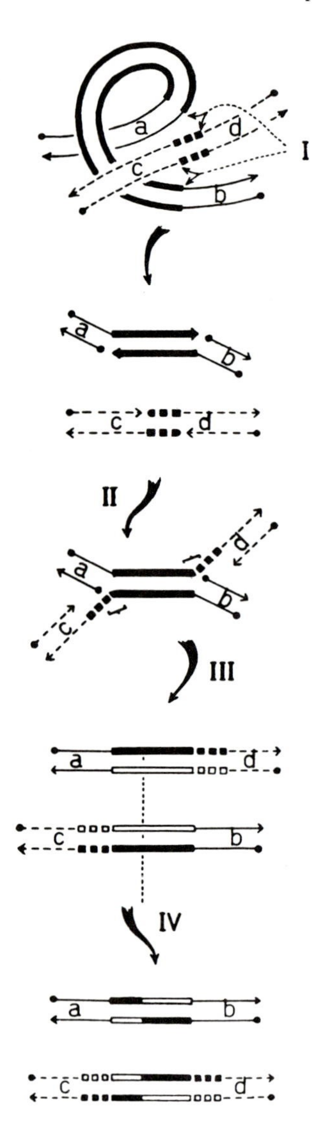

Figure 13-16. Transposition and replication. The top figure illustrates how various regions of double-strand DNA molecules may be brought into close physical proximity for the subsequent cleavage and ligation events. The bottom four drawings show various steps in the transposition process. Solid lines indicate donor DNA; dashed lines indicate target DNA; the heavy bars are parental DNA of the transposable element, and open bars are newly synthesized DNA; the small boxes indicate the oligonucleotide target sequence (filled = parental DNA; open = newly synthesized DNA). The arrowheads indicate 3′-hydroxyl ends of DNA chains, and the dots indicate 5′-phosphate ends. The letters *a, b, c,* and *d* in the duplex arms flanking the transposable elements and target oligonucleotide indicate the genetic structure of the various duplex products. The numbered steps in the model are as follows: **(I)** Introduction of offset nicks (represented by arrowheads). **(II)** Ligation of donor and target strands to give an X-shaped structure centered about the transposable element. **(III)** Filling in of the gaps created by ligation and semiconservative replication of the transposable element. **(IV)** Site-specific reciprocal recombination to regenerate the original donor molecule (but with some newly synthesized material). Note that the target molecule has acquired both the transposable element and a short duplication (the target sequence). From Shapiro, J.A. (1979). Molecular model for the transposition and replication of bacteriophage Mu and other transposable elements. Proceedings of the National Academy of Sciences of the United States of America 76:1933–1937.

region *a* was linked to region *b* and region *c* to region *d*. Those linkages are not affected by the transposition, and so structure IV is then a single circle (cointegrate) containing two copies of Mu. A recombination event between the two copies of Mu then resolves the cointegrate into single circles each containing a recombinant Mu prophage.

In order to derive a simple insertion, only a slight modification must be made in the model. Instead of filling in the gaps of structure II, the remaining links of the transposon DNA to the donor molecule (links to region *a* and region *b*) are degraded. A cointegrate structure cannot then develop. Instead, the donor DNA is left with a gap, as in the case of Tn*10*, and the target DNA (regions *c* and *d*) is left with the transposon plus one gap at each end of the transposon. A simple gap-filling reaction restores molecular integrity of the target.

Summary

Recombination and repair are overlapping processes, but each has its own unique features. Error-free repair is possible by either reversal of the damage to the DNA or excision of the damaged region and insertion of a short patch of about 20 bases. Error-prone repair results from the use of recombination pathways that are more efficient but less precise in their action. It is an inducible system (SOS repair) controlled via the *recA* cistron. The control of *recA* is patterned after the *lac* operon, with *lexA* coding for a protein repressor. A derepressed *recA* cistron produces large quantities of protein that can be activated to destabilize a number of other repressors including the λ repressor and LexA. The destabilization results in self-cleavage of the repressors. A fully activated SOS response leads to long-patch repair (still error-free), inhibition of cell division, and overproduction of RecA and LexA proteins.

Generalized recombination also is dependent on *recA* function, but in this case no induction is necessary. Basal levels of the RecA protein are sufficient for normal needs. RecA coats single-strand DNA and catalyzes the formation of D-loops, which are intermediates in the physical transfer of a strand of DNA from one DNA duplex to another homologous duplex. After the strand has been transferred, structures that are suitable for branch migration develop and produce reciprocal recombinants. Generalized recombination can therefore be viewed as either reciprocal or nonreciprocal, depending on where the genetic markers are located relative to the initial strand exchange. Two pathways for generalized recombination are known, one involving exonuclease V or a substitute and the other involving *recF*.

Site-specific recombination concerns DNA duplexes that have little or no homology. If the recombination event tends always to occur at the same site on both duplexes, as in the case of λ Int, it is classified as double site-specific. If, on the other hand, it tends to occur at a specific site on one DNA duplex but randomly on the other, as in the case of phage Mu or Tn*10*, it is said to be single site-specific. Each type of site-specific recombination has an appropriate protein or proteins associated with it. In the case of transposons, replication may or may not be an integral part of the transposition process.

References

General

Claverys, J-P., Lacks, S.A. (1986). Heteroduplex DNA base mismatch repair in bacteria. Microbiological Reviews 50:133–165.

Derbyshire, K.M., Grindley, N.D.F. (1986). Replicative and conservative transposition in bacteria. Cell 47:325–327.

Friedberg, E.C. (1987). The molecular biology of nucleotide excision repair of DNA: recent progress, pp. 1–23. In: Collins, A., Johnson, R.T., Boyle, J.M. (eds.) Molecular Biology of DNA Repair. Cambridge: Company of Biologists Limited.

Gellert, M., Nash, H. (1987). Communication between segments of DNA during site-specific recombination. Nature 325:401–414.

Kleckner, N. (1983). Transposon Tn*10*, pp. 261–298. In: Shapiro, J.A. (ed.) Mobile Genetic Elements. New York: Academic Press.

Kowalczykowski, S.C. (1987). Mechanistic aspects of the DNA strand exchange activity of *E. coli* RecA protein. Trends in Biochemical Sciences 12:141–145.

Lupski, J.R. (1987). Molecular mechanisms for transposition of drug-resistance genes and other movable genetic elements. Reviews of Infectious Diseases 9:357–368.

Mizuuchi, K., Craigie, R. (1986). Mechanism of bacteriophage Mu transposition. Annual Review of Genetics 20:385–429.

Radding, C.M. (1986). Homologous pairing & strand exchange mediated by RecA nucleoprotein filaments & networks, pp. 77–94. In: Gershowitz, H., Rucknagel, D.L., Tashian, R.E. (eds.) Evolutionary Perspectives and the New Genetics. New York: Liss.

Radman, M., Wagner, R. (1986). Mismatch repair in *Escherichia coli*. Annual Review of Genetics 20:523–538.

Specialized

Amundsen, S.K., Taylor, A.F., Chaudhury, A.M., Smith, G.R. (1986). *recD:* the gene for an essential third subunit of exonuclease V. Proceedings of the National Academy of Sciences of the United States of America 83:5558–5562.

Barbè, J., Villaverde, A., Cairo, J., Guerrero, R. (1986). ATP hydrolysis during SOS induction in *Escherichia coli*. Journal of Bacteriology 167:1055–1057.

Bender, J., Kleckner, N. (1986). Genetic evidence that Tn*10* transposes by a non-replicative mechanism. Cell 45:801–815.

Blanco, M., Herrera, G., Aleixandre, V. (1986). Different efficiency of UmuDC and MucAB proteins in UV light induced mutagenesis in *Escherichia coli*. Molecular and General Genetics 205:234–239.

Chow, S.A., Honigberg, S.M., Bainton, R.J., Radding, C.M. (1986). Patterns of nuclease protection during strand exchange. *recA* protein forms heteroduplex DNA by binding to strands of the same polarity. The Journal of Biological Chemistry 261:6961–6971.

Ennis, D.G., Amundsen, S.K., Smith, G.R. (1987). Genetic functions promoting homologous recombination in *Escherichia coli:* a study of inversions in phage λ. Genetics 115:11–24.

Fazakerley, G.V., Quignard, E., Woisard, A., Guschlbauer, W., van der Marel, G.A., van Boom, J.H., Jones, M., Radman, M. (1986). Structures of mismatched base pairs in DNA and their recognition by the *Escherichia coli* mismatch repair system. EMBO Journal 5:3697–3703.

Hatfull, G.F., Noble, S.M., Grindley, N.D.F. (1987). The gamma-delta resolvase induces an unusual DNA structure at the recombinational crossover point. Cell 49:103–110.

Konforti, B.B., Davis, R.W. (1987). 3′ Homologous free ends are required for stable joint molecule formation by the RecA and single-stranded binding proteins of *Escherichia coli*. Proceedings of the National Academy of Sciences of the United States of America 84:690–694.

Lahue, R.S., Su, S-S., Modrich, P. (1987). Requirements for d(GATC) sequences in *Escherichia coli mutHLS* mismatch correction. Proceedings of the National Academy of Sciences of the United States of America 84:1482–1486.

Lu, C., Scheuermann, R.H., Echols, H. (1986). Capacity of RecA protein to bind preferentially to UV lesions and inhibit the editing subunit (ε) of DNA polymerase III: a possible mechanism for SOS-induced targeted mutagenesis. Proceedings of the National Academy of Sciences of the United States of America 83:619–623.

Maxwell, A., Craigie, R., Mizuuchi, K. (1987). B protein of bacteriophage Mu is an ATPase that preferentially stimulates intermolecular DNA strand transfer. Proceedings of the National Academy of Sciences of the United States of America 84:699–703.

Sargentini, N.J., Smith, K.C. (1986). Quantitation of the involvement of the *recA, recB, recC, recF, recJ, recN, lexA, radA, radB, uvrD,* and *umuC* genes in the repair of x-ray-induced DNA double-strand breaks in *Escherichia coli*. Radiation Research 107:58–72.

Shen, M.M., Raleigh, E.A., Kleckner, N. (1987). Physical analysis of Tn*10*- and IS*10*-promoted transpositions and rearrangements. Genetics 116:359-369.

Surette, M.G., Buch, S.J., Chaconas, G. (1987). Transpososomes: stable protein-DNA complexes involved in the in vitro transposition of bacteriophage Mu DNA. Cell 49:253–262.

Tessman, E.S., Tessman, I., Peterson, P.K., Forestal, J.D. (1986). Roles of RecA protease and recombinase activities of *Escherichia coli* in spontaneous and UV-induced mutagenesis and in Weigle repair. Journal of Bacteriology 168:1159–1164.

Thompson, J.F., Waechter-Brulla, D., Gumport, R.I., Gardner, J.F., Moitoso de Vargas, L., Landy, A. (1986). Mutations in an integration host factor-binding site: effect on lambda site-specific recombination and regulatory implications. Journal of Bacteriology 168:1343–1351.

Wang, T.V., Smith, K.C. (1986). *recA* (Srf) suppression of *recF* deficiency in the postreplication repair of UV-irradiated *Escherichia coli* K-12. Journal of Bacteriology 168:940–946.

Chapter 14
Applying Bacterial Genetic Principles

Bacterial genetics is witnessing an exciting time in its history. For many years this science had been the sole province of the basic researcher. Enormous contributions had been made by bacterial geneticists to our understanding of genetic and molecular biologic processes, but there seemed to be only a few applied uses for the knowledge that had been gained. During the 1980s this picture completely changed, and interest in bacterial genetics suddenly intensified as a result of the development of DNA splicing techniques and their related technologies.

The implications of these methodologies are so great and so far-reaching that it is impossible to estimate the ultimate extent of their effects. Although there has been some exodus from the field of bacterial genetics (begun by Benzer, Delbrück and others who moved into different fields of biology), there is now an influx of new bacterial geneticists who are interested in bacterial genetics as a tool for studying other genetic systems and for developing processes applicable to specific industrial, ecologic, pharmaceutical, or other problems (e.g., fermentation technologists and biochemists).

In this chapter some of the current trends in bacterial genetic research are discussed and some of their implications considered. Old and new technologies are included, for basic research in bacterial genetics does not cease but continues to reveal phenomena important to the understanding of life processes. The topics of this chapter can be viewed as summaries of subjects presently under investigation.

More Information About DNA Splicing Technology

The bare bones outline of DNA splicing was presented in Chapter 2. It is a fundamentally simple process, but as with so many other areas of science the actual practice is more complicated than the theory. The following material provides additional information on the techniques used for DNA splicing and some problems that can arise during the process. The next section considers how vectors can be designed to eliminate or ameliorate the problems.

Restriction Mapping

Restriction enzymes play a central role in DNA splicing, and they can provide some information that contributes to the making of physical maps of DNA molecules. **Restriction maps** show the physical positions of specific recognition sequences on the DNA molecule of interest. They are constructed by analysis of agarose gel photographs such as the one shown in Figure 2-9. The rationale is that because the enzymes are site-specific they should produce a unique set of fragments from a particular DNA molecule. By cutting first with one enzyme and then with another, it is possible to identify those fragments produced by enzyme 1 that carry sequences specific for enzyme 2. Where there is a question as to which smaller fragments correspond to which larger ones, Southern blotting can be performed using the larger fragments as probes. The mobility of the DNA bands is an indication of the fragment size, and therefore the approximate position of the restriction site for enzyme 2 can be determined. A simple restriction map for phage λ is shown in Figure 14-1. More complex examples of restriction maps can be found in Figures 6-12, 11-1, and 11-4.

Restriction mapping is an excellent method for characterizing small DNA molecules quickly and accurately. If two types of DNA yield the same size fragments for all enzymes tested they may be identical, whereas if they yield different-sized fragments, they definitely have some DNA sequence heterogeneity. At the present time a number of laboratories are carrying out macrorestriction experiments in which entire bacterial genomes are cut with restriction enzymes whose recognition sites are comparatively rare to yield some 20 to 80 fragments that can be separated on pulsed field gel electrophoresis. The goal is to prepare a complete physical map of the organism.

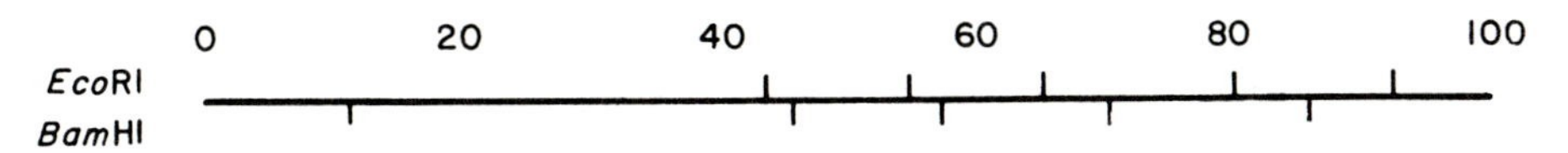

Figure 14-1. Restriction map for phage λ DNA. The phage DNA molecule is represented by the horizontal line. The numerical scale at the top represents distance as a percentage of the total genome. The short vertical lines represent the points of cleavage by either *Eco*RI or *Bam*HI. Note that the cleavages do not overlap. It should be remembered that the cohesive ends of λ may produce a circular molecule that would result in the two fragments at the ends of the DNA molecule merging into a single, larger fragment.

Sources of DNA for Splicing

The discussion in Chapter 2 focused on DNA splicing using simple restriction fragments. As with so much of science, real life is often not that simple, and several problems can arise. There may not be a restriction enzyme that cuts to yield a fragment carrying the entire region of interest. If the DNA is from a eukaryotic organism, it usually has one or more introns within it. Such DNA cannot yield a correctly processed RNA when cloned into a prokaryotic organism, as the only RNA splicing known to occur in prokaryotes is self-splicing. Sometimes the genetic map position for the desired DNA sequence has not yet been determined and therefore it is not possible to predict the nature of the fragment required for the cloning. Each of these problems has a reasonably straightforward solution, although in some cases considerably more work is involved.

When the map position of the appropriate DNA sequence is not known, a common solution is to produce a **library** of DNA clones. A library is prepared by taking the entire genome of the organism and breaking it up into more or less standard-sized pieces. It can be done mechanically by hydrodynamic shear or enzymatically by incomplete digestion with a restriction enzyme that has many recognition sites on the DNA, e.g., *Sau*3A I whose recognition site is 5′GATC. Both techniques generate a series of overlapping DNA fragments that can be sized by various physical techniques. One common method is to use a sucrose gradient, a centrifuge tube in which increasing concentrations of sucrose are present from top to bottom. DNA layered on top of the gradient and subjected to centrifugation moves in the opposite manner as on an agarose gel, i.e., the largest fragments migrate the farthest. By choosing DNA from different regions of the sucrose gradient, it is possible to adjust the average size of the DNA fragment to any desired value.

The overhangs resulting from the *Sau*3A I digestion (GATC) are perfectly complementary to the *Bam*HI overhangs (see Table 2-3). Therefore a vector can be cut with *Bam*HI, and *Sau*3A I-derived fragments ligated into it. One potential problem that can occur here is self-closure of the vector in which the *Bam*HI ends are resealed without the addition of any *Sau*3A I fragment.

If the DNA fragments have been prepared by mechanical means, there are no single-strand ends to match with restriction enzyme overhangs on the vector DNA. A similar problem can arise if the overhangs from the restriction enzyme used to prepare the fragments are not compatible with the vector ends. One way to overcome this difficulty is by using the enzyme **terminal transferase,** which synthesizes single-strand homopolymers (all bases alike) attached to the 3′-ends of DNA strands. By attaching thymidine residues to the fragment and adenine residues to the vector, artificial complementarity can be produced. In this case self-closure of the vector is not possible owing to the identical ends of the vector.

Alternatively, it is possible to obtain **linkers,** small oligonucleotides that contain within their sequences a particular restriction site. These linkers can be **blunt-end ligated** to the fragments. In blunt-end ligation, high concentrations of DNA and ligase are used so that if the ends of DNA molecules should happen to touch they are immediately ligated. There is no formal mechanism for bringing the fragments into apposition. Once ligated, the linkers can be cut with the appropriate restriction enzyme to generate single-strand ends.

When the DNA sequence to be cloned contains introns, it is often simpler to clone a DNA sequence that corresponds to the processed mRNA molecule rather than the actual coding region for the protein. The retroviral enzyme **reverse transcriptase** possesses two enzymatic activities: an RNA-dependent DNA polymerase that makes DNA from an RNA template and a ribonuclease H activity that can degrade RNA:DNA hybrids, mRNA, and poly A tails (Fig. 14-2). The product is **cDNA,** a double-strand DNA

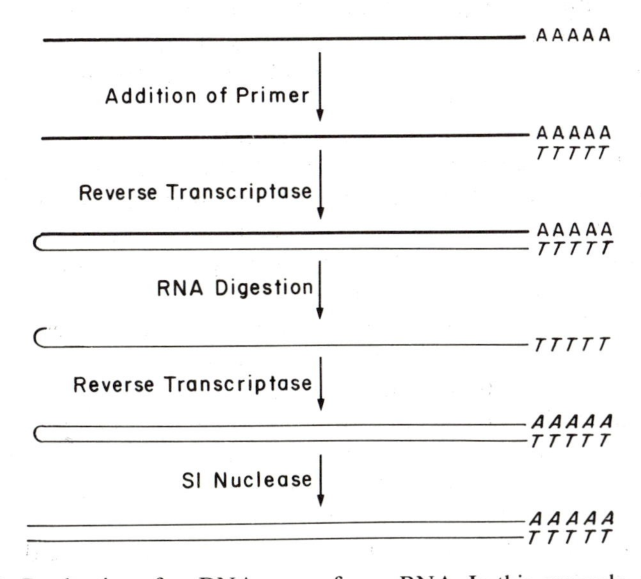

Figure 14-2. Production of a cDNA copy of an mRNA. In this example, the mRNA (thick line) is assumed to be a eukaryotic mRNA molecule that has a poly A tail at its 3′-end. To begin DNA synthesis, a short primer of polydeoxythymidine (italic letters) is hydrogen-bonded to the poly A tail. Reverse transcriptase then uses this primer to initiate and carry out the synthesis of a DNA strand (thin line) complementary to the mRNA base sequence. It is a characteristic of this enzyme that as it reaches the end of the mRNA molecule it loops back on itself to form a continuous hairpin-like structure. The RNA that comprises one leg of the hairpin is removed by the RNase H activity, and the reverse transcriptase can complete synthesis of a double-strand DNA molecule. The loop of the hairpin is opened by the action of S1 nuclease to leave a double-strand DNA molecule that contains all of the sequences present in the original RNA molecule, including the poly A tail.

molecule containing one strand that is complementary to the mRNA. This DNA can then be used directly in cloning experiments if tails or linkers are attached as above. The major problem with the technique is that the reverse transcriptase does not do well in vitro with long transcripts. It is often difficult to obtain cDNA molecules longer than 600 bp in length.

Safety Considerations

The techniques of gene splicing present interesting scientific and ethical problems because they require unusual standards of experimental control. Any DNA can be spliced to any other DNA, regardless of species barriers or any other genetic impedimenta. Moreover, the spliced DNA can be made compatible with essentially any organism by an appropriate choice of vector plasmid. In theory, then, any organism can be endowed with any genetic attribute. The implications of this analysis were not missed by the researchers involved in the early work on gene splicing, and they convened several meetings to discuss the issue. Finally, after an extensive review, the principal scientists involved in gene splicing research agreed to a moratorium on all future experiments until appropriate safety guidelines could be worked out.

The potential biohazards seemed to fall into two broad categories. The first type of potential hazard was thought to be the host–vector combination. The presence of the spliced plasmid might confer new and surprising properties on the host cell. For example, the plasmid might cause production of a potent toxin (similar to phage conversion), or it might utilize sequences from one of the tumor-inducing viruses to code for proteins that would cause gut cells to transform into tumor cells if the host bacterium managed to colonize the intestinal tract of an individual. If such a bacterium–plasmid combination were to escape the confines of the laboratory, it might be dangerous to the surrounding population as well as to the laboratory workers themselves. The problem posed by this type of potential hazard is then one of physical containment—keeping the plasmid and host inside the laboratory and the appropriate culture vessels.

The second type of potential biohazard was thought to be the possible escape of the spliced DNA from the original host cell into some sort of laboratory contaminant. The plasmid might be perfectly harmless in the bacterium in which it was originally cloned but might confer unwanted antibiotic resistance, pathogenicity, etc., on other bacteria. In this case, then, the problem posed is one of biologic containment—keeping the plasmid in the appropriate host organism.

Guidelines addressing these problems were first formulated during a meeting of interested scientists at Asilomar, California, and later formalized as regulations for research funded by U.S. governmental agencies. The initial guidelines were stringent and conservative in their approach. However, as knowledge of the host cell and the DNA interactions has increased, it has become clear that the potential hazards were overrated, and several revisions of the guidelines have gradually relaxed certain of the restrictions. Despite these changes, the basic philosophy of the guidelines has remained unaltered. Physical containment is rated on a scale of one to four, with a biosafety level 1 (BL1) facility being the normal microbiology laboratory and a BL4 facility being a laboratory suitable for work with the most dangerous pathogens. In such a laboratory all air, liquids, and solids leaving the facility are sterilized, and all personnel must shower and change clothing before leaving. Biologic containment is rated on a scale of one to three, with the viability of the host cell and the safety of the plasmid considered together. An HV1 system provides moderate containment using standard laboratory strains of bacteria and suitable plasmids. An HV2 system provides high-level containment due to weakening of the host strain by the introduction of appropriate mutations that greatly reduce its ability to grow under normally encountered environmental conditions, and due to a deliberate alteration of the vector so it does not replicate well in other host bacteria if it should manage to escape biologic confinement. The general survival rate for either host or vector outside the laboratory must be shown to be less than 10^{-8}. An HV3 system has been proposed but not yet constructed and would provide still higher containment levels. Table 14-1 illustrates the application of the regulations to specific cases. Guidelines

Table 14-1. Examples of the types of containment required for DNA splicing experiments under NIH guidelines[a]

Type of donor DNA	Type of physical containment[b]
Escherichia coli	Exempt (BL1)[c]
Salmonella typhimurium	Exempt (BL1)
Bacillus subtilis	Exempt (BL1)
Drosophila	Exempt (BL1)
Hepatitis B virus	BL2
Marburg virus	
Complete genome	BL4
Noninfectious segment	BL2

[a]In each case the host bacterium is assumed to be *E. coli*.

[b]BL1 containment represents normal safe microbiologic techniques; BL2 containment requires that, in addition to BL1 techniques, all experiments be carried out in a special biohazard laboratory with restricted access; BL3 containment requires all of the above plus double doors on the facility and the use of special containment hoods; BL4 containment requires the use of a special facility equipped with negative air pressure, equipment to prevent direct contact of workers with the material, and sterilization of all materials before removal from the facility.

[c]This notation indicates that this type of experiment is not regulated but that good laboratory practice would be to work at the BL1 level.

for large-scale (i.e., industrial) production of organisms carrying spliced DNA have also been prepared.

For the specific case of *Escherichia coli,* the equivalent of an HV1 system is EK1, and the equivalent of an HV2 system is EK2. Curtiss and co-workers constructed a suitable host strain for an EK2 system and designated it χ1776. Careful tests have shown that this strain does not colonize the intestinal tract of test animals even if given in massive doses. An HV2 system is also available in *Bacillus subtilis*.

There are some indications that the presumed barrier to genetic exchange between prokaryotic and eukaryotic organisms is not absolute. One piece of evidence is the existence of the Ti plasmid discussed in Chapter 10 that transfers T-DNA from bacterium to plant. The seeming identity of the introns from phage T4 with those from eukaryotic mitochondria also suggests a connection between the two kingdoms. Furthermore, it seems that transfer may be able to occur in the opposite direction. For example, Carlson and Chelm reported that glutamine synthetase II from *Bradyrhizobium japonicum,* a bacterium commonly found in association with legumes, has no introns but otherwise corresponds well to eukaryotic enzymes with respect to amino acid sequence homology and lack of posttranscriptional modification. Brownlee suggested that such a situation could arise if a eukaryotic **pseudogene** (a nonfunctional region of DNA with extensive homology to an expressed DNA) were somehow transferred into the bacterium, implying that eukaryotic cells do not have a monopoly on the production of certain types of proteins. Supporting this view is the early observation that a particular strain of *E. coli* produced a protein with insulin-like activity.

Some experimentation has been done in an attempt to determine to what extent bacteria are able to exchange DNA under natural conditions. Transformation has been obtained using sterile soil, and conjugation occurs readily in soil, but transduction has yet to be observed. In aquatic environments, on the other hand, all types of genetic exchange have been reported. In the specific case of transduction in *Pseudomonas aeruginosa,* the amount of transduction seen in a natural freshwater environment is inversely proportional to the number of competing organisms present.

Much of the current controversy about genetically engineered organisms revolves around the deliberate release of such organisms into the environment. The U.S. Environmental Protection Agency is attempting to define suitable precautions for such activities. Part of the problem is that the appropriate ecologic studies that might shed light on the persistence of genetically engineered organisms in the environment are only now beginning. For example, Devanas and Stotzky reported that bacteria carrying cloned *Drosophila* DNA survive well in soil in the laboratory and do not seem to be rendered less fit by their extra DNA. Whether this result is generalizable to nature will not be known until appropriate field tests can be made. However, there is substantial discussion as to what constitutes an appropriate field test and how to prevent the very release that is questionable in the first place.

Another area of controversy concerns the type of alteration in the organism proposed for release. How much of an alteration is necessary to make a particular organism hazardous in the environment? In the specific case of *Pseudomonas syringae,* what is wanted is not a new function but, rather, a loss of function. *P. syringae* codes for a protein that causes water to freeze more readily (at a higher temperature than normal). Wild-type bacteria have been killed and mixed with snow-making solutions used on ski slopes to provide a more solid product. However, the protein is also responsible for extensive crop damage when the bacterium grows on the leaves. A spontaneous deletion mutant has been found that lacks the ability to produce the ice protein and therefore should not damage crops. Considerable discussion ensued over potential hazards to the environment before initial field trials were authorized to see if the mutated bacterium would displace the wild-type bacterium from crop plants on which it was sprayed.

Difficulties with the Expression of Cloned DNA

One reason it has been possible to relax the guidelines is that it has turned out to be much more difficult to express foreign DNA sequences in a bacterial host than was expected. There are no particular problems when DNA is transferred from one species to the same species and often not when DNA is transferred among members of the same genus, barring the presence of distinct restriction and modification systems. However, greater taxonomic distance often leads to difficulties that are not necessarily symmetric. For example, DNA from *Bacillus subtilis* can be readily cloned and expressed in *E. coli*, but DNA from *E. coli* is not well expressed in *Bacillus*. The problem lies in the nature of the *Bacillus* promoter. Whereas *E. coli* does not require a specific base sequence outside of the -10 and -35 regions, *Bacillus subtilis* has a requirement for a specific 8 or 9 base sequence upstream of the -10 site. Oddly enough, a particular fragment of phage T5 DNA carrying a promoter is always expressed in *B. subtilis* but is expressed in *E. coli* only during late infection (i.e., after modification of the host RNA polymerase). Along the same lines, bacterial DNA carried by the Ti plasmid is not well expressed in plants unless an appropriate plant promoter is fused to it.

The introns discussed above provide a major block to expression of eukaryotic DNA in prokaryotes. Additionally it is now known that the genetic code is not universal. There have been numerous reports of exceptions to the genetic code in the DNA isolated from mitochondria taken from a variety of sources. For example, in human mitochondria the codon UGA is translated as tryptophan instead of termination (compare with

Table 3-3), and the codon AUA is translated as methionine. In yeast there is also widespread use of UGA to code for tryptophan. The ciliated protozoans have been shown to use the codons UAA and UAG to code for glutamine. Obviously, cloned DNA carrying any translated terminator codons could be properly expressed only in bacterial cells carrying the appropriate suppressor mutation.

Another potential problem with expression is the opposite of those discussed above. In many cases the eukaryotic protein is detrimental, perhaps even lethal, to the producing bacterial cell. This sort of effect provides a strong selection against maintenance of the desired clone. As discussed below, the obvious solution is to provide some sort of regulatory mechanism so the protein encoded by the clone is not produced until just before the experimenter is ready to harvest the cells from the culture.

Cloning Vectors

Cloning vectors must certainly provide the essential replication functions for the cloned DNA, but they can be used to fulfill other functions as well. They can be constructed so as to provide an additional safety margin for the containment of DNA constructs; and they can be used to express the cloned DNA and to provide specific DNA for sequencing or other purposes. Vectors can be derived from simple plasmids or from phages.

Many cloning vectors are based on the types of plasmid that were discussed in Chapter 10. The minimal requirements for a vector are the replication origin, the partition function, and some sort of selectable trait so transformants can be readily identified regardless of the expression of the cloned DNA. Usable insertion sites are defined by those restriction enzyme recognition sequences that are present in single copy or, if double, remove only a small, nonessential portion of the plasmid. Each plasmid is identified by a lower case "p" followed by two or more capital letters indicating the laboratory in which it was prepared and a unique number.

One particularly well known plasmid vector, pBR322, is depicted in Figure 14-3. This vector is based on a ColE1 plasmid and was constructed by Bolivar and Rodriguez. The replication functions are located in the lower left quadrant of the map in Figure 14-3. The plasmid has the property from the ColE1 plasmid of being amplifiable by adding chloramphenicol to the medium, and its presence is detectable by either of the antibiotic resistance markers it carries. Restriction sites are available within each antibiotic resistance reading frame so that cloned DNA can inactivate the normal proteins. Marker inactivation allows the investigator to identify plasmids that have self-closed and religated from those that contain DNA inserts. Although pBR322 is mobilizable by the F plasmid, it is nonconjugative in isolation and therefore provides additional containment for the cloned DNA. It is one component of an EK2 host–vector system.

Cloning vectors derived from phages offer the investigator an automatic

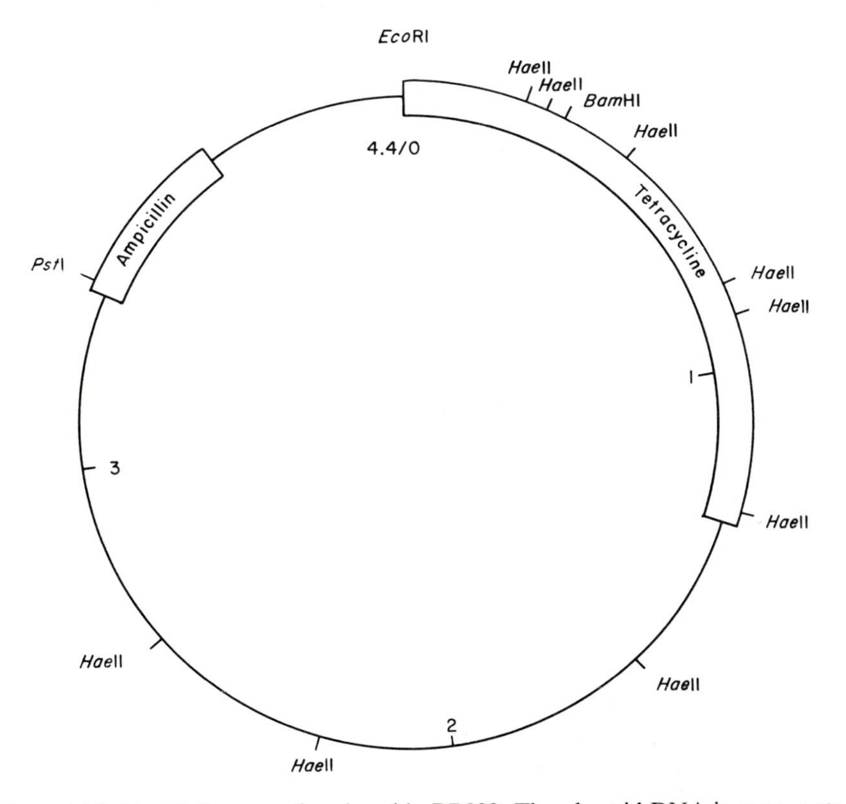

Figure 14-3. Restriction map for plasmid pBR322. The plasmid DNA is represented by the circular structure. Boxed areas represent the sequences coding for the indicated antibiotic resistances. Cleavage sites for a few selected restriction endonucleases are indicated by the short lines on the outer portion of the circle. The single site of cleavage by *Eco*RI is taken as a zero point for the system of coordinates. Numbers along the inner portion of the circle represent the distance from the zero point in kilobases. Total length of the DNA molecule is 4363 bp.

Note that cleavages by *Bam*HI or *Pst*I occur in the middle of the antibiotic resistance sequences, and insertions of DNA at these points result in inactivation of the resistance cistrons. However, *Eco*RI cleaves at a point immediately adjacent to the tetracycline resistance region, and therefore insertions at this site have no effect on antibiotic resistance. The sites for *Hae*II have been included to demonstrate that this enzyme is not suitable for DNA splicing with this plasmid because it cuts the plasmid DNA into too many pieces to be useful. Simultaneous cutting of the plasmid by *Eco*RI and *Bam*HI results in removal of a fragment of the tetracycline DNA and production of a DNA fragment that cannot self-close to form a circular molecule.

delivery system in which the DNA can be isolated already packaged into a phage coat and thereby protected from environmental influences. They circumvent the difficulties that often arise when trying to transform cells with plasmids (see Chapter 8). Two of the most commonly used phages in this regard are λ and M13. One of the major constraints on the λ cloning system is the size of the insert. Like most phages, λ does not have a variable capsid size and therefore must sacrifice some phage DNA in order to accommodate other DNA. Clearly if plaque-forming viruses are to be assembled, the cloned DNA must be kept small. M13 is somewhat less critical in this regard because as a rod-shaped virus its length is determined solely by the length of the DNA to be encapsidated. Practically speaking, DNA that is more than double the normal length is undesirable and tends to spontaneously shorten by deletion. Often the phage is modified to give it a small DNA segment coding for a selectable trait, as in the case of pBR322, so as to allow for rapid detection of lysogens.

In the sections that follow, various specific cloning vectors are described that have been constructed to fulfill perceived needs within the genetic engineering community. The list is by no means exhaustive, and descriptions of new vectors can be found in nearly any issue of journals dealing with bacterial genetics.

Phage λ Vectors

There are numerous cloning vectors derived from λ. Blattner and his co-workers prepared a series of Charon phages (named for the ferryman across the river Styx) that carry deletions from the b2 region through *c*I to allow space for cloned DNA. They also carry the *lacZ* cistron to provide a selectable marker. More recent variations on this theme are the EMBL3 and EMBL4 phages prepared by Murray and co-workers in the European Molecular Biology Laboratory. Like Charon, these phages are **replacement vectors** in which a "stuffer fragment" of DNA is used to make the phage DNA large enough to package. Removal of the fragment then makes any self-closed vectors nonviable. Moreover, the fragment carries *red* and *gam*, thereby conferring the *spi*$^{+}$ phenotype (sensitivity to phage P2 inhibition). If the cloning host is a P2 lysogen, unmodified vectors cannot grow and again are lost. These two properties make it highly probable that any phage that grows on a P2-lysogenic host contains inserted DNA.

Another type of vector is the **cosmid** or cohesive end-containing vector. It is a standard plasmid that includes the λ *cos* site as part of its DNA. After appropriate cutting and ligating experiments have been carried out in vitro, the DNA can be packaged into λ virions by cell extracts prepared from infected cells that contain the appropriate enzymes plus empty heads and tails to yield phage particles that are infectious but contain no λ DNA except for the *cos* site. The virions serve as an efficient delivery system and avoid the complications inherent in the transformation procedure for *E. coli*.

Phage M13 Vectors

Messing and his collaborators developed a series of M13mp phages (more than 19 are available) that are particularly advantageous for cloning DNA that will later be sequenced by the dideoxy technique. The phages are prepared in pairs (e.g., M13mp18 and M13mp19) that contain identical loci. A portion of the amino-terminus of β-galactosidase is included in each phage for identification purposes. When mixed with a host cell protein derived from a particular deletion of the amino-terminus of β-galactosidase, the combination of the peptide from the phage (the α peptide) and the peptide from the bacterium yields a low level of β-galactosidase activity (α-complementation). When such phages are plated on the appropriate indicator strain in the presence of X-gal (5-bromo, 4-chloro, 3-indoyl-β-D-galactoside), they hydrolyze the indicator to yield the intensely blue indole derivative and hence a blue plaque. Located between the promoter for the β-galactosidase fragment and the coding region is a polylinker that contains multiple unique restriction sites. DNA cloned into any one of these sites prevents α-complementation and results in a colorless plaque. Each member of a pair of phages contains the polylinker in opposite orientation.

During infection of the host cell by the phage and its associated insert DNA, the DNA replication process described in Chapter 5 occurs. This process yields single-strand DNA for packaging or extrusion into the medium. The single-strand DNA, whether isolated from the medium or the phage particles, is a suitable substrate for dideoxy DNA sequencing and can be primed with the same DNA sequence that is complementary to the region just upstream of the polylinker (Fig. 14-4). By using a pair of M13mp phages, the experimenter can verify the sequence obtained from one clone by checking its complementarity to the sequence determined from the same clone with the opposite orientation.

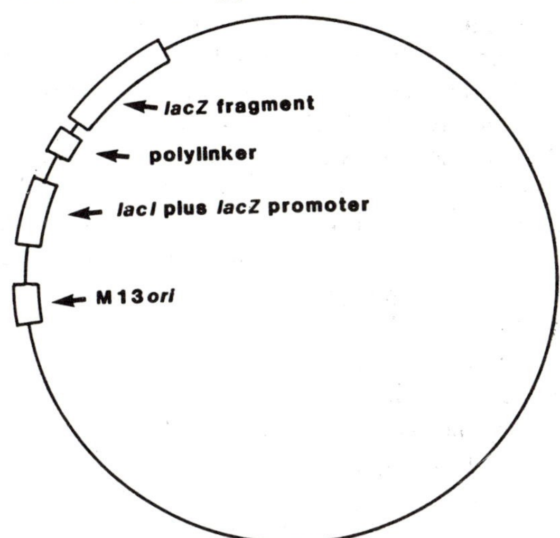

Figure 14-4. General design of the M13mp series phages. Note the presence of the α peptide region of *lacZ* and the location of the primer-complementary region just upstream of the polylinker. The unlabeled region contains the information for the phage proteins. One recent member of this group, M13mp18, is 7250 bp with a 54-bp polylinker.

In terms of the critical component for the M13 cloning vectors, it is the intergenic region (see Figure 5-7) that contains the origin of replication for the rolling circle. Messing has also prepared a series of pUC plasmids that are combinations of the M13 phage origin and *lacZ* region together with portions of pBR322. The plasmid can serve as a standard cloning vector; but when single-strand DNA is needed for sequencing, the M13 origin can be used to provide it. Rolling circle replication is triggered by an M13 infection of the cell carrying the plasmid. A derivative of the pUC plasmids has been prepared by Peeters and co-workers that contains both the replication origins and packaging signals for the single-strand phages Ff and Ike but situated at opposite ends of the polylinker cloning region. Superinfection of a cell carrying this plasmid results in synthesis and packaging of one particular cloned DNA strand because the two phages employ different replication and packaging signals. Thus only one clone is needed to provide both strands for sequencing.

Expression Vectors

In expression vectors the cloning site is located just downstream from a promoter and ribosome-binding site. Therefore the cloned DNA need not include either structure. The promoter may be any one of the standard promoters already discussed or a composite. For example, if it is desired that the cloned DNA be regulated so as to prevent inhibition of host cell growth, the *trp, lac,* or λ promoters might do well. Of course the appropriate repressor molecule must be supplied as a regulator. Its cistron can be located on the same DNA molecule, on another plasmid, or on the bacterial chromosome itself. However unless a low-copy-number plasmid is used for the cloning vector, a chromosomal repressor locus usually cannot synthesize enough product to fully repress the plasmid transcription. High levels of cloned DNA transcription can be obtained by mutating whatever promoter is used so as to bring it closer to the ideal sequence. The *tac* promoter, a combination of the −35 region of the *trp* promoter and the −10 region of the *lac* promoter, has a high efficiency of expression, as do many conventional phage promoters. The phage T5 promoter is capable of outcompeting all *E. coli* promoters in the host cell, thereby yielding large amounts of product.

Particular attention has been paid to the phages that, like T7, code for their own RNA polymerases. They include T3, T7, and the *Salmonella* phage SP6. Each of the polymerases recognizes its own particular promoters and ignores those of other phages or the host bacterium. The differences in the promoters are primarily at base −11. This specificity can be turned to advantage as in the case of a clone prepared by Tabor and Richardson. Here pBR322 was used to carry the T7 polymerase cistron under the control of the λ p_L promoter. The *c*I repressor function was provided in *trans* and prevented synthesis of the T7 polymerase until the repressor was inactivated. At that time the T7 RNA polymerase was synthesized and caused the specific expression of DNA cloned downstream from a T7 promoter. When it is not certain which strand of cloned DNA needs to be transcribed, a combined T7/SP6 vector is often useful. The cloning site is flanked by inwardly directed promoters, T7 at one end and SP6 at the other. The DNA strand that is transcribed is controlled by which phage RNA polymerase is added to the system.

Sometimes a particular mRNA transcript made from cloned DNA is poorly translated in the host cell. This lack of efficiency may be due to an increased susceptibility to degradative enzymes or to the use of rare codons in the mRNA. The only solution for the latter problem is to either change host cells to one that has a more suitable pattern of codon usage or else to use site-specific mutagenesis to change the codons themselves. In the case of the former problem, changing the structure of the mRNA molecule itself can often prevent excessive degradation. In particular, mRNA stability can be enhanced by splicing sequences such as REP onto the end of the cloned DNA so that a suitable hairpin loop is formed. The loop prevents access of the exonuclease to the end of the mRNA.

Broad Host Range Vectors

As discussed in Chapter 10, certain plasmids are capable of existing in a variety of hosts. These plasmids make advantageous vectors for DNA that is intended to be tested in several hosts to observe its effects on metabolism. For best effect, promoters must be supplied that function in all cells to be tested. Timmis and co-workers have used pKT231 as a base to carry the promoters found on the *Pseudomonas* TOL plasmid (toluene degrading). These promoters have been shown to function in many gram-negative bacteria.

A related entity is the **shuttle vector,** which carries replication origins for two different bacteria on the same plasmid. It allows the experimenter to do preliminary manipulations in a well characterized system such as *E. coli* and then transfer the resultant clone to the organism of interest. If the clone is to be expressed in both hosts, it may be necessary to provide two promoters as well.

Runaway Replication Vectors

Cloning vectors are under the same plasmid controls for copy number and partitioning that were discussed in Chapter 11. It is sometimes advantageous to have gene amplification take place so as to augment production of a particular product or provide large amounts of the cloned DNA from a relatively small culture volume. Amplification can be accomplished by suitably mutating the copy number control system so that a conditional mutation, usually temperature-sensitive, is in place. When the culture temperature is raised, all copy number control is lost, and the plasmid can replicate to high levels, perhaps 1000 copies or more per cell. The effect is to kill the host cell, but that is not important because the next step in the procedure is to extract the product from the cells in any case.

Commercial Successes of DNA Cloning

Human Growth Hormone

In the view of the general public, it is not the cloning of bacterial DNA that is important; rather, the subject of importance is the cloning of DNA molecules whose products have a direct impact on man and society. This type of work is still in its early stages, although considerable progress has already been made. The main focus has been on the preparation of hormones, proteins that have a profound biochemical impact on an individual and yet need be present in only trace amounts in order to exert their effects. One such protein, human growth hormone, particularly well illustrates the problems involved in splicing eukaryotic DNA and making it functional in bacteria.

Growth hormone mRNA is synthesized in the pituitary gland. Using pituitary mRNA, a cDNA copy was cloned that codes for a new mRNA consisting of 29 nucleotides at its 5′-end (which are not translated) followed by 651 nucleotides (which are translated) followed by 108 nucleotides at the 3′-end (which again are not translated). The protein produced is a **prehormone** (larger than the active molecule), which must have 26 amino acids removed from the amino-terminus to form the actual hormone. However, the bacteria do not produce the enzyme that would cleave off the amino-terminus to yield the active hormone.

A cDNA fragment was isolated that carried the information for most of the hormone but that lacked the information for the amino-terminus. Then chemically synthesized DNA consisting of the code for methionine (TAC) followed by the 23 bp required to code for the amino-terminus of the growth hormone was prepared. The cloned DNA was removed from its vector by treatment with the same nuclease that had originally been used to produce the fragment, and then the (formerly) cloned DNA was spliced to the synthetic DNA. The new DNA, which now contained all the information necessary to synthesize growth hormone, was spliced into a plasmid that carried two lactose promoter regions. The newly spliced DNA was cloned and shown to cause the *E. coli* that carried it to produce pure human growth hormone.

The resultant protein is methionyl growth hormone, which differs from the natural product by one amino acid. The U.S. Food and Drug Administration has held that any change from the amino acid sequence of a natural protein makes the genetically engineered compound a new drug. Therefore the product from the clone had to undergo appropriate testing for safety and efficacy before being approved for sale. The genetically engineered human growth hormone received its approval for sale in 1986 and immediately raised the ethical issue of whether it was appropriate for parents of normal children who just wanted them to be a little taller to use the drug. The position of the company is that such use is inappropriate, and attempts are being made to restrict distribution of the product to practicing endocrinologists.

Methionyl growth hormone is unusual in that it is readily translated by *E. coli*. Often the cloned eukaryotic DNA yields mRNA molecules whose secondary structures prevent proper translation owing to extensive secondary structure. Schoner and co-workers prepared a system in which the plasmid carries two cistrons downstream from the promoter. The first is a short artificial 31-bp A:T-rich sequence that functions as a spacer to disrupt secondary structure. The second cistron is the actual cloned DNA. If the terminator codon for cistron 1 is located 3′ to the Shine-Dalgarno box of cistron 2 and close to its initiator codon, the translating ribosome unfolds the message, as was seen in the case of the phage MS2 RNA. When methionyl growth hormone was used as cistron 2, about 20% of the cellular protein was found to be growth hormone.

Normally growth hormone is secreted by the producing cell. The gram-negative bacteria are not good secreters of protein but do put certain of their proteins into the periplasmic potential space that lies between the peptidoglycan layer and the cell membrane. Control of this process is via the signal peptide discussed in connection with the Ff phages. From a commercial point of view it is easier to isolate and purify a protein from the periplasm than from the cytoplasm because there are fewer total proteins involved. It was therefore of interest to see if the human signal sequence would function in *E. coli*. Gray and co-workers prepared two plasmids, one using the human signal sequence fused to growth hormone and the other using the *E. coli* alkaline phosphatase signal sequence. In both cases most of the growth hormone was found to be in the periplasm, and the correct disulfide bridges were in place. This finding indicates that at least in this one case the human signal sequence is handled correctly by a bacterium.

Laboratory Reagents

The enzymes involved in the cloning process are specialized and were originally expensive. In many cases their cost has been substantially reduced by the simple expedient of cloning the appropriate DNA sequences. The expression vectors described above can provide large quantities of the protein products of cloned genes. In this way the availability of the Klenow fragment of DNA polymerase I, the RecA protein, and numerous restriction enzymes has been greatly increased. Proteins used in industry are also good candidates for cloning. It has now been reported that the enzyme renin, used in the cheese industry, has been successfully cloned. It is hoped that this new source of supply will free cheese manufacturers from the highly variable present market in which the enzyme must be extracted from calf stomachs.

Other Types of Applied Bacterial Genetics

DNA cloning is not the only possible use for bacterial genetics. Important as cloning is, it is only one component of a broader picture. The discussion that follows presents some ways in which the information from the preceding chapters can be applied to solve specific problems.

Mutagenicity Testing

An interesting new role for bacteria was pioneered by Ames and co-workers. They were concerned with the problem of detecting carcinogenic (cancer-causing) substances within the environment and with the problem of screening chemicals for potential carcinogenic effects prior to commercial use. The traditional method has been to test suspected carcinogens in animals. Massive doses of chemicals are used, long periods of time are required, and maintaining the animals is both expensive and unpopular with a substantial segment of the population. For these reasons, animal testing is not well suited to routine screening, although it is essential as a confirmatory test. What is needed is an inexpensive, quick test that can distinguish those chemicals that have no effect from those that may have an effect. This need is fulfilled by mutagenicity testing, or **Ames testing.**

The theoretical basis for the test rests on the observation that nearly all proved carcinogens that act directly (i.e., attack the DNA, in contrast to indirect hormonal action) are also mutagens. It should be noted, however, that the converse has not been proved true, and not all mutagens are known carcinogens. Therefore when considering how to screen for carcinogens, Ames proposed that initial mutagenicity testing might be helpful. Those chemicals not shown to be mutagenic would be considered noncarcinogenic (unless suspected of hormonal activity), and those chemicals shown to be mutagenic would be subjected to further testing.

Ames and his collaborators set out to construct a bacterial system suitable for rapidly screening chemicals for mutagenic properties. The most convenient bacterial genetic systems involve positive selection for the desired event, so the type of mutation they chose to study was reversion, the reacquisition of a genetically controlled trait. Specifically, they chose to study the reversion of histidine mutations in *Salmonella typhimurium*. No one strain would be suitable for detecting all types of mutations, but by assembling a collection of mutant strains that rarely revert spontaneously they were able to detect frameshifts, amino acid substitutions, and reversion of nonsense mutations. A similar collection of tryptophan mutants has been prepared with *E. coli*. More recently experiments have also been performed using induction of a λ lysogen as an indicator of DNA damage. In all cases increased sensitivity can be obtained by working with DNA repair-deficient strains. Various tissue culture cell lines can be used to test the effect of chemicals on eukaryotic cells. Regardless of the type of system used, the basic technique is the same.

It is easy to demonstrate the effect of a known mutagen. For the *Salmonella* strain, a plate containing all required nutrients except histidine is spread with a lawn of bacteria. A small amount of the mutagen is placed on the center of the plate and allowed to diffuse. Generally the high concentration of the chemical kills all the bacteria in its immediate vicinity; but as the chemical is diluted by diffusion, cells can grow. The mutagen allows more colonies than expected to grow in this zone, and the number of colonies observed is a rough indication of the mutagenicity (and hence potential carcinogenicity) of the tested chemical.

In the formal mutagenicity test, the cells are treated with carefully measured doses of chemicals and then plated on selective medium or tested for induction of the prophage. Many chemicals are not themselves mutagenic but can be converted to proximate mutagens in the course of detoxification in the liver. Therefore the standard screening for mutagenicity also involves treatment of the chemical with a liver extract prior to adding the cells. Any proximate mutagens that may be formed are then further modified by bacterial enzymes such as transacetylases to become direct mutagens. If it is suspected that the chemical to be tested cannot enter the bacterial cell, a transfection system using DNA from specialized transducing phages carrying the same set of mutations can be employed, and the same selection for reversion is possible.

Based on tests with known carcinogens, the Ames mutagenicity testing system shows excellent predictive capabilities. An extensive body of literature exists demonstrating that the *Salmonella* system is about 84% sensitive (finds 84% of known mutagens), and the *E. coli* system is about 91% sensitive. The speed, simplicity, and low cost of the bioassays for mutagenicity have made them a worldwide standard for preliminary experimentation.

Bacteria as a Source of Biochemical Products

Bacteria have traditionally played a role in the biochemical industry. When the entire spectrum of bacterial metabolic activities is considered, a surprising number and variety of commercially important compounds can be produced, especially during the process of anaerobic growth. Such diverse chemicals as ethanol, acetone, and glycerol have a long history of microbial production, but many other compounds can also be synthesized. At present most of these compounds are produced by purely chemical processes; but as energy costs continue to increase, bacterial fermentations can be expected to become cost-competitive in the production of many chemicals. A primary role in the development of biologic technologies will be played by bacterial geneticists working with various fermentation systems to improve the yield and efficiency of the processes.

One example that illustrates the type of research being done is the attempt to produce the amino acid tryptophan on a large scale. This amino acid is an essential human nutrient and frequently is used as a dietary supplement, as it is not prevalent in plant proteins. In bacteria it is synthesized as one of a group of aromatic amino acids that share certain biochemical intermediates. Specifically, one set of enzymes produces the intermediate chorismic acid, which can then be converted, along different pathways, to phenylalanine, tryptophan, and tyrosine. As discussed in Chapter 12, the enzymes specific to tryptophan biosynthesis are coordinately regulated. In addition the tryptophan and chorismic acid pathways are subject to **feedback inhibition,** the inactivation of the first enzyme in a series of reactions by the endproduct of the series.

Pittard and co-workers developed several strains of *E. coli* that produce as much as 80 mg of tryptophan for every gram of cells (dry weight) for each hour spent in the stationary growth phase. In order to do it, transduction was used to remodel the control systems for tryptophan (Figure 14-5). They began by introducing additional mutations, which eliminated the feedback inhibition control of the first enzymes in the chorismic acid and tryptophan pathways, into a strain that already synthesized tryptophan constitutively. This modification permitted tryptophan to accumulate to high levels without slowing down its own synthesis. The next step was

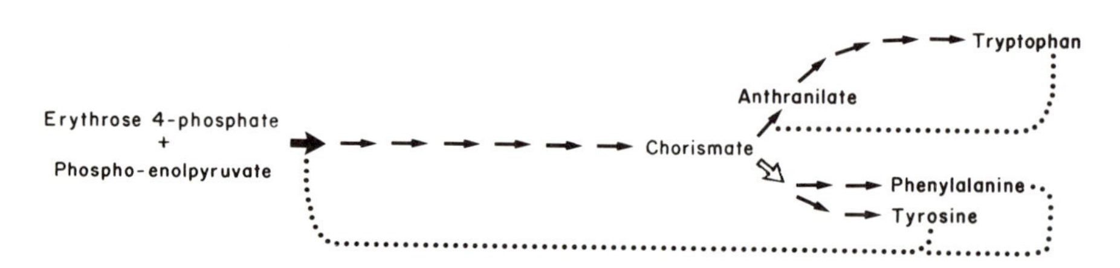

Figure 14-5. Regulatory modifications of the tryptophan operon used by Pittard and co-workers. The biosynthetic pathways used to produce all of the aromatic amino acids are outlined. Each arrow represents an individual enzymatic step. The larger arrows represent steps that are catalyzed by alternative enzymes, each of which is independently regulated. The open arrow represents the step blocked to prevent synthesis of phenylalanine and tyrosine. The dotted lines represent feedback inhibitory pathways that have been blocked by mutation. The result of all of the genetic changes is an organism that requires phenylalanine and tyrosine for growth and that produces large quantities of tryptophan.

to block the enzymatic pathways leading to phenylalanine and tyrosine. If this step were not done, some of the increased amount of chorismic acid would go to produce unwanted aromatic amino acids. The production ability of the mutated cells is so much greater than their biosynthetic needs that, on a molar basis, more tryptophan is recovered from the culture than the amount of phenylalanine and tyrosine added to the medium to permit the culture to grow. The last major alteration was to eliminate the action of the attenuator sequence so that all mRNA molecules that were initiated would be completed. The end result was a strain that uncontrollably synthesized a particular product, in this case tryptophan.

Numerous other fermentative processes have been or are being subjected to similar genetic modifications. The general goal, increased yield per cell, remains the same but other considerations may also enter into the calculations. For the specific case of antibiotic-producing bacterial strains, for example, mutations are sometimes introduced that result in the synthesis of chemically modified antibiotics. The modifications may make the antibiotic more potent or less susceptible to inactivation by R plasmid enzymes. Sometimes it is necessary to transfer the DNA coding for antibiotic production into another bacterium in order to achieve more efficient production. For these reasons, the study of the genetics of the actinomycetes, the principal group of antibiotic producers, has burgeoned.

Bacteria as Degradative Organisms

As the pollution problems of modern society increase, so does interest in bacteria as organisms that promote decomposition. Bacteria always have been the final link in the decomposition chain, but their practical use has been limited to sewage treatment plants and composters. Now, however, intensive research is being carried out to produce bacterial strains capable of degrading particularly troublesome or toxic compounds. Two primary areas of interest are the degradation of crude oil spilled in the environment and the degradation of chlorinated hydrocarbons such as DDT or polychlorinated biphenyls that have been released into the soil or water.

Many of the hydrocarbons normally are stable in nature, but Chakrabarty and co-workers have been able to show that certain strains of *Pseudomonas* can attack and degrade specific hydrocarbons. Moreover, the DNA coding for this ability frequently is located on a plasmid. By producing various combinations of plasmids and by judicious selection of mutants, they have produced strains of *Pseudomonas* that specifically degrade xylenes, toluene, or trimethylbenzenes.

The usual methods for "directing" evolution of bacteria to utilize specific compounds involve gradual introduction of a new substrate into the culture coupled with the gradual removal of normal carbon sources. In one well studied case from *E. coli,* a deletion of the cistron coding for the enzyme β-galactosidase was compensated by the gradual accumulation of mutations in the *ebg* cistron (evolved β-galactosidase) in order to provide a similar function. Five cycles of selective growth were required for the acquisition of optimal activity, including an initial incubation of 25 to 40 days. This type of research is thus a rather lengthy undertaking.

The search for new mutants may be made easier by a device invented by Glaser and co-workers. Their machine, called Cyclops, automatically inoculates plates in a precise array with droplets of liquid containing single bacteria (actually a Poisson distribution), perhaps 80 drops per plate. The plates are then incubated under appropriate conditions and photographed automatically at intervals. A computer can be used to scan the pictures and analyze each colony for growth or lack of it. As many as 100,000 bacteria can be tested per run, which makes possible a kind of brute-force screening never before available to the bacterial geneticist. In theory, all mutants for which a phenotype can be predicted should be capable of isolation, even if no direct selection is possible. Any mutational event occurring at a frequency of 10^{-6} or more should be observed easily in just a few runs of the Glaser machine.

The ultimate goal of all the research of this type is to produce bacteria specific for the degradation of certain compounds. In terms of the hydrocarbon examples presented above, the bacteria could be sprayed on any water or soil contaminated by spilled material. The specialized bacteria would then convert the hydrocarbon into forms suitable for use by the normal bacterial flora of the area. If the special strains were properly constructed, competition from normal bacteria eventually would eliminate them from the environment unless further spills occurred. The commercial potential for such special strains is evident, and considerable progress in this area should be announced now that the U.S. Supreme Court has ruled that it is possible to patent an entire bacterium. Corporations have been understandably reluctant to offer specialized bacteria for sale until they have the means available to prevent unauthorized propagation. Questions about patenting other organisms or individual genes remain unresolved.

Basic Research

Many active research areas have not progressed to the point where there is a direct commercial application or are not truly relevant to product-oriented companies. Nevertheless, the work being done is exciting and important. A few selected areas are discussed below.

Molecular Evolution

Ever since Darwin first postulated the existence of an evolutionary process, geneticists have been looking for information about how the process might work and how the biologic world has arrived at its present state.

Molecular genetics offers some particularly tantalizing pieces of information that bear on this puzzle.

The most obvious area in which new information is available is that of DNA (or RNA) sequences. Every day additional segments of DNA from all kinds of organisms are sequenced. The National Institutes of Health and the European Molecular Biology Organization maintain data banks of sequence information that are accessible by microcomputer. It is therefore possible to examine various genomes for patterns or to compare one kind of organism with another. For example, Blake and Earley have shown that for the nearly 300,000 bp of *E. coli* that have been sequenced, about 51% of the bases are G or C. However, within the coding regions the %(G + C) increases to 53, whereas in the noncoding regions the %(G + C) decreases to 46. It can also be shown that the use of individual codons is not random but is related to the relative abundance of the individual isoaccepting (carrying the same amino acid) tRNA species in nearly all cases.

One common hypothesis to explain how new genetic functions are acquired (originally proposed for tRNA by Jukes and more recently summarized by Grivell) is to assume that cells began with relatively few tRNA coding regions. A process of gene duplication (perhaps mediated by transposons) would lead to tRNA coding regions that could be mutated without harm to the cell because the original sequence was still present. An existing codon might mutate to a form better recognized by a new tRNA molecule. However, unless the number of copies of the new tRNA molecule kept pace with the use of the new codon, genes employing that codon would be at a relative disadvantage. Therefore it is not surprising that rarely synthesized proteins often contain relatively uncommon codons. In support of this theory is the observation that there are striking sequence homologies between certain families of tRNA molecules.

Homology between organisms can also be determined by molecular comparisons. This type of comparison is often used as a measure of the relatedness of two organisms in a new branch of science called molecular taxonomy. Woese and his collaborators have constructed elaborate phylogenetic trees such as the one shown in Figure 14-6 based on similarities in 5S and 16S ribosomal RNAs. A comparison of amino acid and corresponding DNA base sequences for homologous proteins can be particularly fruitful because it permits the investigator to identify silent mutations (same amino acid) in regions that are critical to function for that particular protein.

Hall and his collaborators have continued their investigations into genome evolution by examining the ability of *E. coli* to mutate so as to use β-glucosides such as cellobiose or salicin for growth. They reported that nearly all natural isolates of *E. coli* can spontaneously mutate so as to be able to use one or more β-glucosides. This observation means that nearly all cells carry cryptic or inactivated genes for β-glucoside metabolism. The distinction between the two types of silent gene is that a cryptic gene

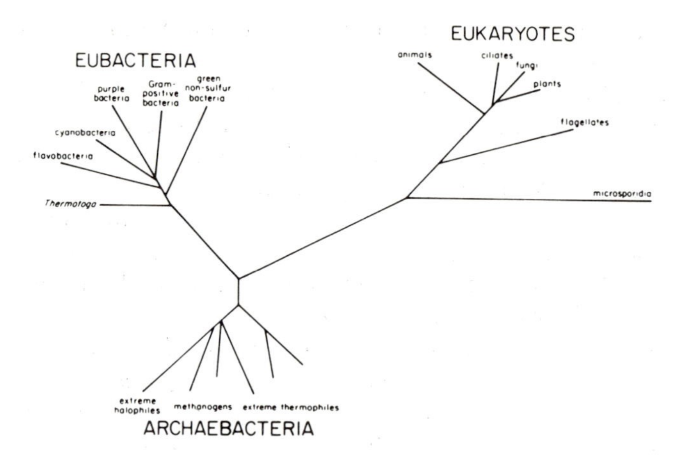

Figure 14-6. Universal phylogenetic tree determined from rRNA sequence comparisons showing the three major kingdoms of organisms (urkingdoms). A matrix of evolutionary distances was calculated from an alignment of representative 16S rRNA sequences from each of the three urkingdoms. It was used to construct a distance tree based on those positions represented in all sequences in the alignment in homologous secondary structural elements. Line lengths of the tree are proportional to calculated distances. Branching order within each kingdom is correct to a first approximation only. From Woese, C. (1987). Bacterial evolution. Microbiological Reviews 51:221–271.

is capable of function but is turned off by a regulator whereas an inactivated gene has a mutation within the coding sequence that prevents any functional product from being made. The question, of course, is why the bacteria maintain these apparently useless regions of DNA. One possible answer is that, according to the model of Jukes, such regions are the very fabric of evolution. Here mutations can occur that have no immediate impact on the cell. Another possibility is that there is some selective disadvantage, e.g., detrimental waste products, to a cell that expresses β-glucoside functions. In that event there is a definite selection for converting the region to silent genes. On the other hand, when selection is applied for β-glucoside utilization, it is comparatively easy to mutate the DNA so it will once again be functional.

Nitrogen Fixation

In terms of energy expenditure, the process of nitrogen fixation (conversion of N_2 to NH_3) is costly, whether performed in a chemistry laboratory by the Haber process or in a bacterial cell such as *Rhizobium* or *Azotobacter*. In terms of its cost to society, however, biologic nitrogen fixation is much to be preferred over industrial nitrogen fixation driven by fossil fuels. Therefore there is considerable interest in developing practical methods for fertilizing commercial crops by natural means.

The principal focus is on adapting one or more of the bacterial nitrogen-fixation systems so it can function as a free-standing system in a plant. Theoretically it should be possible to clone the appropriate genes into the middle of the T-DNA of the *Agrobacterium* plasmid and deliver them directly into a plant chromosome. The practical reality is that the nitrogen-fixation mechanism requires a large number of interrelated functions that have not been fully characterized (see Chapter 12). Although energy analyses indicate that plants can probably provide the necessary input for the process, it seems likely that genetically engineered nitrogen fixation is still years away.

Summary

The applications of bacterial genetics are many and varied, and they lean heavily on the great diversity within the bacterial kingdom. Evolutionary studies are particularly well suited to bacteriology, as the organisms are often easy to grow and many are already well characterized. The model that has evolved postulates that genetic rearrangements, primarily gene duplications, allow a cell the freedom to mutate new functions. Those duplications that have been fixed within the population presumably represent more or less successful adaptations.

Genetic engineering still holds the center stage with respect to the major achievements of bacterial genetics. Many useful proteins have been prepared by cloning the appropriate DNA into *E. coli* or some other bacterium. Although the general public is most familiar with human growth hormone, insulin, and interferon, scientists can point to numerous enzymes and other proteins whose price has been greatly reduced by successful cloning. Numerous specialized cloning vectors have been prepared to facilitate these enterprises along with techniques that allow the preparation of cDNA from purified mRNA.

There are important safety concerns about genetically engineered bacteria. Those concerning laboratory safety have been well studied and suitable precautions taken. At the present time, however, the pressing issue is the extent to which it is appropriate to release organisms of this type into the environment. Careful ecologic studies are required to prevent accidental damage to the ecosystem.

References

General

Brownlee, A.G. (1986). How do bacteria acquire plant genes? Nature 323:585.

Geider, K. (1986). DNA cloning vectors utilizing replication functions of the filamentous phages of *Escherichia coli*. Journal of General Virology 67:2287-2303.

Grivell, L.A. (1986). Molecular evolution: deciphering divergent codes. Nature 324:109-110.

Hall, B.G. (1983). Evolution of new metabolic functions in laboratory organisms, pp. 234-257. In: Nei, M., Koehn, R.K. (eds.) Evolution of Genes and Proteins. Sunderland, Massachusetts: Sinauer Associates.

Hartl, D.L. (1986). Evolution and tinkering: the molecular genetics of bacterial adaptation, pp. 161-175. In: Gershowitz, H., Rucknagel, D.L., Tashian, R.E. (eds.) Evolutionary Perspectives and the New Genetics. New York: Liss.

Hutt, P.B. (1985). Existing regulatory authority to control the processes and products of biotechnology, pp. 3-13. In: Fields, B., Martin, M.A., Kamely, D. (eds.) Genetically Altered Viruses and the Environment. Cold Spring Harbor, N.Y.: Cold Spring Harbor Laboratory.

Jukes, T.H. (1983). Evolution of the amino acid code, pp. 191-207. In: Nei, M., Koehn, R.K. (eds.) Evolution of Genes and Proteins. Sunderland, Massachusetts: Sinauer Associates.

Pouwels, P.H., Enger-Valk, B.E., Brammer, W.J. (1985). Cloning Vectors. Amsterdam: Elsevier.

Silver, S. (ed.) (1985). Biotechnology: Potentials and Limitations. Berlin: Springer-Verlag.

Trevors, J.T., Barkay, T., Bourquin, A.W. (1987). Gene transfer among bacteria in soil and aquatic environments: a review. Canadian Journal of Microbiology 33:191–198.

Specific

Abrahamsén, L., Moks, T., Nilsson, B., Uhlén, M. (1986). Secretion of heterologous gene products to the culture medium of *Escherichia coli*. Nucleic Acids Research 14:7487–7500.

Blake, R.D., Earley, S. (1986). Distribution and evolution of sequence characteristics in the *E. coli* genome. Journal of Biomolecular Structure and Dynamics 4:291–307.

Carlson, T.A., Chelm, B.K. (1986). Apparent eukaryotic origin of glutamine synthase II from the bacterium *Bradyrhizobium japonicum*. Nature 322:568–570.

Devanas, M.A., Stotzky, G. (1986). Fate in soil of a recombinant plasmid carrying a *Drosophila* gene. Current Microbiology 13:279–283.

Gray, G.L., Baldridge, J.S., McKeown, K.S., Heyneker, H.L., Chang, C.N. (1986). Periplasmic production of correctly processed human growth hormone in *Escherichia coli:* natural and bacterial signal sequences are interchangeable. Gene 39:247–254.

Hall, B.G., Betts, P.W. (1987). Cryptic genes for cellobiose utilization in natural isolates of *Escherichia coli*. Genetics 115:431–439.

Mermod, N., Ramos, J.L., Lehrbach, P.R., Timmis, K.N. (1986). Vector for regulated expression of cloned genes in a wide range of gram-negative bacteria. Journal of Bacteriology 167:447–454.

Peeters, B.P.H., Schoenmakers, J.G.G., Konings, R.N.H. (1986). Plasmid pKUN9, a versatile vector for the selective packaging of both DNA strands into single-stranded DNA-containing phage-like particles. Gene 41:39–46.

Saye, D.J., Ogunseitan, O., Sayler, G.S., Miller, R.V. (1987). Potential for transduction of plasmids in a natural freshwater environment: effect of plasmid donor concentration and a natural microbial community on transduction in *Pseudomonas aeruginosa*. Canadian Journal of Microbiology 33:191–198.

Schoner, B.E., Belagaje, R.M., Schoner, R.G. (1986). Translation of a synthetic two-cistron mRNA in *Escherichia coli*. Proceedings of the National Academy of Sciences of the United States of America 83:8506–8510.

Tabor, S., Richardson, C.C. (1985). A bacteriophage T7 RNA polymerase/promoter system for controlled exclusive expression of specific genes. Proceedings of the National Academy of Sciences of the United States of America 82:1074–1078.

Walter, M.V., Porteous, A., Seidler, R.J. (1987). Measuring genetic stability in bacteria of potential use in genetic engineering. Applied and Environmental Microbiology 53:105–109.

Appendix

Laws of Probability and Their Application to Prokaryote Cultures

Chapter 2 includes discussions of some of the problems associated with the analysis of prokaryotic genetic systems, and Chapters 3 and 4 include some additional mathematic analysis. Another theoretical problem that must be considered is that of sampling. As noted earlier, it is not generally possible to recover all of the progeny from a cross because of the necessity of using some sort of selective technique to find a few recombinant individuals among many parental types. Therefore when designing prokaryotic genetic experiments, it is essential to be certain that the sample is representative of the entire population of organisms and that it is of a suitable size to compensate for the random variation observed in any physical procedure. The former concern can be alleviated by good culture agitation techniques, which provide a homogeneous population from which to sample. The latter concern is the subject of this appendix.

Definition of Probability

The concept of probability is, in many respects, an intuitive one. We frequently speak of "the chance" that a particular event will take place. However, what is really being discussed is the probability of the occurrence of the event. Mathematically this probability is expressed either as a proper fraction or a decimal fraction between 0 and 1, although in casual usage it is frequently converted to the corresponding percentage. The numbers 0 and 1 have special meanings, with 0 referring to an event that never can occur and 1 referring to an absolute certainty.

Although other systems are possible, for the purposes of this book all probabilistic events can be characterized in one of two ways, either successes (s) or failures (f). Note that there may be many ways to succeed or fail in a particular system.

Example 1: If success means throwing a die (a cube on which each face bears a unique number from one through six) and having an even number turn up, then there are three ways to succeed and three ways to fail. If all of the numbers on the die are equally likely to turn up (i.e., the die is not "loaded"), we say that all numbers are equally probable and the chance (probability) of success is 50%, or 0.5.

Each time that the die is thrown consitutes a **trial** of the system. According to the analysis in Example 1, on each trial there is a 50% chance of success; yet, as is demonstrated later, a series of five trials, all of which produce even numbers, can be expected to occur 3.1% of the time. Therefore when considering the concept of probability it is necessary to keep in mind that the calculated value for an outcome is really just the proportion of time a particular outcome is expected to occur. The difference between the calculated proportion and unity is the proportion of time an outcome can be expected not to occur. However, expectation does not equate with reality, as anyone who reads the weather forecast knows. The random fluctuations inherent in the physical world can be expected to affect any probabilistic event so that a calculated probability is valid only when a large number of trials is involved. This concept is expressed in the rigorous definition of **probability** by saying that, as the number of trials (t) approaches infinity, the probability of success (p) is the limit of the ratio of the number of ways to succeed (s) divided by the number of ways to succeed plus the number of ways to fail (f), or

$$p = \lim_{t \to \infty} \frac{s}{s + f} \qquad \text{[A-1]}$$

assuming all outcomes occur with the same frequency. In the case of the die above, $s = f$ because all numbers are equally probable; so $p = s/(s + f) = s/(s + s) = s/2s = 0.50 = 50\%$, as previously noted.

In order to develop some feeling for the kinds of data that led to the development of Eq. A-1, look at Table A-1 in which are presented the results from a series of coin tosses. By assuming that a coin can never land on its edge and that a head is equal to a success, these data can be made to fit the same model discussed above. Several points are worth emphasizing. Although it seems intuitively obvious that heads are as likely to occur as tails, the observed proportion is only 46/100 instead of 50/100. This difference between observed and predicted values is not unusual, however, because only 100 trials were involved here instead of a substantially larger number. Note also that, if only subsets of the data are considered, the results can be even more skewed: in the seventh column

Table A-1. Distribution of heads and tails of a coin flipped 100 separate times

T	T	T	T	T	T	H	T	H	T
T	T	H	H	T	H	T	H	T	T
T	T	H	T	H	H	T	H	T	H
T	H	T	H	H	H	H	T	H	T
H	H	H	H	H	T	T	T	H	T
T	H	H	T	H	T	T	H	H	H
H	T	H	T	T	H	T	H	H	H
T	T	T	H	T	T	T	T	T	T
H	H	T	T	T	H	T	T	H	H
H	H	H	T	T	H	T	T	H	T

T means that a tail was observed, and an *H* means that a head was observed. In total there were 46 heads and 54 tails.

there are 80% tails, whereas in the ninth column there are 70% heads. In the short run, wide variations in frequency are possible. Nevertheless, these frequency variations do not change the overall limit function of Eq. A-1, and even after a run of seven consecutive tails the probability of obtaining a head on the next coin toss is still only 50%.

Dependent Versus Independent Events

Thus far the examples used to illustrate the calculations of probability have been chosen for their intuitive clarity. In the real world other cases are far more prevalent. Frequently it is necessary to deal with the results of a series of samples (trials) and to try to infer the nature of the entire population from the composition of this rather limited number of samples.

When dealing with a case involving multiple samples, it is important to distinguish between independent events such as coin tosses or dice throws and dependent events. For a dependent event, the probability of a successful outcome on a subsequent trial is influenced by the results of the preceding trial(s). The same is not true for an independent event. In order to illustrate this point, consider a paper bag containing 20 red marbles and 5 blue ones. It is clear that the probability of drawing out a red marble from the bag is 20/(20 + 5), or 80%. If, before the second trial, the first marble is returned to the bag and the bag shaken well, the process is called sampling with replacement. In such a case, the events are independent and the probability of success is constant at each trial. However, if the first marble is not returned to the bag, it is sampling without replacement, and the events are dependent, because the probability of drawing out a red marble on the second trial is 19/(19 + 5) = 79.2% if a red marble was chosen first but 20/(20 + 4) = 83.3% if a blue marble was chosen first.

It is also necessary to distinguish between outcomes that are mutually exclusive and those that are not. The usual example of a mutually exclusive outcome is the result of drawing a single card from a complete deck of cards. The card cannot be both a seven and an eight, for example, yet it may be that drawing either a seven or an eight would be considered a success. Such a case forces the reconsideration of the method for calculating the probability of a particular outcome for the case of multiple events.

Example 1 reconsidered: Suppose that a single die is to be rolled, and the result is considered a success if the number that appears on top is even. The probability of rolling a two is $1/(1 + 5) = 1/6$. Similarly, the probability of rolling a four is 1/6 and that of rolling a six is also 1/6. The outcomes are mutually exclusive, but any one is acceptable. This proposition is expressed mathematically by saying that the overall probability of success is the sum of the individual probabilities of each mutually exclusive successful outcome, or $1/6 + 1/6 + 1/6 = 1/2$, which is the same result as obtained above.

Example 2: Suppose that the problem of Example 1 is reworded so that, instead of asking for any even number on one role of the die, we want each of the even numbers in turn as the die is rolled three times (first a two, then a four, then a six). The probability of getting a two on the first roll is 1/6, the probability of getting a four on the second roll is 1/6, and the probability of getting a six on the third roll is also 1/6. Each probability is independent of the others, and the outcomes are not mutually exclusive. Therefore one-sixth of the time when the die is rolled, the number that appears will be a two. In the case where a two does appear, only one-sixth of the time will the next number be a four. Therefore the probability of a two and then a four is $1/6 \times 1/6 = 1/36$. Moreover, if we now ask for a six, the probability becomes $1/6 \times 1/6 \times 1/6 = 1/216$. Note that the same result is obtained for the cases where the numbers appear in a different order or where the same preselected number appears three consecutive times.

Example 3: Suppose that the conditions are the same as in Example 1 except that you wish to know the probability of rolling an even number five consecutive times. By the reasoning given in Examples 1 and 2, the probability of rolling an even number is 1/2, so the probability of rolling two consecutive even numbers is $1/2 \times 1/2 = 1/4$, the probability of rolling three consecutive even numbers is $1/2 \times 1/2 \times 1/2 = 1/8$, and the probability of rolling five consecutive even numbers is $1/2 \times 1/2 \times 1/2 \times 1/2 \times 1/2 = 1/32$.

In summary, for the case of multiple trials, the overall probability is the product of the probabilities of a successful outcome at each individual trial. For the case in which the successful outcomes for a single trial are mutually exclusive, the probability of success in that individual trial is the sum of the probabilities for each possible successful outcome.

Application of Binomial Expansion to Probability Theory

Permutation and Combination

Up to this point in the discussion, all of the examples considered have had an easily countable number of outcomes so that the ratio $s/(s + f)$ has had a readily determined numerical value, and the notion of the limit has not been necessary. For the type of probability calculation involved in bacterial genetics, this situation generally does not pertain. Instead, the population of cells, viruses, etc. is so large that exact calculations become more difficult, if not impossible. Nevertheless, it is possible to speak rigorously about certain types of probability by introducing the concept of permutation and combination.

The terms permutation and combination refer to samples taken from a population composed of individually identifiable members. For example, if a bag contains a set of marbles, each of which bears a unique number, there are conceptually two ways to sample the population. One might begin by removing marbles one at a time from the bag and placing them in a row. Suppose that, first, a group of five marbles is removed from the bag and their positions in the line noted. It is clear that, if the marbles were returned to the bag and then again removed one marble at a time, each individual marble might be removed in a different sequence. The sequence of numbers in the first sample might be 12345, and the sequence in the second sample might be 54321. The composition of the two groups is identical, but the sequence in which they were obtained is different. Therefore the two groups represent different permutations of the same combination of items.

When the term **permutation** is used, it refers not only to the composition of a sample but also to the ordering of the items within a sample. On the other hand, the term **combination** refers only to the overall composition of a sample without regard to any sort of internal order within the sample.

If the total size (N) of the population to be sampled is known and the size of the sample (n) is specified, it is possible to calculate precisely how many permutations and combinations can exist. The number of permutations (P) can be calculated from the formula

$$P_{N,n} = \frac{N!}{(N - n)!}$$

$$= \frac{N(N-1)(N-2)\cdots(N-n+1)(N-n)(N-n-1)\cdots(3)(2)(1)}{(N-n)(N-n-1)\cdots(3)(2)(1)}$$

$$= N(N-1)(N-2)\cdots(N-n+2)(N-n+1) \qquad \text{[A-2]}$$

where the sign ! means factorial. The **factorial** of a number is the product of the specified number and each integer less than itself down to and including 1. Thus, $3! = 3 \times 2 \times 1$. The third form of the equation, which is more cumbersome to write but easier to calculate, is obtained by dividing the numerator and denominator by the quantity $(N - n)!$. The number of possible combinations (C) that can occur is, of course, less than the number of permutations and is calculated from the formula

$$C_{N,n} = \frac{N!}{n!(N - n)!} = \frac{N(N-1)(N-2)\cdots(N-n+2)(N-n+1)}{n!} \qquad \text{[A-3]}$$

Example 4: Consider the case of a bag of red and blue marbles, each of which has a unique number on it. There are ten red marbles and five blue marbles, and a sample of three marbles is to be chosen at random. The number of different permutations is

$$P_{N,n} = P_{15,3} = \frac{15!}{12!} = 15 \times 14 \times 13 = 2730$$

whereas the number of different combinations is only

$$C_{N,n} = C_{15,3} = \frac{15!}{3!12!} = \frac{15 \times 14 \times 13}{3 \times 2 \times 1} = 455.$$

Remember that a sample consisting of marbles number 15, 1, and 7 (in that order) is a different permutation from 7, 15, and 1 but represents the same combination. Roughly 26% of the time all of the marbles will be red ($10/15 \times 9/14 \times 8/13 = 0.264$), but it is still possible to distinguish among the marbles because of their numbers.

Binomial Expansion

When endeavoring to deal with a large population that is to be sampled, the theory can be greatly simplified if there are only two possible outcomes, s and f, as above. Each sample can then be categorized as having a particular number of successes and a particular number of failures. In the case of the bag of red and blue marbles, for example, red might be considered a success. Then, if a sample were removed and found to contain only one red marble and two blue marbles, it would have one success and two failures. Such a distribution presumably reflects the actual proportion of red to blue marbles in the entire population. However, if the population is very large and the sample size relatively small, it is difficult to evaluate the significance of the distribution in a single sample due to random fluctuations. In order to obtain more information about the entire population, additional samples would be necessary. One way to determine the ratio of red to blue marbles would be to take a large number of samples, sum the total number of red and blue marbles, and assume that their ratio is the ratio in the larger population. This method can be accurate if enough samples are taken, but it wastes some of the information available from the sampling procedure. A better and more economic way to utilize the information from the samples is to look not only at the red/blue ratio within a sample but also at the frequency with which that sample class appears. The most frequently appearing sample class is presumably most reflective of the general population.

In order to develop a mathematic basis for the procedure just outlined, it is necessary to reverse the problem. Suppose that the population of marbles in the bag is known to have a certain proportion of red marbles (a) and a certain proportion of blue marbles (b). These proportions are determined from Eq. A-1 (e.g., a = number of red marbles/number of red + number of blue marbles). Then, of necessity, $a + b = 1$, as no other color of marbles is possible. Furthermore, assume that the number of marbles involved is so large that removing a few samples does not significantly change a or b. Then, if a sample of five marbles is removed as before, the probability of obtaining a particular arrangement of four red marbles and one blue marble is $a \times a \times a \times a \times b$ or a^4b because the outcomes are not mutually exclusive. The probabilities for other arrangements can be calculated in a similar manner. For instance, the probability would be a^2b^3 for a sample consisting of two red and three blue marbles. However, these probabilities are actually the probabilities only for that particular permutation of marbles, whereas in fact we are interested in the probability of various combinations.

The expression can be corrected to reflect the number of possible combinations by noting that the probability for a particular sample of size n is always equal to $a^r b^{(n-r)}$, where r is the number of successes in the sample. Equation A-3, however, shows that the number of combinations of N things taken n at a time is equal to $C_{N,n}$. Therefore the actual probability of a sample that has a specific amount of success (r) is $C_{N,n}[a^r b^{(n-r)}]$ (where r may range in value from 0 up to n). The term $C_{N,n}[a^r b^{(n-r)}]$ is merely one term of the binomial expansion $(a + b)^n$, however, and therefore the complete probability listing for all possible samples is $(a + b)^n$, which is the binomial expansion.

The binomial expansion thus represents a mathematically exact way of presenting the probabilities of obtaining various samples from a given population. Samples distributed in such a way that they can be described by the binomial expansion are said to be binomially distributed. The binomial distribution has several important advantages. The mean (average) amount of success obtained in a series of trials using a particular population is

$$m = np \qquad \text{[A-4]}$$

where p is the probability of success in a single trial (removal of a single item) and n is the total number of trials made. Sometimes the term m is referred to as the **expectation,** as np represents the expected number of successes in a sample of size n. Random variations result in actual samples with varying amounts of success distributed about the mean value for the amount of success in the population as a whole. The degree of scatter of the amount of success in samples from the same population is usually expressed as the standard deviation (σ). The smaller the numerical value for σ, the more homogeneous are the samples. For the binomial distribution, the standard deviation is easily calculated as

$$\sigma = \sqrt{npq} \qquad [A\text{-}5]$$

where n and p are defined as in Eq. A-4 and $q = 1 - p =$ the probability of failure in a single trial.

Poisson Approximation

Although the binomial distribution is mathematically precise, it is cumbersome to evaluate. Therefore numerous methods that approximate the binomial distribution under certain conditions have been developed. Two such methods are the **normal distribution** and the **Poisson distribution.** The normal distribution is familiar to every student as the famous "bell-shaped curve" so often used to assign grades to classes. Although its use is widespread, it is not as convenient for bacterial genetics as the Poisson distribution, as it functions best when the value for p is near 0.5, and bacterial genetics rarely studies events that are so frequent.

The Poisson approximation to the binomial distribution, however, was specifically developed to deal with rather rare events, cases in which p is considerably less than 0.5. It involves placing limitations on m as defined in Eq. A-4. If n is very large and p is very small, m becomes a number of modest size, on the order of 0.1 to 5.0. Under these conditions, the probability of exactly r successes in a sample of size n is

$$P = \frac{e^{-m} m^r}{r!}. \qquad [A\text{-}6]$$

The number e is an irrational transcendental number, chosen for philosophical reasons that need not concern us, which is the base of the Napierian or natural system of logarithms. Its approximate value is 2.7182828 . . . . Logarithms based on the Napierian system are usually abbreviated ln to distinguish them from logarithms based on the number 10, which are abbreviated log. Some selected values of e^{-m} are given in Table A-2. In addition to the advantages of the binomial distribution, the Poisson approximation is unique because the standard deviation is equal to the square root of the mean, because as p becomes very small q approaches a value of 1. Therefore $\sqrt{npq}$ tends to be approximated by $\sqrt{np}$ for small p, and from Eq. A-4 it is the same as $\sqrt{m}$.

The Poisson approximation is particularly advantageous for bacterial geneticists because in most cases the researcher is dealing with a very large population of organisms and is looking for a rare event by means of selection. An experimental design can be considerably aided by mathematic analysis.

Example 5: A researcher wishes to study the progeny produced by a single phage-infected cell. He begins with 10^8 bacteria and adds to them 10^6 phage particles. If he removes a sample of 100 bacteria from the culture, what is the probability the sample will have no infected cells? One infected cell? Two infected cells? Three or more infected cells?

All that is necessary for the application of Eq. A-6 is the knowledge of m and r. Assume that success is equated with the recovery of a phage-infected cell. By Eq. A-4, $m = np$, and from the statement of the problem we know that $n = 100$ bacteria and $p = 10^6/10^8 = 0.01$ phage/bacterium. Therefore $m = 1$ phage particle per sample. The amount of success is represented by r; and according to the statement of the problem, taking a success to be a phage-infected cell, r is equal to 0, 1, 2, and 3 or more. Using the formula for the case in which $r = 0$, we find

$$P = \frac{e^{-m}m^r}{r!} = \frac{e^{-1} \cdot 1^0}{0!}$$

which is not difficult to evaluate if one can determine a value for 0!. By convention, both 1! and 0! are taken as being equal to 1. Because any number raised to the zero power is equal to 1, for $r = 0$, $P = e^{-1} = 0.367$ using the value of e^{-m} given in Table A-2. By similar reasoning, for $r = 1$,

$$P = \frac{e^{-1} \cdot 1^1}{1!} = e^{-1} = 0.367$$

and for $r = 2$,

$$P = \frac{e^{-1} \cdot 1^2}{2!} = \frac{e^{-1}}{2} = 0.184.$$

Each of the probabilities thus far calculated represents the probability of an exact number of phage-infected cells (successes) per sample. The last part of the problem asks for the case of three or more phage-infected cells. In almost all cases this type of question is best answered by calculating all of the probabilities for the cases that are not included and then subtracting them from unity. For this specific case, the probability of three or more infected cells is equal to $1 - P_0 - P_1 - P_2 = 1 - 0.367 - 0.367 - 0.184 = 0.082$.

Therefore 73.4% of the time the researcher may expect to find either zero or one phage-infected cell in the sample. If this figure is not frequent

Table A-2. Values of e^{-m} for use in Equation A-6[a]

m	e^{-m}	m	e^{-m}	m	e^{-m}	m	e^{-m}	m	e^{-m}
0.1	0.905	0.8	0.449	1.5	0.223	2.4	0.091	3.8	0.022
0.2	0.819	0.9	0.407	1.6	0.202	2.6	0.074	4.0	0.018
0.3	0.741	1.0	0.368	1.7	0.183	2.8	0.061	4.2	0.015
0.4	0.670	1.1	0.333	1.8	0.165	3.0	0.050	4.4	0.012
0.5	0.607	1.2	0.301	1.9	0.150	3.2	0.041	4.6	0.010
0.6	0.549	1.3	0.273	2.0	0.135	3.4	0.033	4.8	0.008
0.7	0.497	1.4	0.247	2.2	0.111	3.6	0.027	5.0	0.007

[a] More extensive values can be found in standard reference works such as the *Handbook of Chemistry and Physics* published by the Chemical Rubber Publishing Company.

enough (i.e., if there are too many cases of multiply infected cells), by changing either n or p the probabilities can be altered to suit the experimental design. For example, if only 50 bacteria are taken per sample, $m = np = 50 \times 0.01 = 0.5$. Then

$$P_0 = \frac{e^{-0.5} \cdot 0.5^0}{0!} = 0.607$$

and

$$P_1 = \frac{e^{-0.5} \cdot 0.5^1}{1!} = 0.303$$

and the researcher has either zero or exactly one phage-infected cell 91% of the time.

When the sample size was 100, the probabilities for the $r = 0$ and $r = 1$ case were identical, but it was not the case when the sample size was 50. This situation is indicative of the way in which the Poisson distribution is skewed for small values of m. Figure A-1 shows the Poisson distribution for several values of m. Only when $m = 5$ does the distribution become symmetric. As m becomes smaller, the zero success case naturally predominates, and the width of the curve becomes smaller because the standard deviation is $\sqrt{m}$. By the time m reaches a value of 0.1, zero successes occur 90.5% of the time.

The problem of inferring distributions within the larger population from the composition and frequency of various samples has already been mentioned. This type of analysis can be done easily with Eq. A-6.

Example 6: The same phage researcher who was laboring in Example 5 has removed a series of samples from a culture. When the samples are tested, 74.1% of them have no phage-infected cells. What is the average number of phage-infected cells in the culture?

In this case it is necessary to solve Eq. A-6 for m. Although it can be

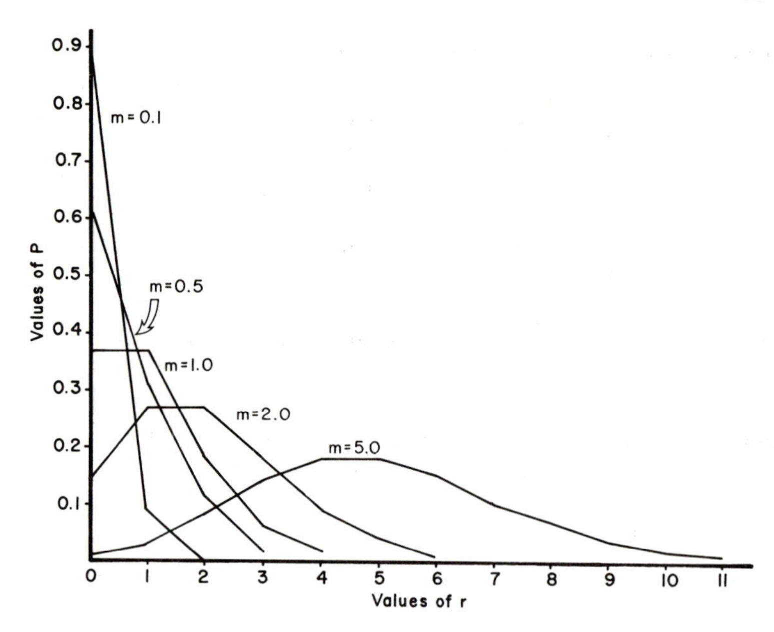

Figure A-1. Poisson distribution. Values of P have been calculated for selected values of m using Eq. A-6.

done for any value of r, the most convenient values are $r = 0$ or $r = 1$. Beginning with $P = e^{-m}m^r/r!$ and substituting $r = 0$, we have $P = e^{-m} \times 1/1$ so that $P = e^{-m}$ or $-m = \ln P$. However, $P = 0.741$, so $\ln P = -0.3 = -m$, so $m = 0.3$. Therefore, on the average, there is three-tenths of an infected cell per sample, or put another way, there are three infected cells per ten samples. Once m is known, it is of course possible to calculate either n or p using Eq. A-4, provided the other value is known.

Another occasional use of the Poisson distribution is to verify that a series of samples, in fact, does reflect a random distribution of success in the population as a whole. This type of analysis is discussed in Chapter 4 in connection with some experiments by Benzer.

Summary

Probability is the likelihood that a particular event will occur. For the case in which there are only two outcomes, successes and failures, probability can be rigorously defined as the

$$\lim_{n \to \infty} \frac{s}{s + f}$$

where n is the total number of trials, s is the number of ways to succeed, and f is the number of ways to fail.

The terms permutation and combination refer to samples that contain more than one item. If the order of the items within the sample is considered, the sample is a specific permutation. If the order of the items is not important, the sample is one type of combination. The number of possible permutations is $P_{N,n} = N!/(N - n)!$, whereas the number of possible combinations is $C_{N,n} = N!/n!(N - n)!$, where N is the number of items in the population and n is the number of items in the sample.

The frequency distribution of all possible combinations of successes and failures within samples of a certain size is given by the binomial expansion $(a + b)^n$, where a is the proportion of successes in the total population, b is the proportion of failures, and n is the sample size. Because this quantity is somewhat difficult to evaluate, an approximation method is generally used. The most useful approximation method for bacterial genetics is that of Poisson, which is $P = (e^{-m}m^r)/r!$, where r is the number of successes in the sample, e is the base of the natural logarithms, and $m = np$. The explicit assumptions of the approximation are that n, the sample size, is very large and p, the proportion of successes in the population, is very small, so that m, the expectation, is a number of modest size.

Problems

Answers to the odd-numbered problems are provided at the end of this section.

1. If the cat goes into the closet one time in five when the door is open, and you forget to check for the cat one time in four before you close the door, what is the probability that you will close the cat in the closet?
2. A child's toy consists of a cone on which five rings of different diameters can be stacked. There is only one way in which the rings can be stacked so that they will all fit on the cone at the same time. If the child chooses the rings randomly, what is the probability the toy will be assembled properly?
3. In a simplified version of roulette, there are 36 numbers on a wheel. A random selector mechanism chooses one number at each trial. If you must pay one dollar for each trial but win 30 dollars if the number on which you bet is selected, what is your expected dollar loss per trial? How many times should you expect to be able to play if you begin with a stake of exactly 60 dollars?
4. If you have a tube containing 20 bacteria and you add to it two phage particles that each infect a separate cell, what is the probability that any particular bacterial cell you select is infected by a phage? If the phages are added so that both could infect the same cell, what is the probability that the bacterial cell you select is infected by both phage

particles? Given that the bacterial cell you select is phage-infected, what is the probability that it is infected by both phage particles?

5. Using a standard deck of cards, what is the probability of drawing out one card and having it be either an ace or a king? What is the probability of drawing two cards and obtaining one king and one ace (assume that you replace the first card before drawing the second)? What is the probability of drawing two cards and obtaining first a king and then an ace (assume that you do not replace the first card)? What is the probability of drawing two cards without replacement and obtaining two kings?
6. A friend has a bag of candy containing 30 jelly beans of a kind you do like and 10 jelly beans of a kind you do not like. What is the probability that you will like the first two jelly beans you select (your friend will be offended if you put any back)? If you select four jelly beans from the bag, how many would you expect to like?
7. After a sudden flood, all the labels on the culture collection have washed off. You have a test tube rack containing 20 bacterial strains, and you know that 10 of the strains are donor cells and 10 are recipients. You have an experiment to do in which it is necessary to mix a donor with a recipient. What is the probability that the first strain you select will be a donor? What is the probability that you will select first a donor and then a recipient? What is the probability that two strains you randomly select will be a donor and a recipient?
8. Suppose that in a particular bacterial strain all mutations occur with a frequency of 10^{-6} and that there are three separately mutable sites on the bacterial genome that give rise to the same phenotype. What is the probability of observing a mutant phenotype? If you require a bacterium that is mutant for both the first phenotypic character discussed above and a second phenotypic character that is determined at a single separate locus, what is the probability of observing it?
9. If you have 15 billiard balls and you are going to choose a group of five, how many permutations could you get? How many combinations could you get?
10. The same flood as in problem 7 washed the labels off six reagent bottles. Three of these bottles are necessary for a particular enzyme assay. How many combinations of three bottles can you make from the six unlabeled bottles? What is the probability of selecting the correct three for your enzyme assay? Assuming that the enzyme assay requires the addition of the reagents in a specific order and assuming that you have correctly selected the three bottles, what is the probability that you will add the reagents in the correct order? What is the probability that you will both select the correct three bottles and add the reagents in the correct order?
11. A somewhat eccentric professor is a fashion plate with a collection of 200 shirts. Unfortunately, 20 of them are grease-stained, and he is

too befuddled to notice. Assuming he selects a shirt at random each morning and hangs it back in the closet each night, what is the expectation that the professor will wear a grease-stained shirt on any given day? What is the probability that he will not wear a stained shirt in 5 days? What is the probability that he will wear exactly one stained shirt in 5 days? What is the probability that he will wear two or more stained shirts in 5 days?

12. A masochistic bicyclist is taking a ride across the desert. If the probability that he will get a thorn in either of his tires is 0.1 for every 100 m he travels, what is the probability of a thornless journey of 100 m? of 500 m? of 1000 m?
13. A culture of bacteria has accidentally been contaminated with a second type of bacterium at a level such that, of every 10^8 bacteria removed from the culture there are 10^7 bacteria that are contaminants. If the culture is streaked on an agar plate to purify it, what is the probability the first colony examined will be the contaminant rather than the correct bacterium? What is the probability that a sample of ten colonies will not have any contaminants? What is the probability of finding at least one contaminating colony in a sample of ten colonies?
14. If you assume that one genetic exchange event will be observed for every 10^5 cells sampled, how many cells should be removed per sample in order for the probability of the sample's having at least one genetic exchange to be 50%? Suppose you wish to limit the frequency of multiple genetic exchanges (two or more) per sample to less than 10%. How should you adjust the size of the sample?
15. In the experiment described in problem 14, you observe that, when samples of 10^5 cells are taken, 67% of the samples show no genetic exchanges. Recalculate the expectation for a genetic exchange event.
16. Culture aliquots of 10^8 cells are tested to see if they contain any cells carrying a particular mutation. It is observed that 5% of the samples contain no mutant cells. What is the average number of mutant cells per sample? What is the frequency with which the mutant cells are observed in the culture?
17. A culture of bacteria is infected with phage at the ratio of 20 bacteria for every phage particle, and aliquots of 100 bacteria are taken. How many phage-infected cells would you expect to find in a sample (assume that all phages infect a cell)? What is the probability of having no phage-infected cells in the sample? What should the sample size be so that the probability of no phage-infected cells becomes 5%?
18. A series of samples has been taken from a bacterial culture, and it has been observed that 67% of the samples have no mutant cells in them and 27% of the samples have exactly one mutant cell. What percentage of the samples would you expect to have two mutant cells? Three mutant cells? At least four mutant cells?

Answers to Odd-Numbered Problems

1. $\frac{1}{5} \times \frac{1}{4} = \frac{1}{20} = 0.05$
3. $\$\frac{1}{6}$; 360 times
5. $\frac{4}{52} + \frac{4}{52} = \frac{2}{13} = 0.154$, $2(\frac{4}{52} \times \frac{4}{52}) = \frac{2}{169} = 0.012$; $\frac{4}{52} \times \frac{4}{51} = \frac{4}{663} = 0.0060$; $\frac{4}{52} \times \frac{3}{51} = \frac{1}{221} = 0.0045$
7. $\frac{10}{20} = 0.5$; $\frac{10}{20} \times \frac{10}{19} = \frac{5}{19} = 0.263$; $2(\frac{10}{20} \times \frac{10}{19}) = \frac{10}{19} = 0.526$
9. $P_{15,5} = \dfrac{15!}{10!} = 15 \times 14 \times 13 \times 12 \times 11 = 360{,}360;$

   $$C_{15,5} = \frac{15!}{5!10!} = \frac{360{,}360}{5 \times 4 \times 3 \times 2 \times 1} = 3003$$

11. $\frac{20}{200} = 0.1$; $P_{5,0} = \dfrac{e^{-0.5} \cdot 0.5^0}{0!} = e^{-0.5} = 0.607$ or $0.9 \times 0.9 \times 0.9 \times 0.9 \times 0.9 = 0.590$

    $P_{5,1} = \dfrac{e^{-0.5} \cdot 0.5^1}{1!} = 0.303$ or $5 \times 0.9^4 \times 0.1 = 0.328$;

    $1 - 0.607 - 0.303 = 0.090$ or $1 - 0.590 - 0.328 = .082$

13. $\dfrac{10^7}{10^8} = 0.1$; $P_{10,0} = \dfrac{e^{-1} \cdot 1^0}{0!} = 0.368$; $1 - 0.368 = 0.632$
15. $0.67 = \dfrac{e^{-m}m^0}{0!}$ or $e^{-m} = 0.67$; and $m = 0.40$; or 4 genetic exchanges/10^6 cells
17. $\frac{1}{20} \times 100 = 5$; $P_{100,0} = \dfrac{e^{-5} \cdot 5^0}{0!} = e^{-5} = 0.007$; $0.05 = e^{-m}$;

    so $m = 3 = np$ and $n = \dfrac{3}{p} = \dfrac{3}{0.05} = 60$ cells/sample

Index

Boldface page numbers indicate the locations of boldface terms. In order to keep the index to a manageable size, references to individuals have generally been omitted.